高等职业院校电力技术类专业系列教材

电气设备检修（三）
——开关设备类（中英双语）

主　编　高　建　张　榆
副主编　杨　熙　廖翔志　王志川
参　编　胡清灵　欧阳仁乐　陈　丽　罗东辉　陈　杰　石　俊

西南交通大学出版社
·成　都·

图书在版编目（CIP）数据

电气设备检修.三,开关设备类：汉、英／高建,张榆主编.--成都：西南交通大学出版社，2023.11
ISBN 978-7-5643-9584-1

Ⅰ.①电… Ⅱ.①高… ②张… Ⅲ.①开关–设备检修–高等职业教育–教材–汉、英 Ⅳ.①TM64

中国国家版本馆 CIP 数据核字（2023）第 229771 号

Dianqi Shebei Jianxiu (San)
—Kaiguan Shebei Lei (Zhong-Ying Shuangyu)

电气设备检修（三）——开关设备类（中英双语）	主编　高　建　　张　榆	策划编辑　李芳芳　张少华 责任编辑　孟媛 封面设计　吴兵

印张：27.75　字数：811千	出版发行：西南交通大学出版社
成品尺寸：185 mm×260 mm	网址：http://www.xnjdcbs.com
版次：2023年11月第1版	地址：四川省成都市二环路北一段111号 西南交通大学创新大厦21楼
印次：2023年11月第1次	邮政编码：610031
印刷：四川玖艺呈现印刷有限公司	营销部电话：028-87600564　028-87600533
书号：ISBN 978-7-5643-9584-1	定价：98.00元

图书如有印装质量问题　本社负责退换
版权所有　盗版必究　举报电话：028-87600562

FOREWORD

为贯彻《国务院关于加快发展现代职业教育的决定》文件精神，更好地满足高等职业教育高质量发展的需要，实现教学内容由知识本位向能力本位的转变，结合人力资源和社会保障部国家职业技能标准《变电设备检修工》的职业能力要求，以提高电气设备检修人员的技术技能水平为目标，为满足市场和企业不断发展的岗位需求，编写了本套教材。本套教材响应"一带一路"建设号召，服务国家战略方针，深化电力行业"一带一路"交流合作。本套教材具备代表性、基础性、针对性及普遍性等特点。本教材的出版有利于拓展国际业务交流，加强电力文化输出，提升国家影响力。

本套教材共分为3本，利用"校企一体，师资互用"机制，践行"双师"结构与"双师"资质，与生产企业共同编制，由专业教师和企业教师一同进行课程内容深度分析。本套教材的编制，旨在以学生为中心，落实立德树人的根本任务，在使学生掌握相关职业资格技能鉴定或教育部"1+X"变电设备检修职业技能等级考试标准制度试点职业技能等级所需的知识和技术技能，满足相关工种的中级工或以上要求的同时，培养学生吃苦耐劳、团结协作、可迁移可转化、注重安全等职业素养和行为习惯，弘扬工匠精神，达到"德技并修、理实并重、手脑并用、讲赛并行、工学结合"的要求。

开关类设备检修是电力系统变电设备检修的重要组成部分，本书是电气设备检修课程的核心组成部分。本书由一线技术专家与专职培训师配合，结合现场工作实际编写完成，兼顾理论和技能操作，对开关类设备检修通用管理规定做了实用化解读，有助于读者的理解和吸收。该书主要包括开关类设备的专业巡视点、检修工艺要求、常见问题以及典型案例内容等。

本书根据理实一体化教学的需要，以项目导向、任务驱动为主线，学习内容遵循由浅入深、循序渐进的原则，采用教室+实训现场的理论实践相结合的教学方法，充分体现了教、学、做一体化。本书在编写过程中突出"工作任务导向、规范作业流程、理论知识够用、突出技能实训"的思想，强调安全作业和标准化作业。全书实操内容较为典型，按教学项目和教学模块设计，突出技能训练，使学员通过对技术工作的任务、过程、环境所进行的整体化感悟和反思，实现知识与技能、过程与方法、情感态度和价值观学习的统一。

本书依托国网四川省电力公司技能培训中心（四川电力职业技术学院）丰富的电气设备检修实训资源，采用了大量源自生产作业现场和技能实训现场的实拍图片，增强了本书的实用性和可读性。

本书由国网四川省电力公司技能培训中心（四川电力职业技术学院）高建、国网四川省电力公司电力科学研究院张榆共同担任主编。全书编写分工如下：项目一电弧及电气接触基本知识由杨熙、廖翔志编写，项目二高压隔离开关检修由高建、王志川编写，项目三高压断路器检修由胡清灵、罗东辉编写，项目四高压开关柜检修由欧阳仁乐、陈杰、石俊编写，项目五组合电器检修由陈丽、张榆编写，高建负责全书内容的审定。

本书由国网四川省电力公司乐山供电公司高级技师李运涛主审，攀枝花供电公司高级技师唐启刚、遂宁供电公司高级技师赵安参与评审。编写过程中得到国网四川省电力公司技能培训中心（四川电力职业技术学院）汤晓青副教授和电网检修部（电力设备技术系）同事的大力支持，在此表示衷心的感谢！

限于编者水平，书中不足和错误之处在所难免，恳请读者批评指正，不胜感激。

编　者

2023 年 5 月

CONTENTS

项目一　电弧及电气接触基本知识 ·· 001
　模块一　电弧的基本知识 ·· 001
　Module 1　Fundamentals of Arc ·· 004
　模块二　电弧的特性和熄灭方法 ·· 008
　Module 2　Characteristics and Extinguishing Methods of Arc ································· 014
　模块三　电气触头的基本知识 ·· 022
　Module 3　Basic Knowledge of Electrical Contact Terminal ···································· 026
　　任务一　认识触头 ··· 032
　　Task 1　Understand Contact Terminals ··· 035

项目二　高压隔离开关检修 ·· 039
　模块一　隔离开关概述 ··· 039
　Module 1　Disconnector Overview ··· 044
　模块二　隔离开关的典型型号 ·· 050
　Module 2　Typical Models of Disconnectors ··· 060
　模块三　隔离开关的操动机构 ·· 072
　Module 3　Operating Mechanism of Disconnector ·· 078
　模块四　隔离开关的运行与维护 ··· 085
　Module 4　Operation and Maintenance of Disconnectors ···································· 090
　模块五　隔离开关检修的基本要求 ·· 098
　Module 5　Basic Maintenance Requirments of Disconnectors ······························· 101
　模块六　隔离开关常见故障分析及处理 ··· 105
　Module 6　Common Fault Analysis and Handling of Disconnectors ······················· 108
　　任务一　GW4-126 型隔离开关的整体调试 ·· 112
　　Task 1　Integral Commissioning of GW4-126 Disconnector ······························ 117
　　任务二　GW7-126 型隔离开关触头过热故障处理 ·· 123
　　Task 2　Response to GW7-126 Disconnector Contact Terminal Overheating ········· 130
　　任务三　CJ6 型电动机操动机构检修 ·· 138
　　Task 3　Maintenance of CJ6 Motor Operating Mechanism ······························· 143

项目三　高压断路器检修 ··· 149
　模块一　高压断路器概述 ·· 149
　Module 1　Overview of HV Circuit Breakers ·· 152

模块二　真空断路器 156
　　Module 2　Vacuum Circuit Breaker 162
　　模块三　SF₆ 断路器 172
　　Module 3　SF₆ Circuit Breaker 181
　　模块四　高压断路器的操动机构 194
　　Module 4　Operating Mechanism of High-voltage Circuit Breaker 216
　　　　任务一　SN10-10 少油断路器灭弧室检修 247
　　　　Task 1　Maintenance of Arc Extinguishing Chamber of SN10-10 Low Oil Circuit Breaker 257
　　　　任务二　断路器二次回路故障处理 270
　　　　Task 2　Fault Treatment of the Secondary Circuit of Circuit Breaker 273
　　　　任务三　断路器机械特性测试 278
　　　　Task 3　Mechanical Characteristic Test of Circuit Breaker 289

项目四　高压开关柜检修 303
　　模块一　配电装置介绍 303
　　Module 1　Power Distribution Unit Introduction 314
　　模块二　高压开关柜基础知识 329
　　Module 2　Fundamentals of HV Switch Cabinet 332
　　模块三　KYN28-12 高压开关柜 336
　　Module 3　KYN28-12 HV Switch Cabinet 341
　　模块四　开关柜的"五防"联锁 347
　　Module 4　"Five-prevention" Interlocking of the Switch Cabinet 354
　　　　任务一　KYN28-12 高压开关柜整体检查与维护 362
　　　　Task 1　Integral Inspection and Maintenance of KYN28-12 HV Switch Cabinet 368
　　　　任务二　KYN28-12 高压开关柜"五防"联锁检查 375
　　　　Task 2　"Five-prevention" Interlocking Inspection of KYN28-12 HV Switch Cabinet 380

项目五　组合电器检修 387
　　模块一　组合电器基础知识 387
　　Module 1　Fundamentals of GIS 400
　　模块二　组合电器的检修 416
　　Module 2　GIS Maintenance 423
　　　　任务一　组合电器指示仪表检查 432
　　　　Task 1　Indicating Instrument Inspection of GIS 435

参考文献 438

项目一　电弧及电气接触基本知识

模块一　电弧的基本知识

一、电弧的特点和危害

电弧是电力系统及电能利用过程中常见的物理现象,它实际上是一种能量集中、温度很高、亮度很大的气体放电现象。电弧对电力系统和电气设备会造成很大的危害。

电弧由阴极区、阳极区和弧柱区三部分组成,如图 1-1 所示。阴极和阳极附近的区域分别称为阴极区和阳极区,在阴极和阳极间的明亮光柱称为弧柱。弧柱中心部位温度最高、电流密度最大,称为弧心;弧柱周围温度较低、亮度明显减弱的部分称为弧焰。

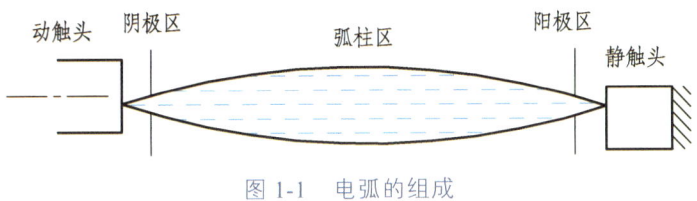

图 1-1　电弧的组成

二、电弧的产生

电弧的产生过程,实际上是气体介质在某些因素作用下,发生强烈游离,产生很多带电质点,由绝缘变为导通的过程。电弧能成为导电通道,是由于电弧的弧柱内存在大量的带电粒子,这些带电粒子的定向运动形成电弧。

1. 自由电子的产生

触头开断的瞬间由阴极通过热电子发射或强电场发射产生少量的自由电子。触头刚分离时,触头间的接触压力和接触面积不断减小,接触电阻迅速增大,使接触处剧烈发热,局部高温使此处电子获得动能,就可能发射出来成为自由电子,这种现象称为热电子发射。另一方面,触头刚分离时,由于触头间的间隙很小,在电压作用下间隙形成很高的电场强度,当电场强度超过 3×10^6 V/m 时,阴极触头表面的电子就可能在强电场力的作用下,被拉出金属表面成为自由电子,这种现象称为强电场发射。

2. 碰撞游离形成电弧

从阴极表面发射出来的自由电子,在触头间电场力的作用下加速运动,不断与间隙中的中性气体质点（原子或分子）撞击，如果电场足够强，自由电子的动能足够大，碰撞时就能将中性原子外层轨道上的电子撞击出来，脱离原子核内正电荷吸引力的束缚，成为新的自由电子。失去自由电子的原子则带正电，称为正离子。新的自由电子又在电场中加速积累动能，去碰撞另外的中性原子，产生新的游离，碰撞游离不断进行、不断加剧，带电质点成倍增加，如图 1-2 所示，此过程愈演愈烈，如雪崩似地进行着，发展成为"电子崩"，在极短促的时间内，大量的自由电子和正离子出现，在触头间隙形成强烈的放电现象，形成了电弧，这种现象称为碰撞游离，又称电场游离。

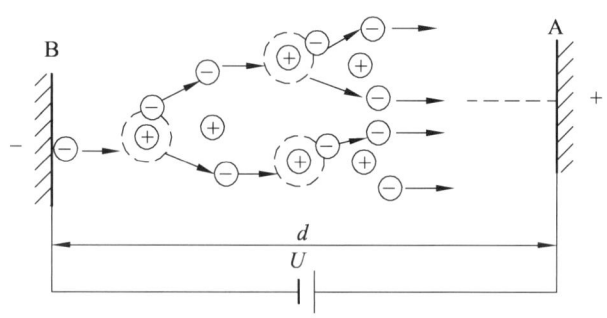

图 1-2　碰撞游离示意图

对于一种气体，能否产生电场游离主要取决于电子运动速度，也就是取决于电场强度、电子的平均自由行程以及气体的性质。触头间电压越高，电场强度也越高，则气体容易被击穿。气体的压力越高，其中自由电子的平均自由行程就越小，因而也就越不容易产生电场游离。不同的气体要从其中性原子外层轨道撞击出自由电子，所需能量值是不同的。

3. 热游离维持电弧

触头间隙在发生了雪崩式碰撞游离后，形成电弧并产生高温。温度增高时，气体中粒子的运动速度也随着增大，就可能使原子外层轨道的电子脱离原子核内正电荷的束缚力（吸引力）成为自由电子，这种游离方式称为热游离。气体温度越高，粒子运动速度越大，原子热游离的可能性也越大，从而供给弧隙大量的电子和正离子，维持电弧稳定燃烧。

一旦触头间隙形成电弧放电后，电弧的电阻很小，导电性很好，触头间隙的电压立刻降至最小，触头间隙的电场强度也大大降低，这时电场游离在间隙中作用不明显。另一方面，由于热平衡，电弧温度达到某一数值后不再上升，电导达到某一值后也不再上升，热游离将在一定强度下稳定下来，达到平衡状态。

综上所述，由于热电子发射或强电场发射在触头间隙中产生少量的自由电子，这些自由电子与中性分子发生碰撞游离并产生大量的带电粒子，从而形成气体导电，即产生电弧，一旦电弧产生后，将由热游离作用来维持电弧燃烧。

三、电弧的熄灭

电弧的熄灭过程，实际上是气体介质由导通又变为截止的过程。电弧中发生游离的同时，还存在着相反的过程，即去游离。去游离使弧隙中正离子和自由电子减少。电弧的熄灭是电弧区域内已电离的质点不断发生去游离的结果。去游离的主要方式包括复合和扩散两种形式。

1. 复合

复合是指正、负带电质点相遇，发生中和成为中性质点的现象。电子的运动速度很快，约为正离子的 1000 倍，所以电子和正离子直接复合的可能性很小。复合的方式是电子先附在中性质点上形成负离子，负离子的运动速度比较小，正负离子的复合就容易进行。目前广泛使用的 SF_6 断路器就利用了 SF_6 气体的强电负性来实现电弧的尽快熄灭。

2. 扩散

扩散是指电弧中的自由电子和正离子散溢到电弧外面，并与周围未被游离的冷介质相混合的现象。扩散是由于带电粒子的无规则热运动，以及电弧内带电粒子的密度远大于弧柱外，电弧的温度远高于周围介质的温度造成的。电弧和周围介质的温度差愈大，带电粒子的密度差愈大，扩散作用就愈强。高压断路器中常采用吹弧的灭弧方法，就是加强了扩散作用。

综上所述，当游离作用大于去游离作用时，电弧电流增加，电弧燃烧加强；当游离作用与去游离作用持平时，电弧维持稳定燃烧；当去游离作用大于游离作用，弧隙中导电质点的数目减少，电导下降，电弧越来越弱，弧温下降，使热游离下降或停止，最终导致电弧熄灭。要使电弧熄灭，必须使去游离作用强于游离作用。

Program 1 Fundamentals of Arc and Electrical Contact

Module 1 Fundamentals of Arc

1.1.1 Characteristics and Hazards of Arc

Arc is a common physical phenomenon in the power system and electric energy utilization process, it is actually a kind of gas discharge phenomenon featuring energy concentration, high temperature and great brightness. Arc can be very harmful to power system and electrical equipment.

The arc consists of three parts: the cathode area, the anode area and the arc column area, as shown in Fig. 1-1. The areas near the cathode and anode are called the cathode area and anode area, respectively, and the bright column of light between the cathode and anode is called the arc column. The center part of the arc column, which has the highest temperature and highest current density, is called the arc center; the part around the arc column with low temperature and significantly reduced brightness is called the arc flame.

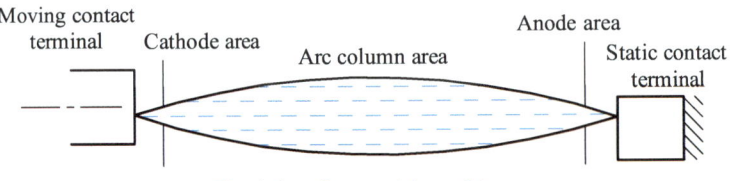

Fig. 1-1 Composition of Arc

1.1.2 Arc Generation

The generation process of arc is actually a process from insulation to conduction in which the gas medium is strongly ionized under the action of some factors, resulting in a lot of charged particles. The arc can become a conductive channel because there are a lot of charged particles in the arc column, and the directional motion of these charged particles leads to an arc.

1. Generation of free electrons

A small number of free electrons are generated by the cathode at the moment when the contact terminal is cut off, either by hot electron emission or by emission from strong electric field. When the contact terminals are just separated, the contact pressure and contact area between the contact terminals continue to decrease, and the contact resistance rapidly increases, causing the contact to heat severely, and the local high temperature allows the electrons to obtain kinetic energy, which may be emitted as free electrons. This phenomenon is called hot electron emission. On the other hand, when the contact terminals are just separated, very high electric field intensity is generated at the gap under the action of voltage because the gap between the contact terminals is very small. When the electric field intensity is more than 3×10^6 V/m, the electrons on the surface of the cathode contact terminal may be pulled out of the metal surface to become free electrons under the action of the strong electric field force. This phenomenon is called the emission from strong electric field.

2. Arc generation by impact ionization

The free electrons emitted from the cathode surface move in an accelerated manner under the action of the electric field force between the contact terminals, and constantly impact with the neutral gas particles (atoms or molecules) in the gap. If the electric field force is strong enough and the kinetic energy of the free electrons is large enough, the electrons in the outer orbit of the neutral atoms can be impacted out and be out of the binding of the attractive force of positive charge in the nucleus to become a new free electron. Atoms that lose their free electrons are positively charged and are called positive ions. The new free electrons accumulate kinetic energy in an accelerated manner in the electric field to impact with other neutral atoms, resulting in new ionization. The impact ionization continues and intensifies continuously, and the charged particles multiply, as shown in Fig. 1-2. This process becomes more and more intense, like an avalanche, and evolves into "electron avalanche". As a large number of free electrons and positive ions appear in a very short period of time, a strong discharge occurs in the gap between contact terminals, resulting in an arc. This phenomenon is called impact ionization, also known as field ionization.

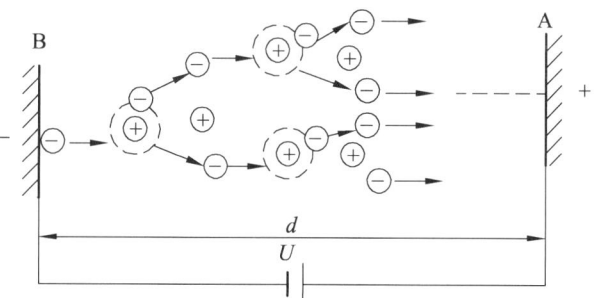

Fig. 1-2　Schematic Diagram of Impact Ionization

For a gas, whether the field ionization can be generated mainly depends on the speed of electron motion. In other words, it depends on the electric field intensity, the average free travel of the electrons, and the properties of the gas. The higher the voltage between the contact terminals is and the higher the electric field intensity is, the easier it is for the gas to be broken down. The higher the pressure of the gas is, the smaller the average free travel of the free electrons in it is, and thus the less likely it is to generate field ionization. Different gases require different values of energy to form free electrons from the outer orbit of the neutral atoms through impact.

3. Arc maintenance by thermal ionization

After the impact ionization like avalanche in the gap between the contact terminals, an arc is formed and high temperature is generated. When the temperature increases, the speed of motion of the particles in the gas also increases, which may make the electrons in the outer orbit of the atoms out of the binding force (attractive force) of positive charge in the nucleus to become free electrons, and this type of ionization is called thermal ionization. The higher the temperature of the gas is, the greater the speed of motion of the particles is, the greater the possibility of atomic thermal ionization is. Thus a large number of electrons and positive ions are supplied to the arc gap to maintain the stable combustion of the arc.

Once the arc discharge is generated in the gap between the contact terminals, the arc resistance is very small, the conductivity is very good, the voltage in the gap between the contact terminals is immediately reduced to the minimum, and the electric field intensity in the gap between the contact terminals is greatly reduced, so the role of field ionization in the gap is not obvious. On the other hand, due to thermal equilibrium, the arc temperature will no longer rise after reaching a certain value, and the conductance will no longer rise after reaching a certain value, and the thermal ionization will be stabilized at a certain intensity and reach equilibrium.

To sum up, because a small number of free electrons are generated in the gap between the contact terminals either by hot electron emission or by emission from strong electric field, these free electrons impact with neutral molecules and are ionized while a large number of charged particles are generated, resulting in gas conduction, i.e. an arc. Once the arc is generated, the arc combustion will be maintained by thermal ionization.

1.1.3 Arc Extinguishing

The arc extinguishing process is actually a process of gas medium from conduction to cut-off. While ionization occurs in the arc, there is also an opposite process, that is, de-ionization. De-ionization reduces the number of positive ions and free electrons in the arc gap. The arc extinguishing is as a result of the continuous de-ionization of the ionized particles in the arc area. The main modes of de-ionization include recombination and diffusion forms.

1. Recombination

Recombination is the phenomenon in which positively and negatively charged particles meet and undergo neutralization to become neutral particles. Electrons move very fast, about 1,000 times faster than positive ions, so direct recombination of electrons and positive ions is very unlikely. The way of recombination is that the electrons are first attached to the neutral particles to form negative ions, which move at a relatively small speed, and the recombination of positive and negative ions is easy to carry out. The SF_6 circuit breaker, which is widely used today, utilizes the strong electronegativity of the SF_6 gas to extinguish the arc as quickly as possible.

2. Diffusion

Diffusion is the phenomenon in which free electrons and positive ions in an arc spill outside the arc and mix with the surrounding cold medium that has not been ionized. Diffusion is caused by the irregular thermal motion of the charged particles, as well as the fact that the density of the charged particles inside the arc is much higher than that outside the arc column, and the temperature of the arc is much higher than the temperature of the surrounding medium. The greater the temperature difference between the arc and the surrounding medium is, the greater the density difference of the charged particles is, the stronger the diffusion effect is. The method of blowing arc is often used to extinguish the arc in high-voltage circuit breakers, which enhances the diffusion effect.

In summary, when the ionization effect is greater than the de-ionization effect, the arc current increases and the arc combustion is enhanced; the arc maintains stable combustion when the ionization effect and de-ionization effect are equalized; when the de-ionization effect is greater than the ionization effect, the number of conductive particles in the arc gap decreases, the conductance decreases, the arc becomes weaker and weaker, and the arc temperature decreases, so that the thermal ionization decreases or stops, which ultimately leads to arc extinguishing. In order to extinguish the arc, the de-ionization effect must be stronger than the ionization effect.

模块二　电弧的特性和熄灭方法

一、交流电弧的特性

在交流电路中产生的电弧称为交流电弧。交流电弧的特性如下：

（1）交流电弧具有动态特性。在交流电路中，电流瞬时值随时间变化，因而电弧的温度、直径以及电弧电压也随时间变化，电弧的这种特性称为动特性。

在一个周期性内交流电弧电流及电压随时间的变化关系如图 1-3 所示，图中 A 点称为燃弧电压，B 点称为熄弧电压，熄弧电压低于燃弧电压。电弧电压呈马鞍形变化，电流小时，电弧电压高，电流大时，电弧电压减小且接近于常数。

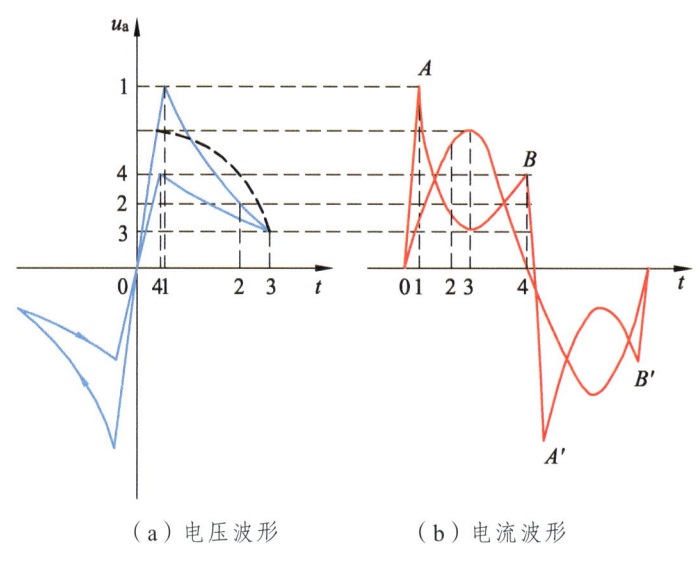

（a）电压波形　　（b）电流波形

图 1-3　交流电弧伏安特性及电压和电流波形

（2）电弧具有热惯性。由于弧柱的受热升温或散热降温都有一定过程，跟不上快速变化的电流，所以电弧温度的变化总滞后于电流的变化，这种现象称为电弧的热惯性。

（3）交流电流每半个周期过零一次，称为"自然过零"。电流过零时，电弧自然熄灭。如果电弧是稳定燃烧的，则电弧电流过零熄灭后，在另半周又会重燃。如果电弧过零后，电弧不发生重燃，电弧就会熄灭。所以，交流电流过零的时刻是熄灭电弧的良好时机，如果在电流过零时采取有效措施使电弧不再重燃，则电弧最终熄灭。

二、交流电弧的熄灭

交流电弧的燃烧过程与直流电弧的基本区别在于交流电弧中电流每半周要经过零点一次，此时电弧自然暂时熄灭。在电流过零时，采取有效措施加强弧隙的冷却，使弧隙介质的绝缘能力达到不会被弧隙外施电压击穿的程度，则在下半周电弧就不会重燃而最终熄灭。交流电流过零后，电弧是否重燃取决于弧隙介质绝缘能力或介电强度和弧隙电压的恢复。

1. 弧隙介质介电强度的恢复

弧隙介质能够承受外加电压作用而不致使弧隙击穿的电压称为弧隙的绝缘能力或介电强度。当电弧电流过零时电弧熄灭,弧隙中去游离作用继续进行,弧隙电阻不断增大,但弧隙介质的介电强度要恢复到正常状态值需要有一个过程,此恢复过程称为弧隙介质介电强度的恢复过程,以能耐受的电压 U_j 表示。

介质介电强度的恢复速度与冷却条件、电流大小、开关电器灭弧装置的结构和灭弧介质的性质有关。图 1-4 所示为不同介质的介电强度恢复过程曲线。从图中可见:在电流过零瞬间($t=0$),介电强度突然出现升高的现象,此现象称为近阴极效应。这是因为电流过零后,弧隙的电极极性发生了改变,弧隙中剩余的带电质点的运动方向也相应改变,质量小的电子迅速向新的阳极运动,而比电子质量大很多倍的正离子由于惯性大,来不及改变运动方向停留在原地未动,导致新的阴极附近形成了一个正电荷的离子层,如图 1-5 所示,正空间电荷层使阴极附近出现了 150～250 V 的起始介电强度。近阴极效应使弧隙在电弧自然熄灭后的极短瞬间能耐受 150～250 V 的外加电压。在低压电器中,常利用近阴极效应这个特性来灭弧。

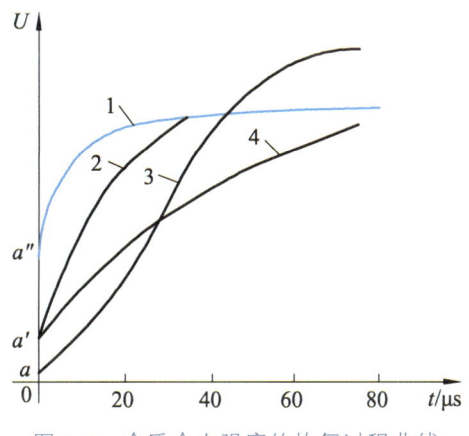

图 1-4 介质介电强度的恢复过程曲线

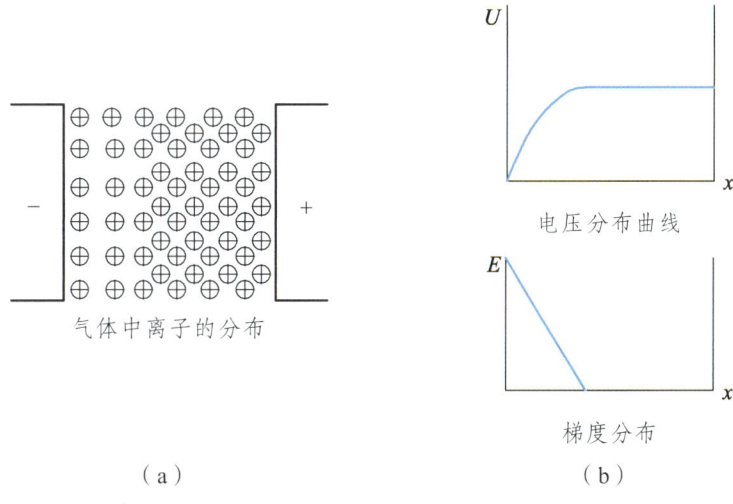

(a) 气体中离子的分布

(b) 电压分布曲线 / 梯度分布

图 1-5 近阴极效应

2. 弧隙电压的恢复过程

电流过零使电弧熄灭后，加在弧隙上的电压称为恢复电压。电弧电流过零前，弧隙电压呈马鞍形变化，电压值很低，电源电压的绝大部分降落在线路和负载阻抗上；电流过零时，弧隙电压等于熄弧电压，正处于马鞍形的后峰值处；电流过零后，弧隙电压从后峰值逐渐增长，一直恢复到电源电压，弧隙电压从熄弧电压变成电源电压的过程称为弧隙电压恢复过程，用 $U_{hf}(t)$ 表示电压恢复过程。电压恢复过程与电路参数、负荷性质等有关。受电路参数等因素的影响，电压恢复过程可能是周期性的变化过程，也可能是非周期性变化过程。图 1-6 所示是弧隙恢复电压按指数规律变化的非周期性过程，图中 U_0 为电弧自然熄灭瞬间的电源相电压，U_{xh} 为熄弧电压，U_{hf} 为弧隙恢复电压，依指数规律上升的恢复电压最大值不会超过 U_0，也就是说不会在电压恢复过程中出现过电压。图 1-7 所示是恢复电压呈现周期性振荡的变化过程，这时弧隙的恢复电压最大值理论上可达到 $2U_0$，实际中由于电阻影响，弧隙恢复电压振荡有衰减，实际最大值为 $(1.3\sim1.6)U_0$。周期性振荡的恢复电压更容易超过弧隙介质强度，造成电弧重燃。

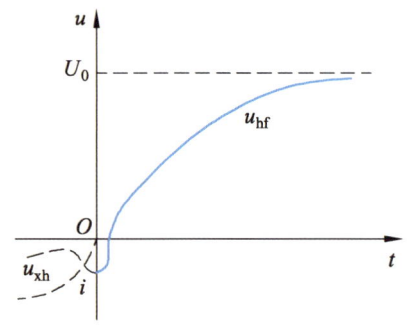

图 1-6　恢复电压非周期性变化

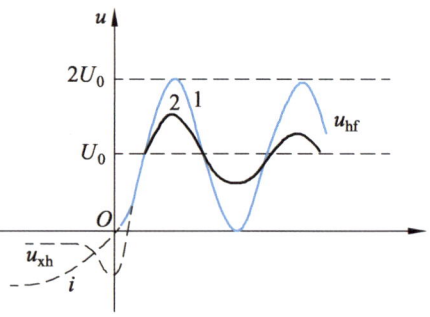

图 1-7　周期性震荡恢复电压变化

3. 交流电弧的熄灭条件

电弧电流过零后，电弧自然熄灭。电流过零后，弧隙中同时存在着两个作用相反的恢复过程，即介质介电强度恢复过程 $U_j(t)$ 和弧隙电压的恢复过程 $U_{hf}(t)$。如果弧隙介质介电强度在任何情况下都高于弧隙恢复电压，则电弧熄灭；反之，如果弧隙恢复电压高于弧隙介质介电强度，弧隙就被击穿，电弧重燃，如图 1-8 所示。

$$U_j(t) > U_{hf}(t)$$

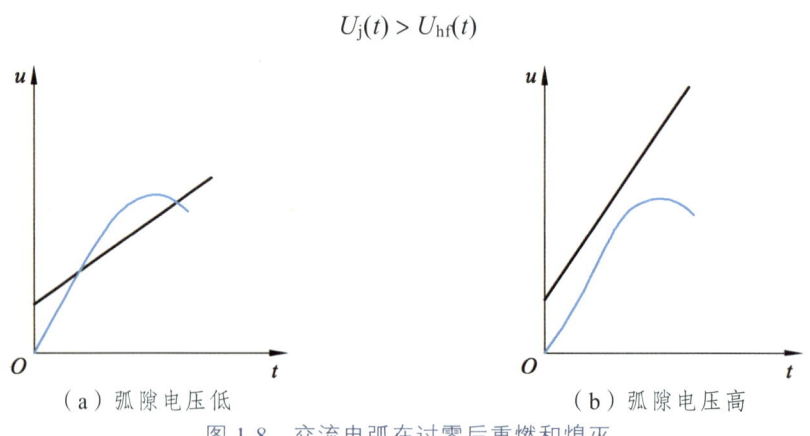

图 1-8　交流电弧在过零后重燃和熄灭

综上所述，在交流电弧的灭弧中，应充分利用交流电流的自然过零点，采取有效的措施，加大弧隙间去游离的强度，使电弧不再重燃，最终熄灭。

三、熄灭交流电弧的基本方法

开断交流电弧时，在电流达到零值以后，加强对弧隙的冷却，抑制热游离，加强去游离。为此，在开关设备中均装设了灭弧装置（灭弧室），灭弧室不断改进，大大提高了开关的灭弧能力。另一方面，为了进一步提高灭弧能力，还可以采用性能更为优越的新型灭弧介质，例如六氟化硫断路器的使用等。目前，在开关电器中广泛采用的灭弧方法有以下几种。

1. 吹弧

利用灭弧介质（气体、油等）在灭弧室中吹动电弧，广泛应用在开关电器中，特别是高压断路器中。

用弧区外新鲜、低温的灭弧介质吹拂电弧，对熄灭电弧起到多方面的作用。它可将电弧中大量正负离子吹到触头间隙以外，代之以绝缘性能高的新鲜介质；它使电弧温度迅速下降，阻止热游离的继续进行，使触头间的绝缘强度提高；被吹走的离子与冷介质接触，加快了复合过程的进行；吹弧使电弧拉长变细，加快了电弧的扩散，弧隙电导下降。按吹弧方向分为：

（1）横吹。

吹弧方向与电弧轴线方向相垂直时，称为横吹，如图 1-9（a）所示。横吹更易于把电弧吹弯拉长，增大电弧表面积，还能加强冷却和增强扩散。

（2）纵吹。

吹动方向与电弧轴线方向一致时，称为纵吹，如图 1-9（b）所示。纵吹能促使弧柱内带电质点向外扩散，使新鲜介质更好地与炽热的电弧相接触，冷却作用加强，并把电弧吹成若干细条，易于电弧熄灭。

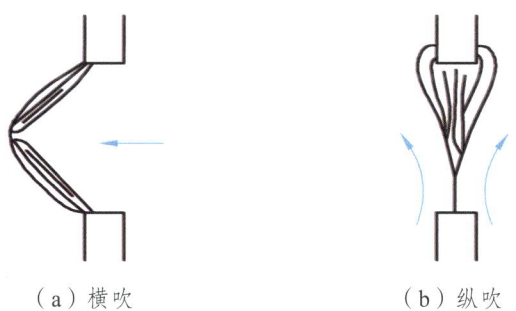

（a）横吹　　　　　（b）纵吹

图 1-9　吹弧方法

（3）纵横吹。

横吹灭弧室在开断小电流时，因灭弧室内压力太小，开断性能差。为了改善开断小电流时的灭弧性能，可将纵吹和横吹结合起来。在开断大电流时主要靠横吹，开断小电流时主要靠纵吹。

2. 采用多断口灭弧

在许多高压断路器中，每相有两个或多个断口相串联，如图 1-10 所示。熄弧时，利用多断口把电弧分解为多个相串联的短电弧，使电弧的总长度加长，弧隙电导下降；在触头行程、分闸速度相同的情况下，电弧被拉长的速度成倍增加，促使弧隙电导迅速下降，提高了介电强度的恢复速度；同时采用多断口时，加在每一断口上的电压会成倍减小，输入电弧的功率和能量也随之减小，弧隙电压的恢复速度降低，缩短了灭弧时间。多断口比单断口具有更好的灭弧性能，便于采用积木式结构（用于110 kV及以上电压的断路器中）。

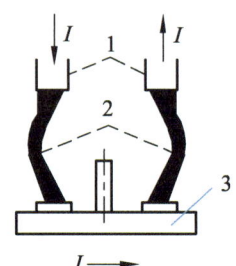

1—多断口；2—电弧；3—均压电容。

图1-10 双断口灭弧

采用多断口的结构后，每个断口上的电压会出现分配不均的现象，这是由于两断口之间的导电部分对地电容的影响而引起的。为了使各个灭弧室的工作条件相接近，通常采用断口并联电容的方法，在每个断口外边并联一个比对地电容大得多的电容 C，称为均压电容，其容量一般为1000~2000 pF。接了均压电容后，只要电容容量足够大，多断口的电压就接近相等了。实际中要做到电压完全均匀，必须装设容量很大的电容，造成投资增大，经济性不好，因此，一般按断口间最大电压不超过均匀分配值10%的要求来选择均压电容的电容量。

3. 提高分闸速度

迅速拉长电弧，有利于迅速减小弧柱内的电位梯度，增加电弧与周围介质的接触面积，加强冷却和扩散作用。现代高压开关中都采取了迅速拉长电弧的措施灭弧，如采用强力分闸弹簧，其分闸速度已达 16 m/s。

4. 用耐高温金属材料制作触头

触头材料对电弧中的去游离也有一定影响，选用熔点高、导热系数和热容量的耐高温金属制作的触头，可以减少热电子发射和电弧中的金属蒸汽，从而减弱游离过程，利于电弧熄灭。常用的材料有铜钨合金和银钨合金。

5. 采用优质灭弧介质

灭弧介质的特性，如导热系数、介电强度、热游离温度、热容量等，对电弧的游离程度有很大影响，这些参数值越大，去游离作用越强。现代高压开关中，广泛采用油、压缩空气、SF_6气体、真空等作为灭弧介质。

6. 短弧原理灭弧

这种灭弧方法常用于低压开关电器中，如自动开关和电磁接触器等。利用一个金属灭弧栅将电弧分为多个短弧，利用近阴极效应的方法灭弧，如图1-11所示。灭弧栅用金属材料制成，触头间产生的电弧被磁吹线圈驱入灭弧栅，每两个栅片间就是一个短弧，每个短弧在电流过零时新阴极产生 150～250 V 的起始介电强度，如果所有串联短弧的起始介电强度总和始终大于触头间的外加电压，电弧就不会重燃而熄灭。在低压电路中，电源电压远小于起始介质强度之和，因而电弧不能重燃。

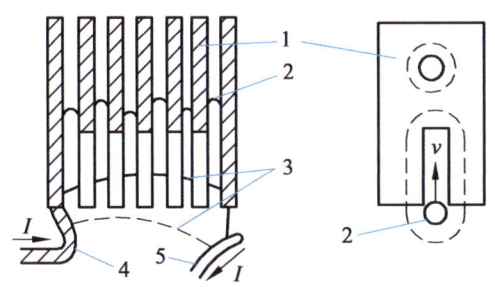

1—金属灭弧栅；2—触头；3—灭弧介质；4—动触头；5—灭弧片。

图 1-11　电弧在灭弧栅内熄灭

7. 利用固体介质的狭缝灭弧装置灭弧

低压开关中也广泛应用狭缝灭弧装置灭弧。狭缝由耐高温的绝缘材料（如陶土或石棉水泥）制作，通常称为灭弧罩。电弧形成后，用磁吹线圈产生的磁场作用于电弧，电弧受电动力作用吹入狭缝中，把电弧迅速拉长的同时，电弧与灭弧罩内壁紧密接触，热量被冷的灭弧罩吸收，电弧温度下降，电弧表面被冷却和吸附；又因窄缝中的气体被加热使压力很大，加强了电弧中的复合过程。图 1-12 是狭缝灭弧装置的工作原理示意图。

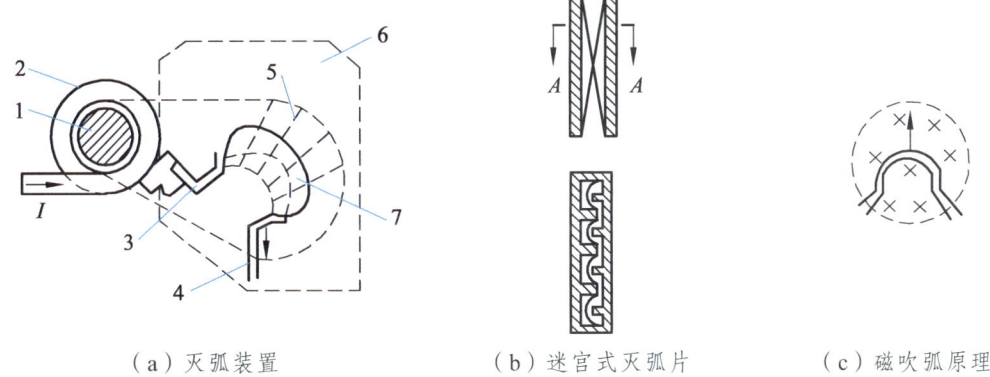

（a）灭弧装置　　　　（b）迷宫式灭弧片　　　　（c）磁吹弧原理

1—磁吹铁心；2—磁吹线圈；3—静触头；4—动触头；5—灭弧片；6—灭弧罩；7—电弧移动位置。

图 1-12　狭缝灭弧装置的工作原理

磁吹力的产生靠外加磁场，使电弧在磁场中受力向灭弧室狭缝中移动。产生磁场的方法有三种：

（1）磁吹线圈与电路串联。其特点是吸力的方向不随电流方向的改变而变化，磁吹力的大小与电弧电流的平方成正比。当切断小电流时，可能磁吹力太小，电弧不能被拉入狭缝中。

（2）磁吹线圈与电路并联。磁吹力不受电弧电流影响，可以获得恒定的磁场强度，开断小电流时不会降低它的开断能力。但磁吹力具有方向性，在使用中必须注意磁吹线圈的极性。

（3）永久磁铁式。其工作原理与并联磁吹相同，但它不需要线圈，结构简单。它同样具有方向性，一般只应用于直流电路中。

Module 2　Characteristics and Extinguishing Methods of Arc

1.2.1　Characteristics of the AC Arc

An arc generated in an AC circuit is called an AC arc. AC arc is characterized as follows:

(1) AC arcs have dynamic characteristics. In the AC circuit, the instantaneous value of the current varies with time, and thus the temperature and diameter of the arc and the arc voltage also vary with time, and this characteristic of the arc is called the dynamic characteristic.

The current and voltage variation of AC arc with time in a cycle is shown in Fig. 1-3 where point A is called arc combustion voltage and point B is called arc extinguishing voltage, and extinguishing voltage is lower than arc combustion voltage. The arc voltage shows a saddle-shaped change. When the current is small, the arc voltage is high; when the current is large, the arc voltage decreases and is close to constant.

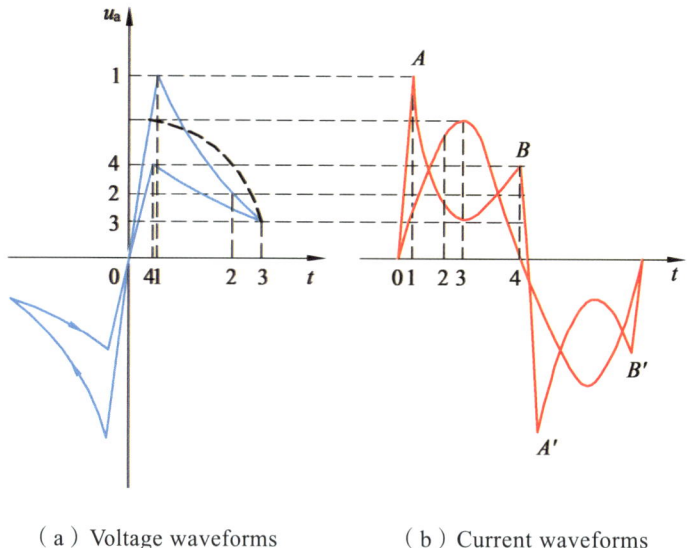

(a) Voltage waveforms　　(b) Current waveforms

Fig. 1-3　Volt-ampere Characteristics and Voltage and Current Waveforms of AC Arc

(2) The arc has thermal inertia. Because the heating or cooling of the arc column takes a certain time, which can not keep up with the rapid change of current. Therefore, the change of arc temperature always lags behind the change of current, and this phenomenon is called thermal inertia of arc.

(3) The value of AC turns to zero once every half cycle, which is called "natural zero crossing". The arc is naturally extinguished at the zero crossing of current. If the arc is combusting steadily, the arc is extinguished at zero crossing and then rekindled during the other half cycle. If the arc is not rekindled after zero crossing, the arc is extinguished. Therefore, the moment of the zero

crossing of AC is a good time to extinguish the arc, and if effective measures are taken at the time of zero crossing of current so that the arc is not rekindled, the arc will eventually be extinguished.

1.2.2 Extinguishing of AC Arc

The basic difference between the combustion process of the AC arc and the DC arc is that the current in the AC arc turns to zero once every half cycle, at which time the arc is naturally extinguished temporarily. At the time of zero crossing of current, if effective measures are taken to strengthen the cooling of the arc gap so that the insulation capacity of the arc gap medium will be adequate to avoid breakdown by the voltage applied outside the arc gap, the arc will not be rekindled and finally extinguished in the second half cycle. After zero crossing of AC, whether the arc rekindling depends on the insulation capacity or dielectric strength of arc gap medium and arc gap voltage recovery.

1. Recovery of the dielectric strength of the arc gap medium

The insulation capacity or dielectric strength of the arc gap refers to that arc gap medium can withstand the action of the applied voltage without causing the arc gap breakdown. When the arc is extinguished at the time of zero crossing of the current, the de-ionization in the arc gap continues, and the arc gap resistance increases continuously, but it takes certain time for the dielectric strength of the arc gap medium to recover to the normal state value. This process is called the recovery process of the dielectric strength of the arc gap medium, which is expressed by the withstand voltage U_j.

The recovery rate of dielectric strength is related to the cooling conditions, the magnitude of current, the structure of the arc extinguishing device and the nature of the arc extinguishing medium of the switching device. Fig. 1-4 shows the recovery process curves of dielectric strength for different media. As can be seen from the figure: At the instant when the current turns to zero ($t = 0$), there is a sudden increase in dielectric strength, a phenomenon known as the near-cathode effect. This is because after the current value turns to zero, the polarity of the electrode of the arc gap changes, and the direction of motion of the remaining charged particles in the arc gap also changes accordingly. The electrons with small mass quickly move toward the new anode, while the positive ions with a mass many times larger than that of the electrons stay in place since it is too late to change the direction of motion due to the large inertia, which results in the formation of a positively charged ion layer near the new cathode. As shown in Fig. 1-5, the positive space charge layer gives rise to a starting dielectric strength of about 150 to 250 V near the cathode. The near-cathode effect allows the arc gap to withstand an applied voltage of 150–250 V for a very short period of time after the arc is naturally extinguished. In low-voltage electrical appliances, the characteristic of near-cathode effect is often utilized to extinguish arcs.

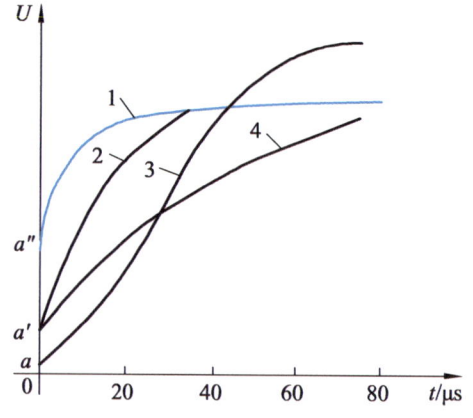

Fig. 1-4　Recovery Curve of Dielectric Strength

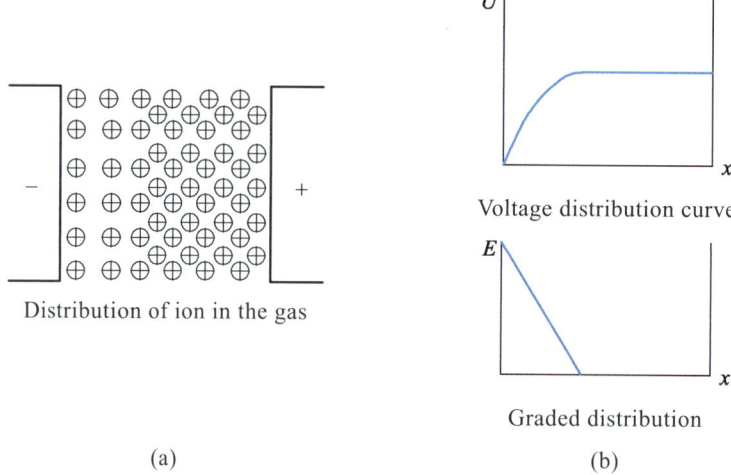

(a) Distribution of ion in the gas
(b) Voltage distribution curve / Graded distribution

Fig. 1-5　Near-cathode Effect

2. Recovery process of arc gap voltage

After arc extinguishing caused by current zero crossing, the voltage exerted on the arc gap is called recovery voltage. Before arc current zero crossing, the arc gap voltage shows saddle-shaped change and the voltage is very low. Most part of the power voltage falls on the line and load impedance. During current zero crossing, the arc gap voltage is equal to the arc extinguishing voltage, which is at the post-peak of the saddle. After current zero crossing, the arc gap voltage gradually increases from the post-peak value until it recovers to the power voltage. The process in which arc gap voltage becomes arc extinguishing voltage is called the recovery process of arc gap voltage, represented by $U_{hf}(t)$. The voltage recovery process is related with circuit parameters and load characteristics. Due to the influence of factors such as circuit parameters, the voltage recovery process may be a process of periodic change or a process of aperiodic change. Fig. 1-6 shows the aperiodic process of exponential change of arc gap recovery voltage. In the figure, U_0 represents the instantaneous phase voltage of power supply when the arc naturally extinguishes, U_{xh} represents arc-extinguishing voltage, and U_{hf} represents arc gap recovery voltage. The maximum recovery voltage which shows an exponential increase will not exceed U_0. That is, overvoltage will not occur during the process of voltage recovery. Fig. 1-7 shows the process of periodic oscillation of recovery voltage. At this moment, the maximum theoretical recovery voltage of arc gap may reach $2U_0$. In practice, due to the impact of resistance, there is attenuation in the oscillation of arc gap recovery voltage. The actual maximum value is (1.3–1.6) U_0. The recovery voltage showing periodic oscillation will be more likely to exceed the dielectric strength of arc gap, causing arc to reignite.

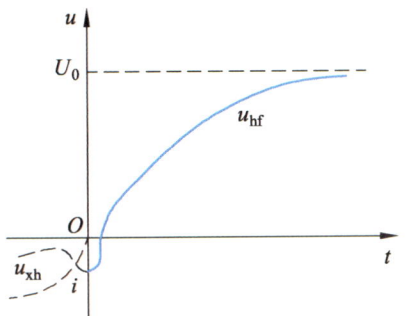 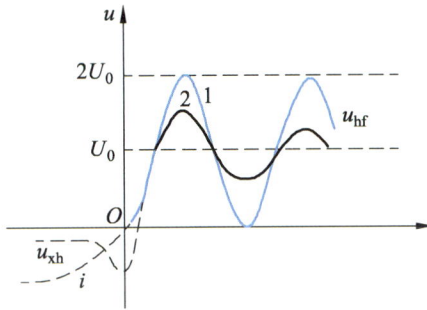

Fig. 1-6　Aperiodic Change of Recovery Voltage

Fig. 1-7　Change of Recovery Voltage with Periodic Oscillation

3. Extinguishing condition for AC arc

After arc current zero crossing, the arc will naturally quench. After current zero crossing, there exist two recovery processes with reverse action in the arc gap simultaneously, i.e. the recovery process $U_j(t)$ of dielectric strength and the recovery process $U_{hf}(t)$ of arc gap voltage. If the dielectric strength of arc gap medium is higher than the recovery voltage of arc gap in any case, the arc will quench; if the recovery voltage of arc gap is higher than the dielectric strength of arc gap medium, the arc gap will be broken down and the arc will reignite, as shown in Fig. 1-8.

$$U_j(t) > U_{hf}(t)$$

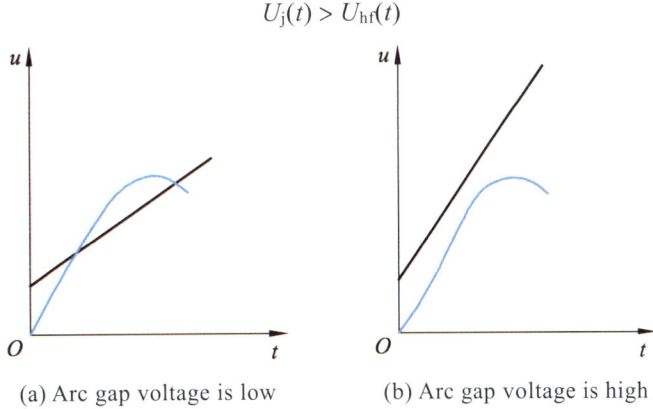

(a) Arc gap voltage is low　　(b) Arc gap voltage is high

Fig. 1-8　Re-ignition and Extinguishing of AC Arc after Zero Crossing

In conclusion, during arc extinguishing of AC arc, natural zero crossing of AC current shall be fully used, and effective measures shall be taken to increase the strength of de-ionization in the arc gap, so that the arc will not reignite and will finally extinguish.

1.2.3 Basic Methods for AC Arc Extinguishing

Break the AC arc, and when the current reaches zero, strengthen cooling for the arc gap, suppress thermal ionization and enhance de-ionization. For this purpose, an arc extinguishing device (arc extinguishing chamber) has been installed in each switchgear, which greatly improves the arc extinguishing ability of switchgear. In addition, to further improve arc extinguishing ability, new arc extinguishing media with better performance such as SF6 circuit breaker may be used. At present, several arc extinguishing methods are widely used in switching devices, as shown below.

1. Arc blowing

Arc extinguishing media (such as gas and oil) are used to blow the arc in the arc extinguishing chamber. This method is widely used in switching devices, especially high-voltage circuit breaker.

The fresh and low-temperature arc extinguishing medium outside the arc area is used to blow the arc, which plays multiple roles in extinguishing the arc. It is able to blow a large number of positive and negative ions in the arc out of the clearance between contact terminals and the media will be replaced with fresh medium with higher insulating property. It quickly lowers the arc temperature, inhibits thermal ionization, and improves the insulating strength of gap between contact terminals. Ions which have been blown away contact the cold medium, which accelerates the process of ionization recombination. Blowing makes the arc become longer and thinner, which accelerates arc diffusion and reduces the conductivity of arc gap. Arc blowing is divided into the following types according to the blowing direction:

(1) Horizontal blowing.

When the blowing direction is perpendicular to the arc axis, such method is called horizontal blowing, as shown in Fig. 1-9(a). Horizontal blowing can easily bend and stretch the arc, increase the surface area of arc, and improve the effect of cooling and diffusion.

(2) Vertical blowing.

When the blowing direction is consistent with the arc axis, such method is called vertical blowing, as shown in Fig. 1-9(b). Vertical blowing can diffuse charged particles in the arc column outward, so that the fresh medium can better contact the burning arc. Such method improves the cooling method and blow the arc into several thin strips, which makes it easy for the arc to extinguish.

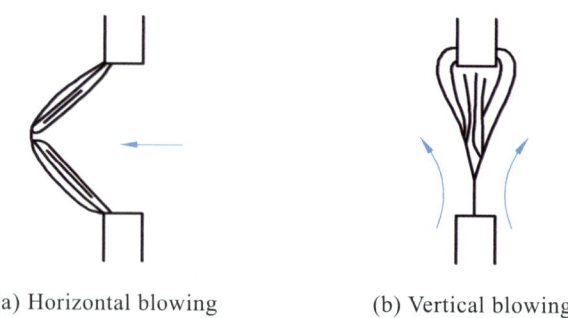

(a) Horizontal blowing   (b) Vertical blowing

Fig. 1-9 Methods for Arc Blowing

(3) Horizontal and vertical blowing.

When a small current is broken in the arc extinguishing chamber for horizontal blowing, since the pressure in the arc extinguishing chamber is too small, the breaking performance is poor. To improve the arc extinguishing performance during the breaking process of small current, horizontal blowing may be combined with vertical blowing. Horizontal blowing is mainly used for breaking large current, while vertical blowing is mainly used for breaking small current.

2. Multi-break arc extinguishing

For many high-voltage circuit breakers, two or more breaks are connected in series in each phase, as shown in Fig. 1-10. During arc extinguishing, multiple breaks divide the arc into several short arcs connected in series, which increases the overall length of arc and reduces the conductivity of arc gap. Assume the stroke and opening speed of contact terminal remain unchanged. The speed at which the arc is stretched doubles, which quickly reduces the conductivity of arc gap and improves the recovery rate of dielectric strength. Meanwhile, if multiple breaks are used, the voltage exerted on each break will decrease several times. As a result, the power and energy of the arc will decrease, the

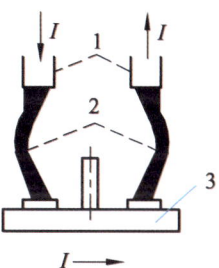

1-multiple breaks; 2-arc; 3-voltage-sharing capacitor.

Fig. 1-10 Double-break Arc Extinguishing

recovery rate of arc gap voltage will go down, and the arc extinguishing time will be reduced. Multi-break circuit breaker has better arc extinguishing performance than single-break circuit breaker, which is suitable for modular construction (used in circuit breakers with a voltage of 110 kV and above).

If multi-break structure is adopted, voltage will be distributed unevenly on different breaks. This is caused by the ground capacitance of the conductive part between two breaks. To provide similar operating conditions for different arc extinguishing chambers, the break is usually connected with a capacitor in parallel. Specifically, a capacitor C is connected in parallel outside each break, the capacitance of which is much larger than ground capacitance. Such capacitor is called voltage-sharing capacitor and the capacitance is usually 1,000−2,000 pF. After the voltage-sharing capacitor is connected, as long as the capacitance is large enough, the voltage on each break will be almost equal to each other. In practice, to ensure that voltage is completely distributed uniformly on each break, a capacitor with large capacitance must be installed, which will increase investment and cause poor economic performance. Therefore, when we select the capacitance of voltage-sharing capacitor, the maximum voltage between breaks shall not exceed 10% of the uniformly distributed voltage.

3. Improve the opening speed

Quickly stretching the arc can quickly decrease the electrical potential gradient in the arc column, increase the contact area between the arc and surrounding medium, and promote cooling and diffusion. Modern high-voltage switches quickly stretch the arc to realize arc extinguishing. For example, a forced opening spring is used and the opening speed can reach 16 m/s.

4. Contact terminal made of high-temperature resistant metal material

The material of contact terminal will have a certain impact on arc de-ionization. The contact terminal made of high-temperature-resistant metal with high melting point, thermal conductivity and thermal capacity can reduce thermionic emission and metal vapor in the arc, so as to weaken ionization and promote arc extinguishing. The commonly used materials include copper-tungsten alloy and silver-tungsten alloy.

5. The use of high-quality arc extinguishing media

The properties of arc extinguishing medium such as thermal conductivity, dielectric strength, thermal ionization temperature and thermal capacity has a great impact on the degree of ionization of arc. The higher these parameters are, the stronger effect of de-ionization they will cause. In modern high-voltage switches, oil, compressed air, SF_6 gas and vacuum are widely used as arc extinguishing media.

6. Short-arc extinguishing

Such arc extinguishing method is usually used in low-voltage switching devices such as automatic switch and electromagnetic contactor. A metal arc chute is used to divide the arc into several short arcs, and near-cathode effect is used for arc extinguishing, as shown in Fig. 1-11. The arc chute is made of some metal material. The arc generated between contact termianls is driven into the arc chute by magnetic blow-out coil. There is a short arc between two arc chutes. During current zero crossing, the new cathode in each short arc produces 150–250 V initial dielectric strength. If the sum of initial dielectric strength of all the short arcs connected in series is always larger than the applied voltage between contact terminals, the arc will not reignite. Instead, it will extinguish. In a low-voltage circuit, the power voltage is much less than the sum of initial dielectric strength. Therefore, the arc will not reignite.

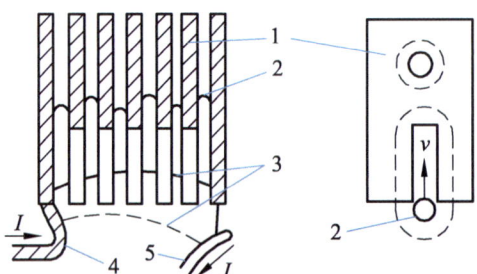

1-metal arc chute; 2-contact terminal; 3-arc-extinguishing medium; 4-moving contact; 5-arc chute.

Fig. 1-11　Arc Extinguishing in the Arc Chute

7. Slit arc extinguishing device in solid medium is used for arc extinguishing

Slit arc extinguishing device is also widely used for arc extinguishing in low-voltage switches. The slit is made of high-temperature-resistant insulating materials (such as pottery clay or asbestos cement), which is usually called arc extinguishing shield. Once an arc forms, the magnetic field generated by magnetic blow-out coil will act on the arc. The arc is driven by electromotive force into the slit and is quickly stretched, Meanwhile, the arc contacts closely with the inner wall of arc extinguishing shield. Since the heat is absorbed by the cold arc extinguishing shield, the arc temperature falls, and the arc surface is cooled down. Besides, the gas in the narrow slit is heated and the pressure becomes very large, which promotes the recombination process of arc. Fig. 1-12 is the schematic diagram of the operating principle of slit arc extinguishing device.

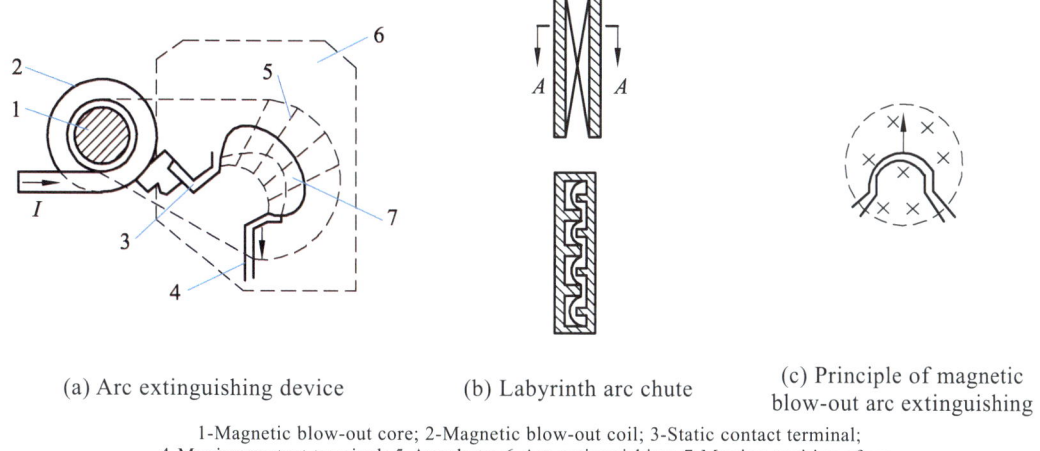

(a) Arc extinguishing device   (b) Labyrinth arc chute   (c) Principle of magnetic blow-out arc extinguishing

1-Magnetic blow-out core; 2-Magnetic blow-out coil; 3-Static contact terminal;
4-Moving contact terminal; 5-Arc chute; 6-Arc extinguishing; 7-Moving position of arc.

Fig. 1-12   Operating Principle of Slit Arc Extinguishing Device

The external magnetic field produces a magnetic blow-out force. Under the action of such force in the magnetic field, the arc moves towards the slit in the arc extinguishing chamber. Three methods can be used to generate the magnetic field:

(1) The magnetic blow-out coil is connected with the circuit in series. The direction of the attraction will not change with the direction of current, and the magnetic blow-out force is in direct proportion to the square of arc current. When a small current is cut off, the arc cannot be pulled into the slit maybe because the magnetic blow-out force is too small.

(2) The magnetic blow-out coil is connected with the circuit in parallel. The magnetic blow-out force will not be influenced by arc current and a constant magnetic field intensity can be obtained. When a small current is cut off, the breaking capacity will not be reduced. However, the magnetic blow-out force has a certain direction. Attention shall be paid to the polarity of magnetic blow-out coil during use.

(3) Permanent magnet. It has the same operating principle as the magnetic blow-out coil in parallel, but it does not need any coil and the structure is simple. It also has a certain direction. Generally, it is only used in DC circuit.

模块三　电气触头的基本知识

一、触头的接触电阻

下面分析影响接触电阻的主要因素。

1. 触头间的压力

即使精细加工的触头表面，从微观上看也是凹凸不平的，触头接触面积的大小受施加压力大小的影响，如图 1-13 所示。在不加外力情况下，将两个触头对接放置时，触头间仅有一点 a 接触，如图 1-13（a）所示。施加外力 F_1 时，a 点被压平形成接触面；若施加比 F_1 更大的外力 F_2，则 a 接触面增大，同时又将 b 点接触并形成新的接触面，总的接触面增大了，接触电阻就小了。故压力是影响接触电阻的重要因素。

在开关电器中，一般在触头上附加钢性弹簧，目的是增大并保持触头间的接触压力，使触头接触可靠，减小接触电阻并保持稳定。

2. 触头材料及预防氧化的措施

一些电气设备，如变压器、电机等采用铜制引出端头，如果是在屋外和潮湿的场所中，就不能将铝导体用螺栓与铜端头连接。因为铜铝直接接触会形成电位差（约为 1.86 V），当含有溶解盐的水分渗入接触面的缝隙时，会产生电解反应，铝被强烈地电腐蚀，导致触头损坏，并可能酿成重大事故。为了避免出现这种情况，通常采用铜铝过渡接头，其结构是一端为铝，一端为铜，如图 1-13 所示。对于具有触头的转动式摩擦接触电器，检修后转鼓上需要薄薄的涂敷一层凡士林，以抗氧化和腐蚀，在投入正常运用后就不用频繁涂抹凡士林。

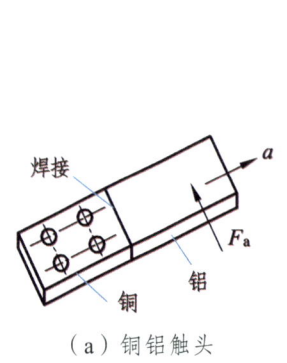

（a）铜铝触头

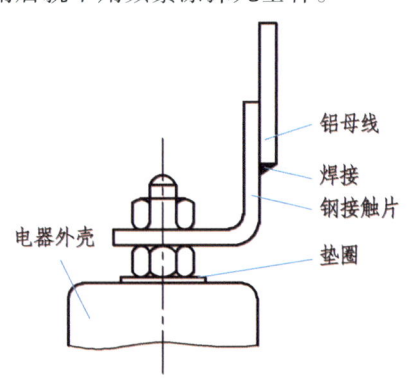

（b）铝母线接到电器铜端头上用的接头

图 1-13　触头材料

二、触头的分类及其结构

（一）按接触面的形式分类

1. 点触头

点触头是指两个触头间的接触面为点状的触头，如球面和平面接触、两个球面接触等都是点接触。这种接触形式的优点是压强较大、接触点较固定、接触电阻稳定、触头结构简单、自净作用较强；缺点是接触面积小、不宜通过较大电流、热稳定性差。因此，这种触头通常只用在工作电流和短路电流较小的情况下，如继电器和开关电器的辅助触点等。图 1-14（a）所示为点接触示意图。

2. 线触头

线触头是指两个触头的接触面为线状的触头，如柱面与平面接触，或两个圆柱面间的接触等都属于线接触，图 1-14（b）所示为线接触示意图。线触头的压力强度较大，在同样压力下，线触头比面接触触头的实际接触点要多。线触头在接通或断开时，触头间的运动形式是一个触头沿另一个触头的表面滑动。由于触头的压强很大，滑动时很容易把触头表面的金属氧化层破坏掉（这种效应也被称为自洁作用），从而可减小接触电阻，铜制线触头的接触电阻是平面触头的 1/2～1/3。线触头的接触面积比较稳定，广泛应用于高、低压开关电器中。

3. 面触头

面触头是指两个平面或两个曲面的接触触头，触头容量较大。在受到较大压力时，接触点数和实际接触面积仍比较小，所以，为保证触头的动稳定，减小接触电阻，就必须对触头施加更大的压力。图 1-14（c）所示为面接触示意图。

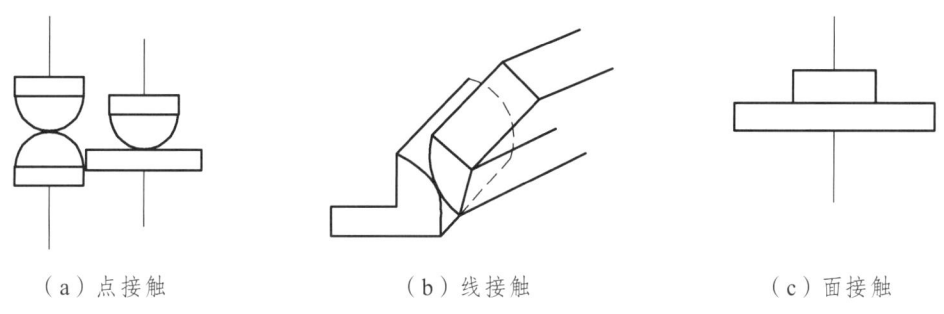

(a) 点接触　　　　　　　　(b) 线接触　　　　　　　　(c) 面接触

图 1-14　触点的三种接触形式

（二）按结构形式分类

常见触头的结构形式主要有三种：固定触头、可断触头、滑动触头，如图 1-15 所示。各种触头均需满足接触性能、动热稳定性、抗熔焊、耐电弧烧伤等各种要求，同时还要尽可能地便于安装、维修，降低造价。

1. 固定触头

固定触头是指连接导体之间不能相对移动的触头，如母线之间，母线与电器引出端头的连接等。如图 1-15（a）、（b）、（c）、（d）所示为常见的固定触头形式。

固定触头按其连接方式可分为可拆卸和不可拆卸两类。

（1）可拆卸的连接：采用螺栓连接方式，以方便安装和维修。

（2）不可拆卸的连接：采用铆接或压接方式，触头连接后便不可拆卸。压接时，使用专用的压接模具，由压接工具施压成形。

固定触头的接触表面应采取适当的防腐措施，以防止外界的侵蚀，保证接触可靠、耐用。防腐的方法一般是在触头连接后，在外面涂以绝缘漆、瓷釉或凡士林油等。

2. 可断触头

可断触头是在工作过程可以分开的触头，广泛应用于高低压开关电器中，按其结构可分为以下几种：

（1）对接式触头。如图 1-15（e）、（f）、（g）所示。这种触头优点是结构简单，分断速度快；缺点是接触面不够稳定，关合时易发生触头弹跳，由于触头间无相对运动，故基本上没有自净作用，触头容易被电弧烧伤、动热稳定性较差。因此，对接式触头只适用于 1000 A 以下的断路器中。

（2）插入式触头。如图 1-15（h）~（k）所示。其结构特点是所需接触压力较小，有自洁作用，无弹跳现象，触头磨损小，动热稳定性好。缺点是除了刀形触头外，结构复杂，分断时间长。

（3）刀形触头。如图 1-15（h）、（i）所示，其结构简单，广泛用于手动操作的高低压电器，如刀开关、隔离开关等。

（4）瓣形触头，又称插座式或梅花形触头，如图 1-15（j）所示，其静触头是由多瓣独立的触指组成一个圆环，如同插座状，动触头是圆形导电杆。接通时导电杆插入插座内，由强力弹簧或弹簧钢片把触指压向导电杆，静触指与动触头间形成线接触。插座式触头接触面工作可靠，接触电阻稳定，结构复杂，断开时间较长，广泛用于少油断路器中作为主触头和灭弧触头。为了使触头具有抗电弧烧伤能力，常在外套的端部加装铜钨合金保护环，在动触头的端部镶嵌铜钨合金制成的耐弧端。

（5）指形触头。如图 1-15（k）所示，它由成对的装在载流体两侧的接触指、楔形触头和夹紧弹簧组成。其优点是动稳定性好，有自洁作用；缺点是不易与灭弧室配合，工作表面易被电弧烧伤。用在少油断路器中作工作触头，在一些隔离开关中也有应用。

3. 滑动触头

滑动触头也叫中间触头，又称可动触头，是指在工作中被连接的导体总是保持接触，能由一个接触面沿着另一个接触面滑动的触头，其结构形式如图 1-15（l）~（n）所示。这种触头的作用是给移动的受电器供电，如电机的滑环碳刷、行车的滑线装置、断路器的滑动触头等。

（1）豆形触头。如图 1-15（l）所示，它的静触指分上、下两层，均匀分布在上、下触头座的圆周上，每一触指配有小弹簧作缓冲，以减少摩擦力和防止动触杆卡涩，动触杆从其中心孔通过。这种触头接触点多，在较小的接触压力下，具有良好的导电能力，而且结构紧凑。缺点是通用性差。

（2）"Z"形滑动触头。如图 1-15（m），"Z"形触头的结构与插座式触头相近。它是把"Z"形触指（静触头）装在导电座里面，用弹簧保持触指的位置，并将触指紧压在圆形导电座和动触杆上。这种触头结构简单、工作可靠，没有导电片，高度低，接触稳定而有自洁作用。

（3）滚动式滑动触头。如图 1-15（n）滚动式滑动触头是在工作中，导体由一个接触面沿着另一个接触面滑动的触头。它由圆形导电杆、成对的滚轮、固定导电杆以及弹簧等组成。弹簧的作用是保持滚轮和可动导电杆以及固定导电杆的接触压力。在接通和断开过程中，滚轮沿着导电杆上、下滚动。滚动式滑动触头接触面的摩擦力小，自洁作用较差。

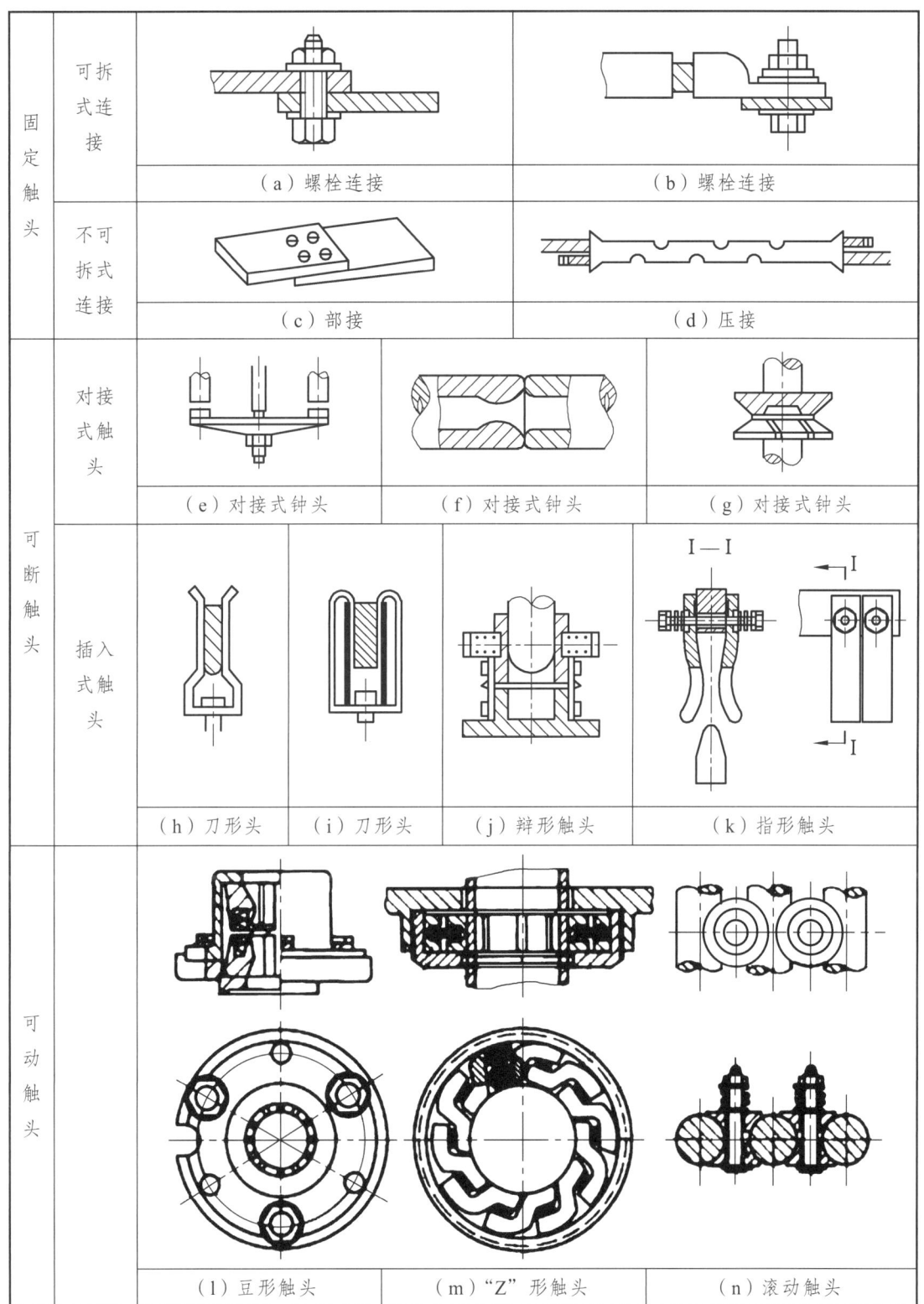

图 1-15 触头的结构与分类

Module 3 Basic Knowledge of Electrical Contact Terminal

1.3.1 Contact Resistance between Contact Terminals

The main factors that influence the contact resistance are analyzed as follows.

1. Pressure between contact terminals

Even the surface of contact terminal made by fine processing is rough and uneven from a micro perspective. The contact area between contact terminals is influenced by the pressure applied, as shown in Fig. 1-13. If no external force is applied, put two contact terminals together, they contact only at point a, as shown in Fig. 1-13(a). When an external force F_1 is applied, point a is flattened and forms a contact surface. If an external force F_2 which is larger than F_1 is applied, the contact area a will increase. At the same time, they will contact at point b and point b will form a new contact surface. As a result, the total contact area increases, while the contact resistance becomes smaller. Therefore, pressure is an important factor that affects contact resistance.

For switching devices, a rigid spring is usually attached to the contact terminal, with the purpose of increasing and keeping the contact pressure between contact terminals, ensuring reliable contact between contact terminals, reducing the contact resistance and keeping it stable.

2. Materials and anti-oxidation measures for contact terminal

Copper leading-out terminals are used for some electrical equipment such as transformer and motor. In outdoor environment and damp place, bolts cannot be used to connect the aluminium conductor to the copper terminal. This is because when copper and aluminium contact each other directly, a potential difference (about 1.86 V) will form. When the water containing dissolved salt penetrates into the slit on the contact surface, electrolytic reaction will occur. As a result, aluminium is subject to strong electrical corrosion, causing damage to the contact terminal and even major accident. To avoid this situation, a copper-aluminium transition joint is usually used, with one end made of aluminium and the other end made of copper, as shown in Fig. 1-13. For electrical equipment with a contact terminal which works depending on rotary frictional contact, a thin layer of vaseline shall be applied to the drum after maintenance, in order to prevent oxidation and corrosion. Frequent application of vaseline is not required after normal operation.

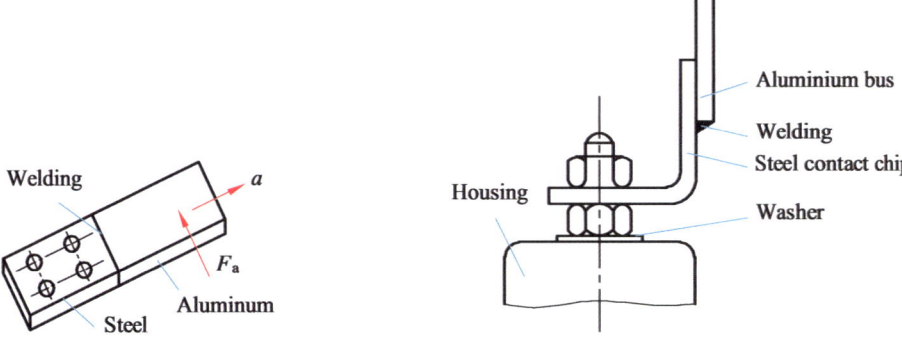

(a) Aluminum contact terminnal

(b) Connector used to connect the aluminium bus to the copper terminal of electrical appliance

Fig. 1-13　Materials for Contact Terminal

1.3.2　Classification and Structure of Contact Terminal

1.3.2.1　Classification according to the form of contact surface

1. Point contact terminal

For point contact terminal, the contact surface between two contact terminals is a point-like one. Point contact can be found between a spherical surface and a plane and between two spherical surfaces. The advantages of such form of contact include high pressure, fixed contact point, stable contact resistance, simple structure of contact terminal, and strong effect of self-purification. There are also some shortcomings such as small contact surface, small current, and poor thermal stability. Therefore, such contact terminal is only used when the working current and short-circuit current are small. For example, it is used as an auxiliary contact for relays and switching devices. Fig. 1-14 (a) is the schematic diagram of point contact.

2. Linear contact terminal

For linear contact terminal, the contact surface between two contact terminals is linear. The contact between a cylindrical surface and a plane and the contact between two cylindrical surfaces are linear contact. Fig. 1-14 (b) is the schematic diagram of linear contact. Linear contact terminal has a high intensity of pressure. Under the same pressure, a linear contact terminal has more contact points than a surface contact terminal. When a linear contact terminal is connected or disconnected, the mode of motion between contact terminals is that one contact terminal slides on the surface of another contact terminal. Due to the high intensity of pressure between contact terminals, when the contact terminal is sliding, the metal oxide layer on the surface of contact terminal can be easily damaged (such effect is also called self-cleaning effect), so as to reduce the contact resistance. The contact resistance between copper linear contact terminals is 1/2–1/3 of that between plane contact terminals. Depending on the stable contact surface, linear contact terminal is widely used in high-voltage and low-voltage switching devices.

3. Plane contact terminal

Plane contact terminal refers to the contact terminal between two planes or two curved surfaces. Such contact terminal has a large capacity. Under a large pressure, the number of contacts and the actual contact area are quite small. Therefore, to ensure dynamic stability of the contact terminal and reduce the contact resistance, a larger pressure should be applied to the contact terminal. Fig. 1-14 (c) is the schematic diagram of surface contact.

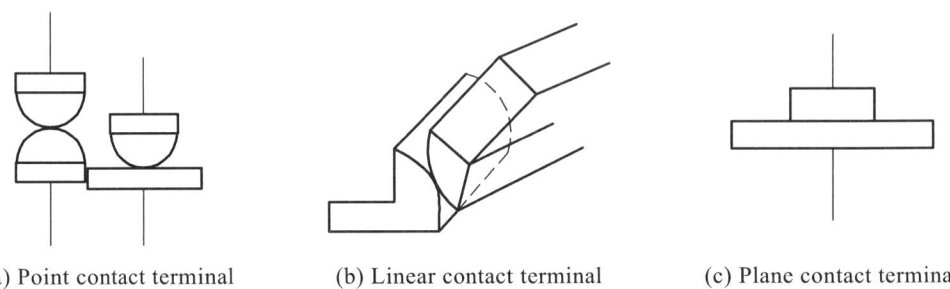

(a) Point contact terminal　　　(b) Linear contact terminal　　　(c) Plane contact terminal

Fig. 1-14　Three Forms of Contact between Contacts

1.3.2.2　Classification according to the structural form

According to the structural form, there are mainly three types of common contact terminals: fixed contact terminal, break-away contact terminal, and sliding contact terminal, as shown in Fig. 1-15. All types of contact terminals shall meet the requirements for contact performance, thermal dynamic stability, fusion welding resistance, and arc burn resistance. Meanwhile, they shall facilitate installation and maintenance and reduce the cost.

1. Fixed contact terminal

Fixed contact terminal refers to the contact terminal which cannot move relatively between conductors. They include connectors between buses and between a bus and a leading-out terminal of electrical appliance. The common types of fixed contact terminals are shown in Fig. 1-15 (a), (b), (c), and (d).

Fixed contact terminals can be divided into detachable and non-detachable contact terminals according to the type of connection.

(1) Detachable connection. Bolts are used for connection, in order to facilitate installation and maintenance.

(2) Non-detachable connection. Riveting or crimping is used for connection, and the contact terminal is non-detachable after connection. A special crimping die is used for crimping and a crimping tool is used to apply pressure.

Appropriate anti-corrosion measures shall be taken for the contact surface between fixed contact terminals, so as to prevent external corrosion and ensure reliable and durable contact. In general, insulating paint, porcelain glaze or vaseline shall be applied to the external surface of contact terminal after connection, so as to achieve the goal of corrosion prevention.

2. Break-away contact terminal

Break-away contact terminal can be separated during operation. It is widely used in low-voltage switching devices. It is divided into the following types according to the structural form:

(1) Butt-joint contact terminal. They are shown in Fig. 1-15 (e), (f), and (g). The advantages of such contact terminals are simple structure and high breaking speed. They also have some disadvantages. For example, the contact surface is not stable enough and the contact terminal may bounce during closing. Besides, since there is no relative motion between contact terminals, they can hardly provide any self-cleaning effect. As a result, the contact terminal may be burnt by arc and the thermal dynamic stability is poor. Therefore, butt-joint contact terminal is only suitable for circuit breakers with a current of below 1,000 A.

(2) Plug-in contact terminal. They are shown in Fig. 1-15 (h) through (k). They are characterized by small contact pressure, self-cleaning effect, no bounce, small wear between contact terminals, and good thermal dynamic stability. The shortcomings include complex structure and long breaking time (except knife contact terminal).

(3) Knife contact terminal. As shown in Fig. 1-15 (h) and (i), with a simple structure, they are widely used in manually operated high-voltage and low-voltage electrical appliances such as knife switch and disconnector.

(4) Disc-type contact terminal is also called socket-type or quincunx contact terminal, as shown in Fig. 1-15 (j). The static contact terminal is a ring composed of multi-disc independent contact fingers, like a socket. The moving contact terminal is a circular conductive rod. Once connected, the conductive rod is inserted into the socket. A strong spring or steel spring presses contact fingers to the conductive rod. Linear contact forms between static contact fingers and the moving contact terminal. Socket-type contact terminal is characterized by reliable contact surface, stable contact resistance, complex structure, and long breaking time. It is used as the main contact terminal and arc extinguishing contact terminal in the low oil circuit breaker. To enable the contact terminal to prevent arc burn, the end of the shield is usually provided with a guard ring made of copper-tungsten alloy, and the end of the moving contact terminal is inlaid with an arc-proof terminal made of copper-tungsten alloy.

(5) Finger-type contact terminal. As shown in Fig. 1-15 (k), it is composed of a pair of contact finger on both sides of current carrier, wedge-shaped contact terminal and clamping spring. It has advantages of good dynamic stability and self-cleaning effect. The disadvantages are that it cannot cooperate with the arc extinguishing chamber easily and the working surface is susceptible to arc burn. It is used as a working contact terminal in low oil circuit breakers. It is also used in some disconnectors.

3. Sliding contact terminal

Sliding contact terminal is also called middle contact terminal or movable contact terminal. The conductors connected can always keep contact during operation, and the contact terminal can move from one contact surface to another. The structural form is shown in Fig. 2-15 (l) through (n). Such contact terminal is used to supply power for moving current collectors such as the slip ring and carbon brush of motor, the slide wire device of bridge crane, and the sliding contact terminal of circuit breaker.

(1) Bean-shaped contact terminal. As shown in Fig. 1-15 (l), the static contact finger includes upper layer and lower layer, which are uniformly distributed along the circumference of upper and lower contact terminal seat. Each contact finger is provided with a small spring as a buffer, so as to reduce friction and prevent jamming of the moving contact rod. The moving contact rod passes through the center hole of the contact terminal. Such contact terminal has many contacts. Under a small contact pressure, it has good electrical conductivity and compact structure. The shortcoming is poor universality.

(2) Z-shaped sliding contact terminal. As shown in Fig. 1-15 (m), Z-shaped contact terminal has similar structure as socket-type contact terminal. Z-shaped contact finger (static contact terminal) is installed in the conductive seat. Spring is used to maintain the position of the contact finger and press the contact finger on circular conductive seat and moving contact rod. Such contact terminal features simple structure, reliable operation, no conducting strip, low height, stable contact, and self-cleaning effect.

(3) Rolling-type sliding contact terminal. As shown in Fig. 1-15(n), when the rolling-type sliding contact terminal is working, the conductor slides from one contact surface to another. It is composed of circular conductive rod, a pair of roller, fixed conductive rod, and spring. The spring is used to maintain the contact pressure between the roller and the moving conductive rod and the fixed conductive rod. During closing and breaking process, the roller rolls up and down along the conductive rod. The rolling-type sliding contact terminal has a small friction on the contact surface and a poor self-cleaning effect.

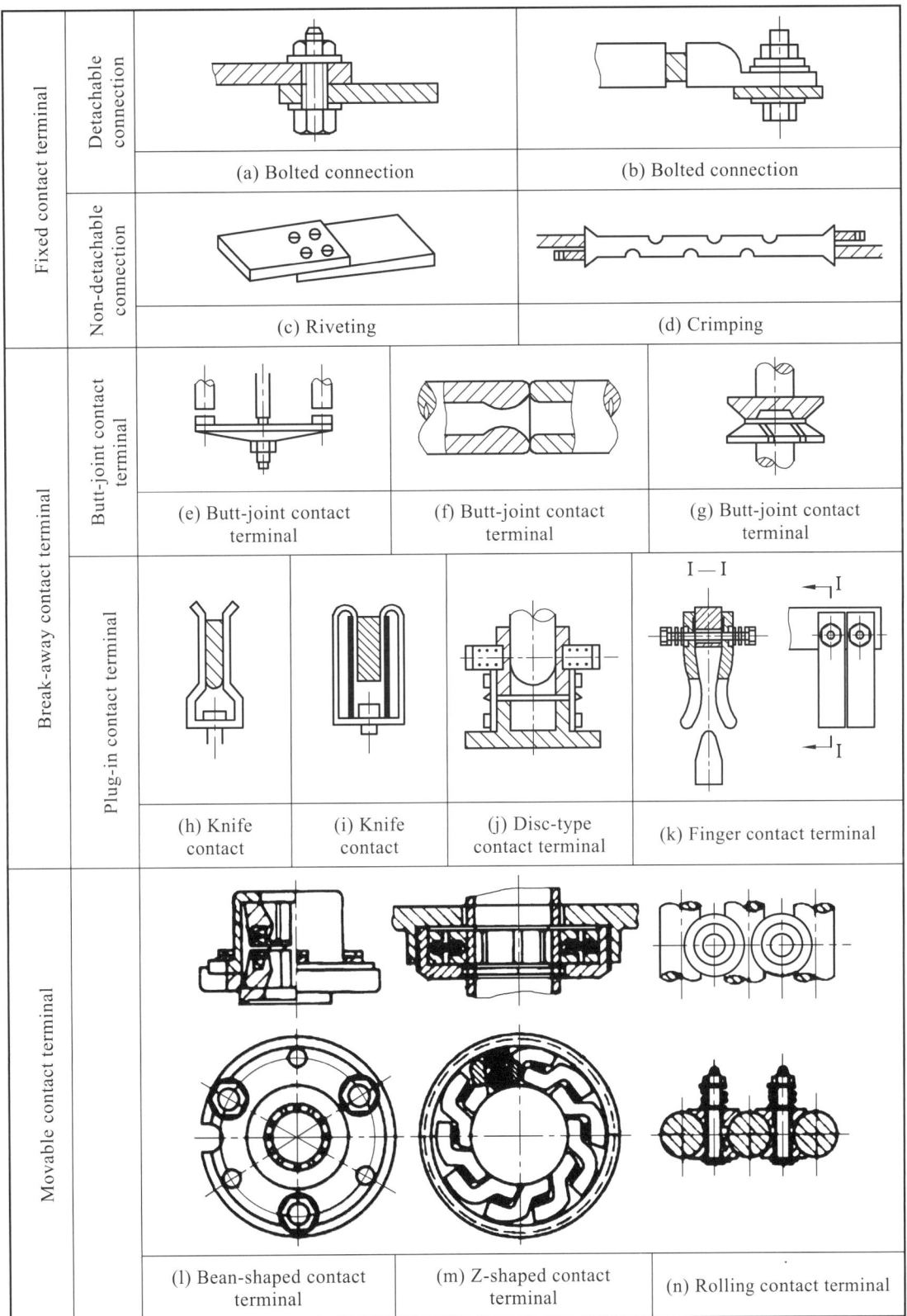

Fig. 1-15 Structure and Classification of Contact Terminals

任务一 认识触头

一、工作任务

按照规范的作业流程以及工艺要求,在实训大厅寻找、收集各类触头,熟悉各类触头的基本结构、动作原理。

二、引用标准

(1)《国家电网公司变电运维管理规定》。
(2)《国家电网公司变电检修管理规定》。
(3)《国家电网公司电力安全工作规程》(变电部分)。

三、工作要求

(1)作业人员精神状态良好,接受工作负责人交底,并签字后开展工作。
(2)作业人员应规范穿戴工作服和安全帽,做好安全防护。
(3)作业为室内备用设备作业,不带电,无天气要求。

四、工作准备

(一)危险点及预控措施(见表1-1)

表1-1 危险点及预控措施

序号	防范类型	危险点	预防控制措施
1	误碰、误动、误登运行设备与高压带电设备距离不够	意外伤人	观察设备时应与带电设备保持足够的安全距离。110 kV不小于 1.5 m,35 kV不小于 1 m,10 kV不小于 0.7 m
2	擅自打开设备网门,擅自移动临时安全围栏,擅自跨越设备固定围栏,误入带电间隔	观察设备低压触电	观察时不得直接触碰裸露设备,作业过程中禁止野蛮操作
3	误碰误触机械传动机构,造成机械伤害	观察人员误碰、误登带电设备	观察设备禁止变更检修现场安全措施,禁止改变检修设备状态,身体禁止进入储能装置内
4	进入高压室为随手关门,造成小动物进入	小动物进入导致短路	进入高压室,必须随手将门锁好
5	不戴安全帽,不按规定着装,在突发事件时失去保护	高空落物伤人	进入设备区必须戴安全帽
6	现场安全措施不规范,如警告标志不齐全,孔洞封闭不良带电,设备隔离不符合要求原因,易造成人身伤害	工作所需各类生产安全工器具资料等,无法满足现场工作需求,导致人员触电	使用前进行全面检查,不合格的工器具禁止带入工作现场

(二)工器具及材料

工器具及材料见表1-2。

表1-2 工器具及材料表

类别	名称	规格型号	数量	备注
通用工具	内六角扳手		1套	
	平口钳		1把	
	尖嘴钳		1把	
	一字螺丝刀	150 mm	1把	
	十字螺丝刀	150 mm	1把	
	梅花扳手	整套	一套	
	呆扳手	整套	一套	
	活动扳手	150 mm、250 mm	各一	

(三)作业人员分工

人员分工见表1-3。

表1-3 人员分工表

序号	工作岗位	数量	职责
1	工作负责人	1	负责本次工作的人员分工、现场查勘、作业方案制定、召开班前会、作业过程中安全监督、工作中突发状况的处理、工作质量的监督、班后会总结
2	操作人员	1	负责拆除触头的主要操作
3	辅助操作人员(可无)	1	辅助操作人员进行触头拆除

五、作业程序

作业流程见表1-4。

表1-4 作业流程表

序号	作业内容	作业步骤及标准	安全措施及注意事项
1	工作前准备工作	1. 检查工器具是否齐全,检查工器具外观和试验合格; 2. 工作负责人同工作许可人巡视待检修设备,确认工作票所列安全措施已经正确执行,安全措施是否完备,现场是否具备开工条件,必要时进行补充; 3. 执行工作许可手续; 4. 对工作班成员召开班前会; 5. 抄写设备铭牌参数	1. 工器具无损伤、变形、失灵现象,需要试验的工器具合格证在有效期内; 2. 巡视现场时禁止无关人员进入现场; 3. 班前会应包含工作地点双重名称、工作时间与工作任务、人员分工、工作危险点及预控措施、停电范围及工作现场安全措施; 4. 全体工作成员应当正确穿戴安全帽、工作服、工作鞋、劳保手套等劳动保护用品

续表

序号	作业内容	作业步骤及标准	安全措施及注意事项
2	确认设备状态	负责人带领工作班成员确认设备处于断电状态	
3	寻找各类触头	收集各类触头，观察、对比相同和不同点，学习触头结构	
4	现场恢复至初始状态	1. 关闭所有柜门； 2. 断开控制回路、合闸回路电源； 3. 将"禁止合闸，有人工作"标识牌挂回控制回路、合闸回路处	
5	工作终结	1. 清理作业现场，做到"工完料尽场地清"； 2. 召开班后会； 3. 终结工作票	严禁负责人前去终结工作票的同时清理场地

Task 1   Understand Contact Terminals

1.1   Work Tasks

According to the specified operation procedures and technological requirements, look for and collect various contact terminals in the training hall, and get familiar with the basic structure and operating principle of various contact terminals.

1.2   References

(1) *Regulations of State Grid Corporation of China on Management of Substation Operation and Maintenance.*

(2) *Substation Maintenance Management Regulations of State Grid Corporation of China.*

(3) *Electric Power Safety Working Regulations (Power Transformation) of State Grid Corporation.*

1.3   Work Requirements

(1) The operators shall keep a good mental state, receive the disclosure made by the person in charge of work, and carry out work after signing.

(2) The operators shall wear working clothes and safety helmet and take safety protection measures according to the specification.

(3) The operation will be carried out on indoor de-energized standby equipment and there is no requirement for weather.

1.4   Preparation for Work

1. Hazards and preventive and control measures (see Table 1-1)

Table 1-1   Hazards and Preventive and Control Measures

S/N	Type of precautions	Hazard	Prevention and control measures
1	Accidental touch with, maloperation of or climbing on the running equipment, or insufficient distance from high-voltage live equipment	Accidental personal injury	Keep enough safe distance from live equipment when observing the equipment. The distance shall be no less than 1.50 m for 110 kV equipment, no less than 1.00 m for 35 kV equipment, and no less than 0.70 m for 10 kV equipment
2	Open the equipment door, move temporary safety fence, or stride over the fixed equipment fence without authorization, or enter the energized partition accidentally	Low-voltage electric shock when the operator is observing the equipment	Do not touch exposed equipment directly during observation, and avoid rough handling during operation
3	Accidental touch with mechanical transmission mechanism, causing mechanical injury	The observer touches or climbs on live equipment accidentally	Do not change safety measures on the maintenance site and the maintenance state of equipment when the observer is observing the equipment, and do not enter the energy storage device

Continued

S/N	Type of precautions	Hazard	Prevention and control measures
4	Fail to close the door after entering and going out from the high-voltage room, causing small animals to enter the room	Intrusion of small animals into the high-voltage room, causing short circuit	Close the door after entering and going out from the high-voltage room
5	Fail to wear safety helmet and working clothes as specified, so that protective measures cannot be provided in case of emergency	Personal injury caused by falling objects	The operator shall wear safety helmet when entering the equipment area
6	The safety measures on site do not meet the requirements. For example, warning signs are not complete, holes are poorly closed, and the isolation of live equipment does not meet the requirements, which may easily cause personal injury	The production and safety tools and instruments and data required for work cannot meet the job requirements on site, causing electric shock	Conduct a comprehensive inspection before use and do not bring non-conforming tools and instruments to the work site

2. Tools, instruments and materials

Tools, instruments and materials are shown in Table 1-2.

Table 1-2 Tools, Instruments and Materials

Category	Name	Specification and model	Quantity	Remarks
General tools	Allen wrench		1 set	
	Flat-nose pliers		1	
	Long-nose pliers		1	
	Slotted screwdriver	150 mm	1	
	Cross screwdriver	150 mm	1	
	Spline end wrench	A complete set	One set	
	Open spanner	A complete set	One set	
	Monkey wrench	150 mm、250 mm	One for each	

3. Division of labor among operators

The division of labor is shown in Table 1-3.

Table 1-3 Division of Labor

S/N	Job	Quantity	Responsibilities
1	Person in charge of work	1	Be responsible for work division of working staffs, site survey, stipulation of operation scheme, pre-shift meeting, safety supervision during operation, handling of emergencies during work, supervision of work quality and summary of post-shift meeting
2	Operator	1	They shall be responsible for the main operation of contact terminal removal and shall get familiar with the structure
3	Auxiliary operators (optional)	1	Auxiliary operators shall remove contact terminals

1.5 Operation Procedures

The operation procedures are shown in Table 1-4.

Table 1-4 Operation Procedures

S/N	Scope of work	Operational steps and standards	Safety measures and precautions
1	Preparations before work	1. Check if all instruments and tools are complete, and if their appearance and test are acceptable; 2. The person in charge of work and work permitter shall make an tour inspection for equipments under maintenance, and confirm all safety measures as listed by work ticket have been properly implemented. It is also necessary to check if safety measures are complete, and if the site is provided with conditions for commencement of work, and make supplements as required; 3. Implement work permit procedures; 4. Call in toolbox meeting participated by members of work team; 5. Record parameters on the equipment nameplate	1. Instruments and tools are free of damage, deformation and malfunction. Qualification certificates for instruments and tools to be tested are within the term of validity; 2. Prevent other persons from entering the site during tour inspection; 3. The content of pre-shift meeting shall include double names of workplace, working time and work task, division of labor, hazards and preventive and control measures, scope of power failure, and safety measures on the work site; 4. All work members shall properly wear such labor protection appliances as safety helmet, working clothes, working shoes and protective gloves
2	Confirm equipment status	The person in charge shall lead members of the shift team to confirm that the equipment has been powered off	
3	Seek various contact terminals	Collect various contact terminals, observe and compare their similarities and differences, and study the structure of contact terminals	
4	Restore the site to its initial state	1. Close all the cabinet doors; 2. Cut off the power supply for control circuit and closing circuit; 3. Hang the sign board "No Closing, Work in Progress" back to the control circuit and closing circuit	
5	End of work	1. Clean up the work site, and make sure that elements and materials are fully utilized and the site is cleaned; 2. Convene a post-shift meeting; 3. Terminate the work ticket	It is strictly forbidden for the person in charge to clean up the site while terminating the work ticket

项目二　高压隔离开关检修

模块一　隔离开关概述

在电力系统中,为了安全生产运行,需要将带电运行的电气设备停电检修或转为备用,设备与电源之间、设备与设备之间必须有明显可见且足够大的断开点。隔离开关正是在电路中设置的这种断开点,以确保运行和检修的安全。

隔离开关主要用来隔离电路。在分闸状态下有明显可见的断口,在关合状态下,导电系统中可以通过正常的工作电流和故障下的短路电流。隔离开关没有灭弧装置,除了能开断很小的电流外,不能用来开断负荷电流,更不能开断短路电流,但隔离开关必须具备一定的动、热稳定性。

一、隔离开关专业术语

1. 隔离开关

在分闸位置时,触头间符合规定要求的绝缘距离和明显的断开标志;在合闸位置时,能承载正常回路条件下的电流及规定时间内异常条件(例如短路)下的电流的开关设备。

2. 接地开关

释放被检修设备和回路的静电荷以及为保证停电检修时检修人员人身安全的一种机械接地装置。它可以在异常情况下(例如短路)耐受一定时间的电流,但在正常情况下不通过负载电流,它通常是隔离开关的一部分。

3. 断口距离

隔离开关的主隔离开关在正常分闸位置时,同相两极触头之间的最短距离,对多断口隔离开关而言,最短距离是指全部断口最短绝缘距离之和。

4. 合闸不同期性

合闸不同期性是指两相或多相隔离开关的主隔离开关不同时接触时的差异,通常以距离表示。

二、隔离开关的作用和基本要求

（一）隔离开关的作用

（1）隔离电源。电气设备检修时，在开断电流以后，用隔离开关将需要检修的电气设备与带电的电源隔离，形成明显可见的断开点，以保证检修人员和设备的安全。

（2）倒换线路或母线。利用等电位间没有电流通过的原理，用隔离开关将电气设备或线路从一组母线切换到另一组母线上，改变运行方式。

（3）关合与开断小电流电路。可以用隔离开关关合和开断正常工作的电压互感器、避雷器电路；关合和开断母线和直接与母线相连接的电容电流；关合和开断电容电流不超过 5 A 的空载电力线路；关合和开断激磁电流不超过 2 A 的空载变压器等。

（二）对隔离开关的基本要求

按照隔离开关在电网中担负的任务及使用条件，其基本要求如下。

（1）有明显的断开点。

在隔离开关分开状态下，应具有明显可见的断开点，以便清楚地确定被检修的设备或线路是否已与电网隔离，从而能更好地保证检修工作人员的安全。

（2）有可靠的绝缘。

隔离开关断开点之间应有可靠的绝缘，以保证在恶劣的气候条件下也能可靠工作，并在过电压及相间闪络的情况下，不致从断开点击穿而危及人身安全。

（3）有一定破冰能力。

隔离开关的触头敞露在大气中，因此对户外式隔离开关，要求在分开时能破碎覆盖在触头上的一定厚度的冰层。对于用在气候寒冷地区的户外型隔离开关应具有设计要求的破冰能力，在冰冻的环境里应能可靠地分合闸。

（4）有足够的热稳定性和动稳定性。

隔离开关应具有足够的热稳定性和动稳定性，尤其不能因电动力的作用而自动断开，否则将引起严重事故。

（5）隔离开关的结构应简单，动作要可靠，有一定的机械强度；金属制件应能耐受氧化而不腐蚀；户外隔离开关在恶劣的气候条件下必须能可靠地分、合闸。

（6）带有接地开关的隔离开关必须有可靠的机械闭锁，以保证先断开隔离开关后，才能合接地开关；先拉开接地开关后，才能合隔离开关的操作顺序。

（7）有可靠的联锁装置。

在隔离开关本身或其操动机构上应有锁扣装置，以防其在通过短路电流时由于电动力作用而自动分开。通过辅助触点，隔离开关与断路器之间应有电气闭锁，保证正确的操作顺序，以防带负荷误拉、合隔离开关。

（8）对于一般户外非 GIS 或 HGIS 用隔离开关，其外绝缘爬电距离应满足安装地点的污秽等级要求，同时应根据设计要求留有一定裕度。

三、隔离开关的基本结构和技术参数

(一)隔离开关的基本结构

隔离开关的结构示意图如图 2-1 所示,主要由以下几个部分组成。

(1)导电部分:主要起传导电路中的电流,关合和开断电路的作用,包括触头、闸刀、接线座。在合闸状态能可靠的通过正常工作电流和短路电流,在分闸状态有明显的间隙,并具有可靠的绝缘。

(2)绝缘部分:主要起绝缘作用,实现带电部分和接地部分的绝缘。包括支持绝缘子和操作绝缘子。

(3)传动机构:它的作用是接受操动机构的力矩,并通过拐臂、联杆、轴齿或是操作绝缘子,将运动传动给触头,以完成隔离开关的分、合闸动作。

(4)操动机构:与断路器操动机构一样,通过手动、电动、气动、液压向隔离开关的动作提供能源。

(5)支持底座:该部分的作用是起支持和固定作用,其将导电部分、绝缘子、传动机构、操动机构等固定为一体,并使其固定在基础上。

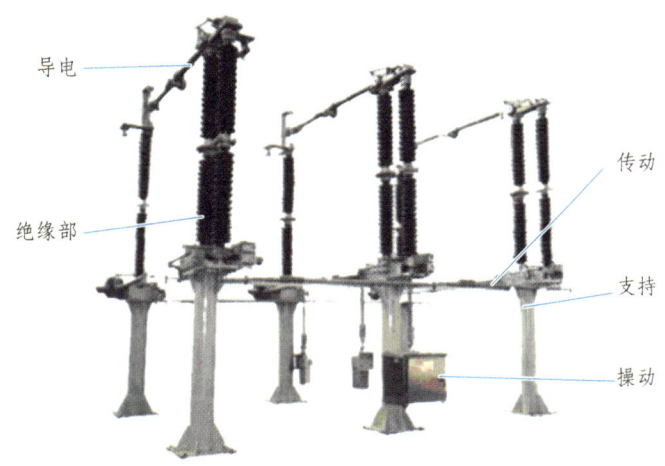

图 2-1 隔离开关结构示意图

(二)隔离开关的技术参数

(1)额定电压(kV):指隔离开关长期运行时承受的工作电压,是线电压。

(2)最高工作电压(kV):隔离开关所能承受的超过额定电压的电压。它不仅决定了隔离开关的绝缘要求,而且在相当程度上决定了隔离开关的外部尺寸。

(3)额定电流(A):指隔离开关可以长期通过的工作电流,即长期通过该电流,隔离开关各部分的发热不超过允许值。

(4)热稳定电流(kA):指隔离开关在某一规定的时间内,允许通过的最大电流。它表明了隔离开关承受短路电流热稳定的能力。

(5)极限通过电流峰值(kA):指隔离开关所能承受的瞬时冲击短路电流,这个值与隔离开关各部分的机械强度有关。

(6)回路接触电阻(μΩ):指隔离开关导电回路在各种电接触形式下的导电性能,是检验及设计、制造工艺装配的技术能力。

四、隔离开关的分类、型号和使用条件

(一)隔离开关的分类

隔离开关种类很多,可根据装设地点、电压等级、极数和构造进行分类。隔离开关分类见表 2-1。

表 2-1　隔离开关分类表

序号	分类方式	类别	
1	按安装场所	户内式,户外式	
2	按极数	单极、三极	
3	按有无接地开关	不接地,单接地,双接地	
4	按用途	(1)一般输配电用　　　　(2)快速分闸用 (3)变压器中性点接地用　(4)大电流母线用	
5	按动作方式	闸刀式、旋转式、插入式	
6	按结构形式	按照支柱绝缘子数量划分	(1)单柱式隔离开关 (2)双柱式隔离开关 (3)双柱式隔离开关
		按照支柱绝缘子的数量和导电活动臂开启方式划分	(1)单柱垂直缩式 (2)双柱水平旋转式 (3)双柱水平伸缩式 (4)三柱水平旋转式
7	按所配操动机构	手动式、电动式、气动式和液压式	
8	按使用环境	普通型和防污型	

(二)隔离开关的型号

高压隔离开关的型号主要由以下六个单元组成:

| 1 | 2 | 3 | - | 4 | 5 | / | 6 |

1:名称;G-隔离开关。

2:安装地点;N-户内型,W-户外型。

3:设计序号;由行业管理部门根据鉴定及申领型号的先后顺序确定,用阿拉伯数字"1、2、3"等表示。

4:额定电压(kV)。

5:补充特性:C-瓷套管出线,D-带接地刀闸;K-快分型,G-改进型,T-统一设计。

6:额定电流(A)。

例如:型号 GW16-252D/3150 中,G 表示隔离开关,W 表示户外,16 是设计序号,额定

电压是 252 kV，D 是表示有接地开关，额定电流是 3150 A。

目前，国内使用比较多的有 GN19、GW4、GW5、GW6、GW7、GW16、GW17 等系列隔离开关。

（三）隔离开关的使用条件

隔离开关的额定使用条件如下：

（1）海拔高度不大于 1000 m 为普通型，海拔高度大于 1000 m 为高原型。

（2）地震裂度不超过 8 度。

（3）环境温度不高于 +40 ℃，户内产品环境温度不低于 -10 ℃，户外产品温度不低于 -30 ℃。

（4）户内产品空气相对湿度在 +25 ℃ 时，其日平均值不大于 95%，月平均值不大于 90%（有些产品要求空气相对湿度不大于 85%）。

（5）户外产品的覆冰厚度分为 5 mm 和 10 mm。

（6）户内产品周围空气不受腐蚀性或可燃气体、水蒸气等显著污秽的污染，无经常性的剧烈振动。户外产品的使用环境为普通型用于Ⅰ级污秽区、防污型用于Ⅱ级（中污型）、Ⅲ级（重污型）污秽区。

Program 2 Maintenance of High-voltage Disconnector

Module 1 Disconnector Overview

In the power system, for work safety, interruption maintenance or switch of live electrical equipment to standby is required, and there must be clearly visible and large enough disconnection points between the equipment and the power supply and between the equipment. The disconnector is just this disconnection point set in the circuit to ensure the safety of operation and maintenance.

The disconnector is mainly used to isolate circuits. There is an obvious break in the open status, and in the closed status, normal working current and short-circuit current under fault can pass through the conductive system. The disconnector has no arc extinguishing device, which cannot be used to disconnect the load current, let alone the short-circuit current, except for disconnecting very small current. However, the disconnector must have certain dynamic and thermal stability.

2.1.1 Terminology of Disconnector

1. Disconnector

In the open position, the insulation distance between contact terminals meets the specified requirements and there is an obvious disconnection sign. In the closed position, it can bear the current under normal circuit conditions and the current under abnormal conditions (such as short circuit) within a specified time.

2. Grounding switch

Grounding switch is a mechanical grounding device which releases the static charge of the equipment and circuit under maintenance and ensures the personal safety of the maintainer during interruption maintenance. It can withstand the current for a certain time under abnormal conditions (such as short circuit), but it does not pass through the load current under normal conditions. It is usually a part of the disconnector.

3. Break distance

When the main disconnector of the disconnector is in the normal open position, the break distance is the shortest distance between the same-phase two-pole contact terminals. For the multi-fracture disconnector, the shortest distance refers to the sum of the shortest insulation distances of all the breaks.

4. Closing non-synchronism

Closing non-synchronism refers to the difference when the main disconnectors of two-phase or multi-phase disconnectors do not contact at the same time, usually expressed by distance.

2.1.2 Function and Basic Requirements of Disconnectors

1. Function of disconnectors

(1) Isolate the power supply. During the electrical equipment maintenance, the electrical equipment to be maintained is isolated from the live power supply by an disconnector after breaking current, forming visible disconnected points, so as to ensure the safety of maintainers and equipment.

(2) Switch the line or bus. Based on the principle of no current between equipotentials, electrical equipment or lines are switched by a disconnector from one set of buses to another set, changing the operation mode.

(3) Close and break small current circuits. The circuits of voltage transformer and lightning arrester under normal operation can be closed and disconnected by a disconnector. Besides, the disconnector can be used to close and disconnect the bus and the capacitor current directly connected to the bus; close and disconnect the no-load power line with the capacitor current no more than 5 A; close and disconnect the no-load transformer with excitation current no more than 2 A.

2. Basic requirements for disconnectors

According to the tasks and service conditions of the disconnector in power grid, its basic requirements are as below.

(1) There are obvious disconnection points.

The disconnector should have an obvious disconnection point under the open status, so as to clearly determine whether the equipment or line under maintenance has been isolated from the power grid, thus better ensuring the safety of the maintainer.

(2) There is reliable insulation.

There should be reliable insulation between the disconnection points of the disconnector to ensure reliable operation under harsh weather conditions, and in the case of overvoltage and interphase flashover, it is not possible to endanger personal safety due to breakdown from the disconnection point.

(3) There is a certain ability to break ice.

The contact terminal of the disconnector is exposed in the atmosphere, so it is required that the outdoor disconnector can break a certain thickness of ice covering the contact terminal when it is separated. For the outdoor disconnector used in cold climate areas, it should have the ice-breaking ability required in the design, and it should be able to close and disconnect reliably in the frozen environment.

(4) There is sufficient thermal and dynamic stability.

The disconnector should have sufficient thermal and dynamic stability. Especially, it cannot be automatically disconnected due to the action of electrodynamic force, or else it will cause serious accidents.

(5) The disconnector should have a simple structure, and the action should be reliable, with certain mechanical strength. Metal parts should be able to withstand oxidation without corrosion. The outdoor disconnector must be able to open and close reliably under bad weather conditions.

(6) The disconnector with grounding switch must have reliable mechanical locking to ensure that the following operation sequence: the grounding switch can be closed after the disconnector is disconnected; the disconnector can be closed after the grounding switch is opened.

(7) There is a reliable interlocking device.

There should be a locker on the disconnector itself or its operating mechanism to prevent it from automatically separating due to electrodynamic action when passing through short-circuit current. Through the auxiliary contact terminal, there should be electrical locking between the disconnector and the circuit breaker to ensure the correct operation sequence and prevent the disconnector from being opened and closed with load.

(8) For general outdoor non-GIS or HGIS disconnectors, the external insulation creepage distance should meet the pollution level requirements of the installation site, and a certain margin should be left according to the design requirements.

2.1.3　Basic Structure and Technical Parameters of Disconnectors

1. Basic structure of disconnectors

The schematic diagram of structure of the disconnector is shown in Fig. 2-1, and it is mainly composed of the following parts.

(1) Conduction part. It is mainly used to conduct the current in the circuit, close and disconnect the circuits, including contact terminal, knife switch and wire holder. Normal working current and short-circuit current can reliably pass through it in the closed status, and there is an obvious gap in the open status, with reliable insulation.

(2) Insulation part. It mainly plays a role of insulation and insulates the live parts and grounding parts, including supporting insulator and operating insulator.

(3) Transmission mechanism. It receives the torque of the operating mechanism and transmits the motion to the contact terminal through the crank arm, link rod, gear shaft or operating insulator to complete the opening and closing of the disconnector.

(4) Operating mechanism. Like the operating mechanism of the circuit breaker, it supplies the energy to the actions of the disconnector manually, electrically, pneumatically and hydraulically.

(5) Support base. This part mainly plays a role of supporting and fixing. It integrates and fixes the conduction part, insulator, transmission mechanism and operating mechanism on the foundation.

Fig. 2-1　Schematic Diagram of Structure of the Disconnector

2. Technical parameters of disconnectors

(1) Rated voltage (kV). It refers to the working voltage of the disconnector when it runs for a long time, which is the line voltage.

(2) Maximum operating voltage (kV). Maximum working voltage (kV). It is the voltage that the disconnector can bear and exceeds the rated voltage. It not only determines the insulation requirements of the disconnector, but also determines the external dimensions of the disconnector to a large extent.

(3) Rated current (A). It refers to the long-term working current of the disconnector, that is, the heating of each part of the disconnector does not exceed the allowable value when passing through the current for a long time.

(4) Thermal stable current (kA). It refers to the maximum allowable current of the disconnector within a specified time. It indicates the thermal stability of the disconnector under short-circuit current.

(5) Peak let-through current limit (kA). It refers to the instantaneous impact short-circuit current that the disconnector can withstand. This value is related to the mechanical strength of each part of the disconnector.

(6) Circuit contact resistance ($\mu\Omega$). It refers to the conductivity of the conductive circuit of disconnector under various forms of electrical contact, and it is the technical ability of inspection, design, manufacturing process and assembly.

2.1.4　Classification, Model and Service Conditions of Disconnectors

1. Classification of disconnectors

There are many kinds of disconnectors, and they can be classified according to installation location, voltage class, number of poles and structure. See Table 2-1 for the classification of disconnectors.

Table 2-1   Classification of Disconnectors

S/N	Classification method	Category	
1	By the installation location	Indoor, outdoor	
2	By the number of poles	Single pole, three poles	
3	By the grounding switch	No grounding, single grounding, double grounding	
4	By purpose	(1) General transmission and distribution; (2) Quick opening; (3) Transformer neutral grounding; (4) Heavy current bus	
5	By the action mode	Knife switch, rotary and plug-in types	
6	By the structure type	By the number of post insulators	(1) Single column disconnector; (2) Double column disconnector; (3) Three-column disconnector
6	By the structure type	By the number of post insulators and opening mode of conductive movable arm	(1) Single-column vertical contraction type; (2) Double-column horizontal rotary type; (3) Double-column horizontal telescopic type; (4) Three-column horizontal rotary type
7	By the operating mechanism	Manual, electric, pneumatic and hydraulic types	
8	By the service environment	General and anti-pollution type	

2. Model of disconnectors

The model of HV disconnectors is mainly composed of the following six units:

| 1 | 2 | 3 | - | 4 | 5 | / | 6 |

1: Name; G-disconnector.

2: installation site; N-indoor, W-outdoor.

3: Design number; It is determined by the industry management department according to the sequence of identification and model applied, which is expressed by Arabic numerals "1, 2, 3", etc.

4: Rated voltage (kV).

5: Supplementary properties: C-outgoing line of porcelain bushing, D-knife switch with grounding; K-quick opening, G-improved, T-unified design.

6: Rated current (A).

For example, in the model GW16-252D/3150, G stands for disconnector; W refers to outdoor; 16 is the design serial number; the rated voltage is 252 kV; D stands for grounding switch; the rated current is 3150A.

At present, GN19, GW4, GW5, GW6, GW7, GW16 and GW17 series of disconnectors are widely used in China.

3. Service conditions of disconnectors

The rated service conditions of disconnectors are as below:

(1) It is an ordinary type when the disconnector is applied where the altitude is not more than 1000 m, and it is a plateau type when the disconnector is applied where the altitude is more than 1000 m.

(2) Seismic intensity is not more than 8.

(3) The ambient temperature is not higher than +40 °C; the ambient temperature of indoor products is not lower than −10 °C; the outdoor product temperature is not lower than −30 °C.

(4) When the relative air humidity of indoor products is +25 °C, the daily average is not more than 95%, and the monthly average is not more than 90% (some products require that the relative air humidity is not more than 85%).

(5) The icing thickness of outdoor products is divided into 5 mm and 10 mm.

(6) The air around indoor products is not polluted by corrosive or combustible gases, water vapor and other obvious pollutants, and there is no frequent violent vibration. The service environment of outdoor products is as below: ordinary type for Grade I polluted areas, anti-pollution type for Grade II (medium polluted type) and Grade III (heavy polluted type) polluted areas.

模块二　隔离开关的典型型号

一、户内式隔离开关

户内式隔离开关的主要结构类型有插入式和转动式。

1. GN19-10 系列户内高压隔离开关

（1）结构特点。GN19-10 系列插入式户内高压隔离开关，其外形如图 2-2 所示，实物图如 2-3 所示。GN19-10 系列隔离开关采用三相共底架结构，主要由静触头、基座、支柱绝缘子、拉杆绝缘子、动触头组成。隔离开关的导电部分由动触头、静触头、触座组成，每相导电部分通过两个支柱绝缘子固定在基座上，三相平行安装。每相动触头为两片槽型铜片，它不仅增大了动触头的散热面积，对降低温度有利，而且提高了动触头的机械强度，使隔离开关的动稳定性提高。隔离开关动触头一端通过轴销安装在触座上，另一端与静触头相连接，其接触压力是靠两端接触弹簧维持的，每相动触头中间均连有拉杆绝缘子，拉杆绝缘子与安装在基座上的主轴相连，主轴转动，带头拉杆绝缘子操动动触头完成分、合闸。主轴两端伸出基座，其任何一端均可与所配用的手动操动机构相连。

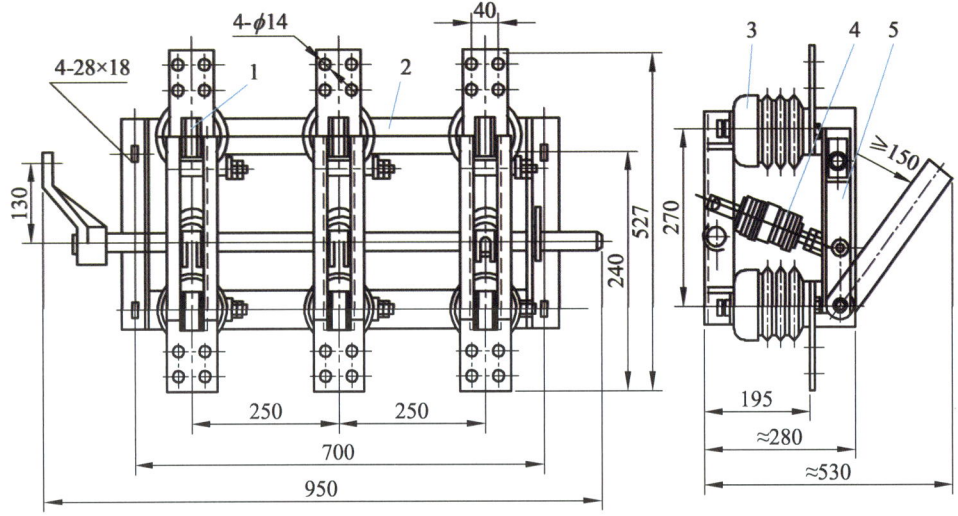

1—静触头；2—基座；3—支柱绝缘子；4—拉杆绝缘子；5—动触头。

图 2-2　GN19-10 型户内高压隔离开关

图 2-3　GN19-10 型隔离开关实物图

GN19-10/1000 型及 GN19-10/1250 型在动静触头接触处装有两件磁锁压板，当很大的短路电流通过时，磁锁压板相互间产生的吸引电磁力增加了动静触头的接触压力，从而增大了触头的动热稳定性。

（2）动作原理。分闸时由操动机构通过拉杆拐臂带动主轴旋转，操作绝缘子向上推动闸刀，使闸刀和静触头分开，闸刀绕触座旋转，动触头也在闸刀的带动下向上移动至分闸位置，完成分闸动作。

合闸时由操动机构通过拉杆拐臂带动主轴旋转，使操作绝缘子拉着闸刀向下绕触座旋转，在动触头和静触头相遇后，动触头继续运动插入静触头至合闸位置，完成合闸动作。

2. GN30-10 系列户内高压隔离开关

GN30-10 型旋转式户内高压隔离开关基本尺寸如图 2-4 所示，实物如图 2-5 所示。GN30-10 型隔离开关是一种旋转触刀式的新型隔离开关，特别适用于安装高压开关柜内，使高压开关柜结构紧凑、简单、占用空间小，提高其安全可靠性。开关主体是通过两组绝缘子固定在开关底架上下两个面上，上下两个面之间由固定在开关架上的隔板完全分开，通过旋转触刀，从而实现开关的合闸与分闸。由于静触头分别安装在开关柜的上下两个面上，使其带电部分与不带电部分在开关柜内完全隔开，从而保证维修人员的安全。GN30-10D 型带接地刀闸，以满足接地的需要。

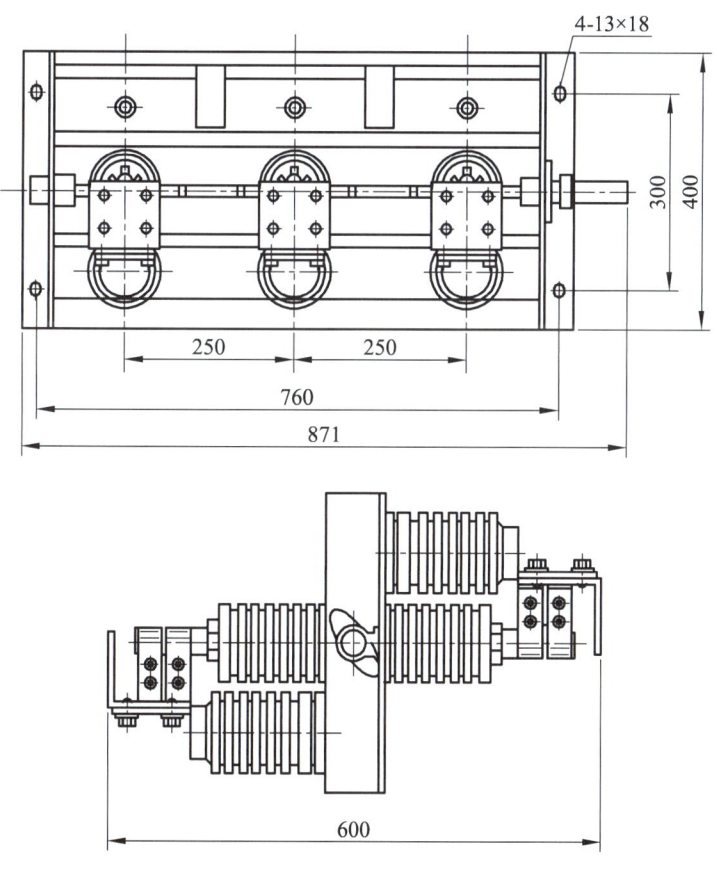

图 2-4　GN30-10 型外形及安装尺寸

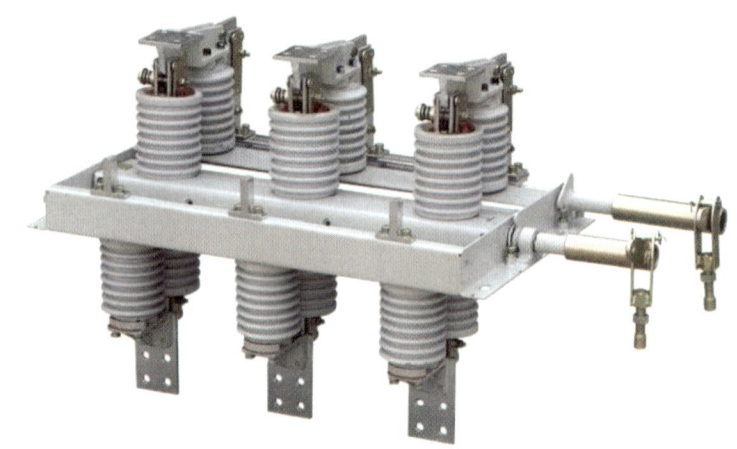

图 2-5 GN30-10 型隔离开关实物图

二、户外隔离开关的结构类型

(一)户外隔离开关的结构形式

按照支柱绝缘子的数量和导电活动臂的开启方式划分,一般有单柱垂直缩式、双柱水平开启式、双柱水平伸缩式、三柱水平旋转式四种型式。

(1)单柱垂直伸缩式。上半部为折叠式导电杆,下半部为一个垂直绝缘支柱,导电杆能上下活动,开断时形成一个垂直方向的空气间隙。它可以直接布置在母线下方,按照不同母线设计方案布置,三相隔离开关可以垂直于母线中心线布置,也可以错开布置。

(2)双柱水平旋转式。两个支柱分开并垂直布置,开断时其顶部活动杆在水平面内分别旋转 90°,形成一个水平方向的空气间隙。也有的两个支柱成 V 形布置,即 V 型隔离开关。

(3)双柱水平伸缩式。在一个支柱上安装静触头,另一个支柱上安装折叠式导电杆,导电杆能水平伸缩,开断时可形成一个水平方向的空气间隙。

(4)三柱水平旋转式。三个绝缘支柱分开并垂直布置,两边侧支柱固定不动,中间支柱上部安装一个水平式活动导电杆,开断时它在水平面内旋转约 60°,形成两个水平方向的空气间隙。

(二)常见型号的户外式隔离开关

1. 双柱式隔离开关

(1)GW4-110 型隔离开关。

GW4-110 型隔离开关为双柱单断口水平开启式结构,由底座、绝缘支柱、导电部分和操动机构组成,实物图如图 2-6 所示。每极有两个实心棒式绝缘支柱,分别装在底座两端的轴承座上,用交叉连杆连接,可以水平转动。导电闸刀分成两段,分别固定在两个绝缘支柱的顶端。触头接触的地方在两个绝缘支柱的正中位置,指形触头上装有防护罩,用以防雨、雪及灰尘。GW4-110 型隔离开关的单极结构,如图 2-7 所示。

隔离开关分合闸操作时,由操作机构通过垂直拉杆和主拐臂旋转 180°,使主拉杆带动首相的左侧绝缘子旋转 90°,并通过同期拉杆使另一侧绝缘子反向旋转 90°,再通过相间水平拉杆联动使 B、C 两相同步实现断开或闭合。为使引出线不随支柱的转动而扭曲，在闸刀与出线接线端子之间装有挠性连接的导体。

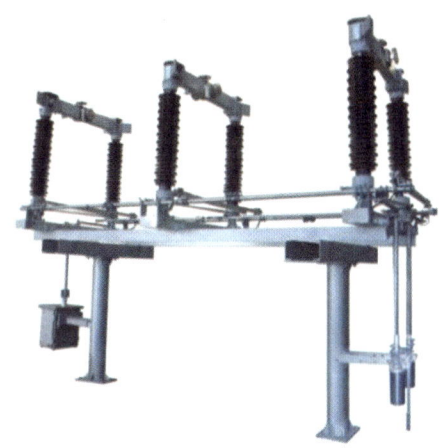

图 2-6　GW4-110 型隔离开关实物图

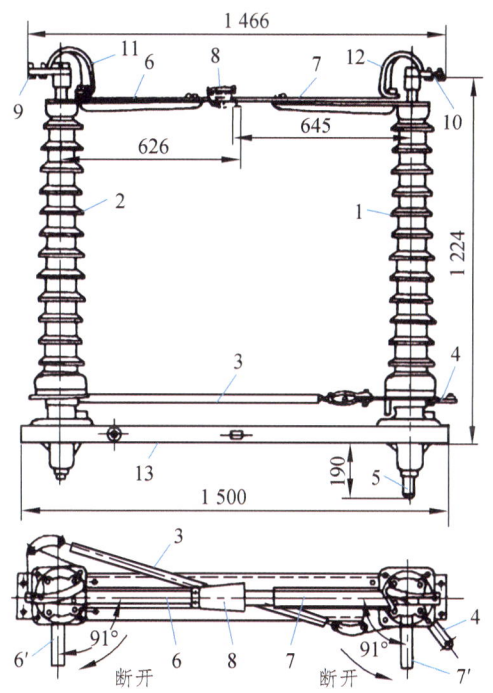

1，2—绝缘支柱；3—连杆；4—操动机构的牵引杆；5—绝缘支柱的轴；6，7—闸刀；
8—触头；9，10—接线端子；11，12—挠性连接的导体；13—底座。

图 2-7　GW4-110 型双柱式隔离开关

GW4 型隔离开关可配用手动、电动和气动操动机构，三相联动操作，电动和气动操作可实现远方控制。根据需要还可配装接地开关。该型隔离开关结构简单紧凑，尺寸小，质量轻，广泛用于 10~110 kV 配电装置中。由于闸刀在水平面内转动，因而对相间距离要求大。

GW4 本体的动作原理：当操动机构操作时，带动底架中部的传动轴旋转 180°，通过水平拉杆，带动一侧的支持绝缘子（安装于转动杠杆上）旋转 90°，并借同期拉杆使另一侧支持绝缘子反向旋转 90°，于是两主闸刀及其中间触头实现合闸或分闸。接地刀闸的操动机构合分时，借助传动轴及水平固定杆，使地刀转轴旋转一角度，从而使接地闸刀合闸或分闸。接地刀转轴上有扇形板与紧固在瓷柱法兰上的弧形板组成联锁，能确保主刀分—地刀合、地刀分—主刀合的正确动作顺序。

（2）GW5-110D 隔离开关。

GW5-110D 隔离开关是由双柱隔离开关改进而成，开关由底座、棒式支柱绝缘子、导电闸刀、左右触头和传动部分等组成，隔离开关每相的两个棒式绝缘子成 V 型布置，交角 50°，固定在一个底座上，故也称为 V 型隔离开关。棒式绝缘子上装有闸刀，可动触头成楔形连接。GW5-110D 隔离开关单极外形，如图 2-8 所示，一台三相隔离开关由三个单极组成。

操动机构可配用手动、电动或气动操动机构。根据需要该隔离开关可配装接地闸刀。进行操作时，两个棒式绝缘子以相同速度作相反方向（一个顺时针，另一个逆时针）的转动，两半闸刀同时绕绝缘子轴线转动 90°，使隔离开关接通或断开。

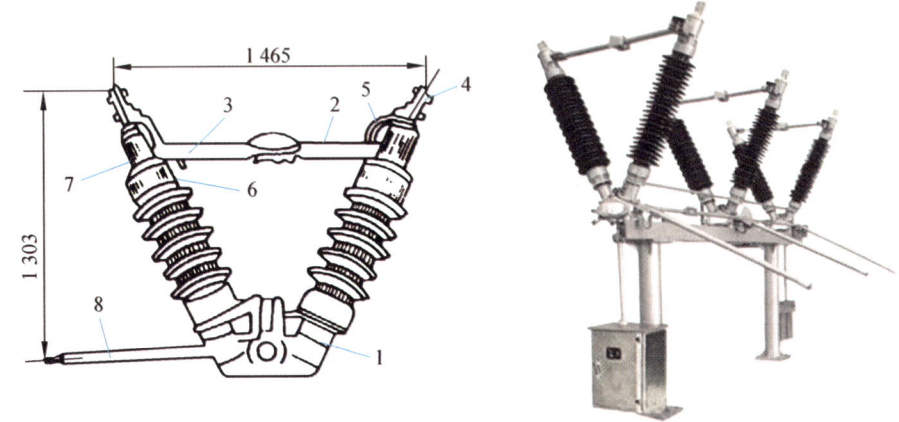

1—底座；2、3—闸刀；4—接线端子；5—挠性连接导体；
6—棒式绝缘子；7—支承座；8—接地刀闸。

图 2-8　GW5-110D 型隔离开关

GW5 型隔离开关广泛用于 35～110 kV 电压等级中。其主要特点是结构简单，尺寸小，质量轻；闸刀分成两半，以减少闸刀导电杆长度，操作时闸刀水平等速运动，使冰层受到很大剪力易于破除；合闸时支持绝缘子受弯折力，因而要求绝缘子具有较高强度；因闸刀水平转动，极间距离要求较大。

（3）GW11-252 型隔离开关。

① 结构特点。GW11-252 型隔离开关采用双柱水平伸缩式结构，实物图如 2-9 所示，闸刀的动作方式为水平伸缩式，分闸后形成水平方向的绝缘单断口，分合状态清晰，便于巡视。隔离开关制成单极形式，由三个单极组成一台三相隔离开关。其结构包括底座、绝缘支柱、传动装置、导电闸刀、静触头和操动机构等。三极隔离开关由一台电动操动机构联动操作，三极接地开关由一台人力操动机构联动操作（也可由一台电动操动机构联动操作）。每极隔离开关动、静触头侧均可配装一个接地开关静触头供接地用。GW11-252 型隔离开关的结构如图 2-10 所示。

图 2-9 GW11-252 型隔离开关实物图

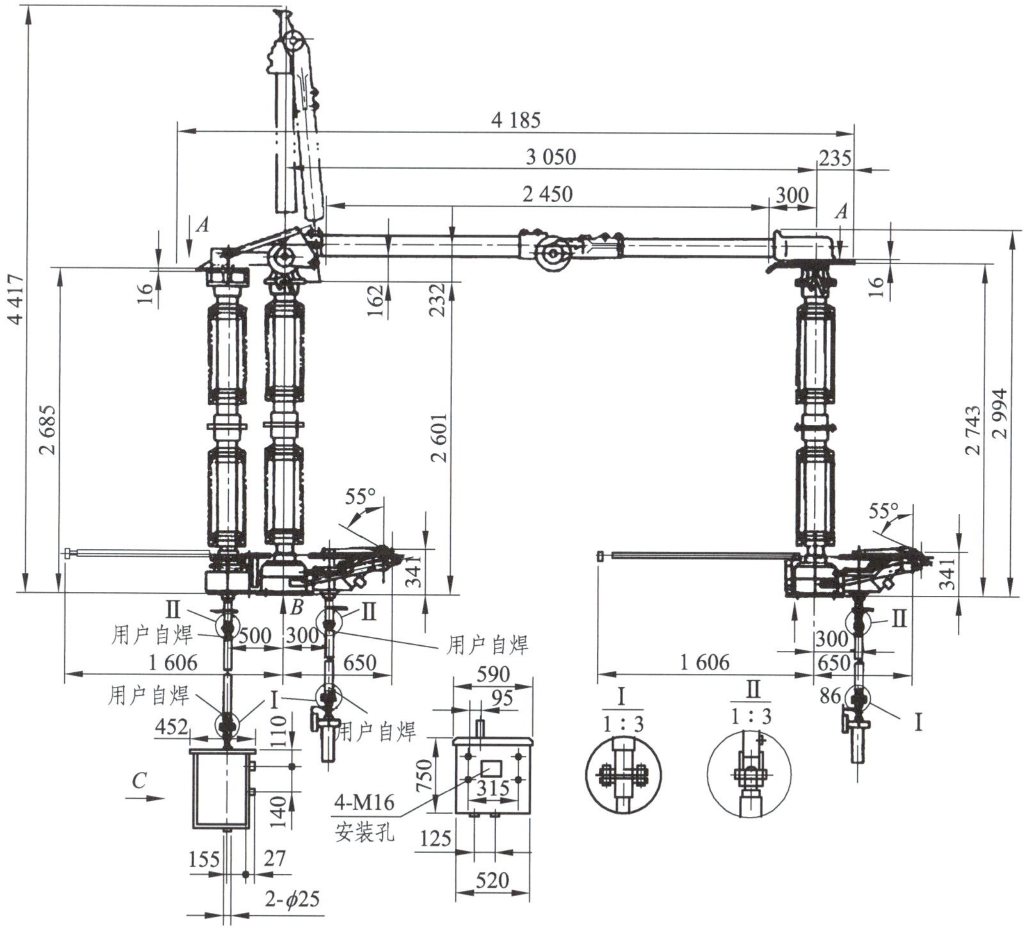

图 2-10 GW11-252 型双柱水平伸缩式隔离开关及其操动机构外形

② 动作原理。电动机操动机构由异步电动机驱动，通过机械减速装置将力矩传递给机构主轴，再借助连接钢管将力矩传给隔离开关操作支柱，操作支柱动作一次约转 90°，操作支柱的顶部通过连杆传动装置带动导电闸刀，合闸过程导电闸刀下导电杆向下转动，上导电杆以联轴节为圆心作圆周运动，使上下导电杆串成直线，动触头插入静触头，完成合闸动作。分闸过程与此相反，使上下导电杆折叠竖立在绝缘支柱顶部，保证断口的安全绝缘距离。隔离开关分闸动作完成后再操作接地闸刀，顺时针方向摇动手动机构手柄，通过蜗轮蜗杆传动，机构主轴通过连杆将力矩传递给接地开关转轴，使导电杆向上旋转约 75°，当动静触头相接触后，动触头向上运动，插入静触头，完成合闸动作。分闸过程与此相反。隔离开关、接地开关的三级联动通过极间拉杆实现。

在动触头侧，通过机械连锁装置使隔离开关与接地开关实现主分—地合、地分—主合，在静触头侧，采用电磁锁来保证操作顺序的正确。

2. 三柱式隔离开关

（1）GW7-220 型隔离开关结构特点。GW7-220 型的高压隔离开关采用三柱双断口水平旋转开启式结构，如图 2-11 所示。由底座、绝缘支柱、导电闸刀、静触头、传动装置和操动机构等组成。静触头分别在两边的棒形支柱绝缘子上端，中间棒形支柱绝缘子用以支持闸刀，并可带动闸刀作水平转动。图 2-12 是 GW7-220 型隔离开关单级结构图。

（2）GW7 动作原理。

① 隔离开关。由电动机构带动设在主极底座中的转动轴，旋转 180°，通过拐臂、连杆组成的四连杆机构驱动中间瓷柱转动，带动导电闸刀在水平面上旋转约 70°，即可完成分、合闸动作。主极通过相间水平连接管，带动两个边极同步完成分、合闸动作。

② 接地开关。合闸时，主闸刀应在分闸位置，操作地刀操作机构的主轴旋转 180°，通过拐臂及连杆使接地开关向上运动，插入静触头中，完成接地开关的合闸过程，分闸过程与此相反。

图 2-11　GW7-220 型隔离开关实物图

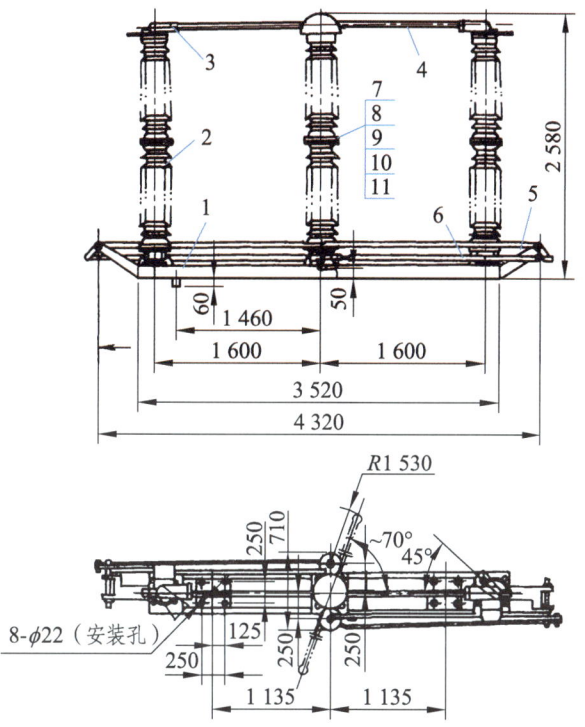

1—底座；2—绝缘支柱；3—静触头；4—主闸刀；5—接地开关；6—拉杆。

图 2-12　GW7-220 型隔离开关单极外形图

Ⅱ型接地开关为两步动作。合闸时，接地闸刀向上运动（约 80°），与静触头相碰后变为上伸运动，动触头插入静触头中。分闸过程与此相反，接地闸刀先下缩一定距离，使动触头从静触头中拔出，然后向下摆落到水平位置。三极接地开关通过水平连接管达到同步动作。

隔离开关与接地开关之间设有机械联锁，以保证隔离开关在合闸时，接地开关不能合闸；接地开关在合闸位置时，隔离开关不能合闸。机械联锁是利用设在主极中间轴承座上的一对月牙板来实现的。

3. 单柱式隔离开关

（1）GW6-220GD 型隔离开关。

GW6-220GD 型隔离开关每极具有两个瓷柱，即支持瓷柱和操作瓷柱。由于只有一个支持瓷柱，故称为单柱式。静触头固定在架空硬母线或悬挂在架空软母线上。

动触头固定在导电折架上。通过操动机构使操作瓷柱转动，带动传动装置去操作导电折架上下运动，从而使动触头垂直上下运动，夹住或释放静触头，即可实现合、分闸，形成电气绝缘断口。结构如图 2-13 所示，图中虚线部分是合闸位置时可动部分的位置。

（2）GW10-252 型隔离开关。

GW10-252 型隔离开关为单柱立式伸缩室外交流高压隔离开关，如图 2-14 所示，包括底座、绝缘支柱、传动装置、导电闸刀、静触头、操动机构等。三相隔离开关由三个单极组成。

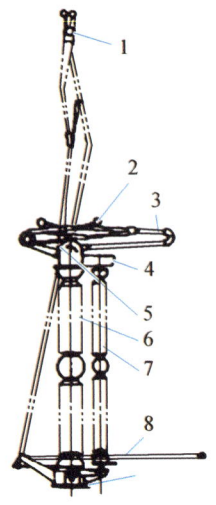

1—静触头；2—动触头；3—导电折架；4—传动装置；5—接线板；6—支持瓷柱；
4—传动装置；5—接线板；6—支持瓷柱；7 操作瓷柱；8—接地开关；9—底座。

图 2-13　GW6-220GD 隔离开关

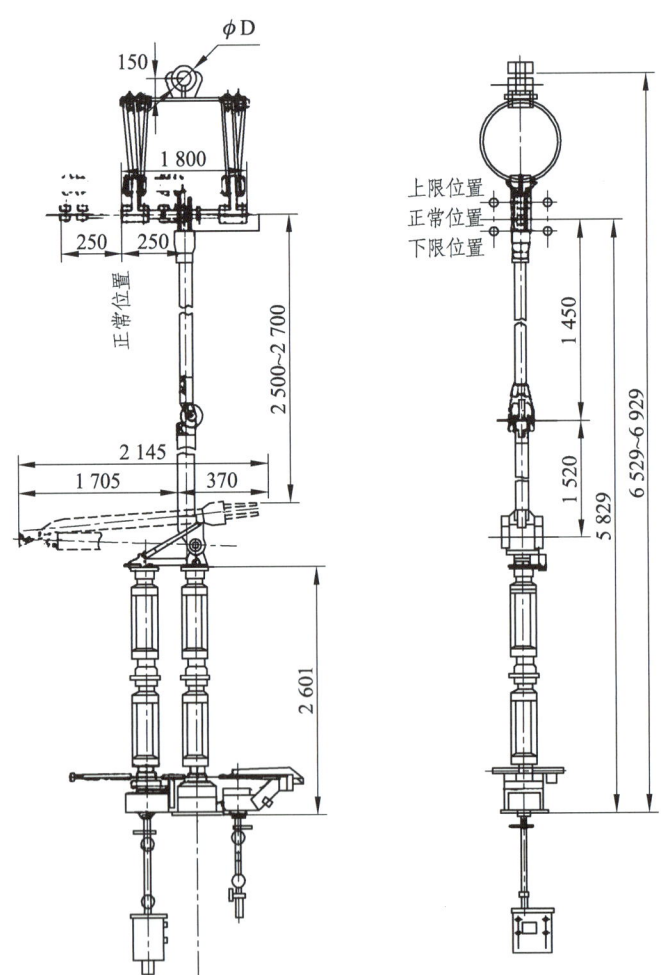

图 2-14　GW10-252 型单柱垂直伸缩式隔离开关及其操动机构外形

GW10-252 型隔离开关的结构和动作原理与 GW11-252 型隔离开关相致，主要的区别在于导电闸刀与静触头的结构。

隔离开关的静触头是由镀有银层的铜管与接线夹板组成，两侧的接线夹板上有线孔，以便和母线相连。静触头安装于母线上，在分闸后形成垂直方向的绝缘单断口，闸刀的动作方式为垂直伸缩。导电闸刀采用伸缩式（半折架）结构，由动触头、上导电杆、联轴节（即导电关节）、齿轮、齿条、操作杆、弹簧、滚轮触头等组成，在分闸位置时，上下导电杆通过联轴节折叠在水平位置，当机构带动下段铝管向上传动时，上段铝管以联轴节（装在下段铝管的顶部，它本身也以传动装置的底座为轴心作圆周运动）为圆心作圆周运动。在合闸位置时上下导电杆串接成一直线，动触头夹紧静触头。动触头装于上导电杆的顶部，触片有较长的接触面，接触压力由传动件的弹性装置（在上导电杆的内部）产生并保持稳定的数值。

每极隔离开关装配一个接地开关供断口下端接地使用，接地开关为单杆分步动作式。

Module 2 Typical Models of Disconnectors

2.2.1 Indoor Disconnector

The main structural types of indoor disconnectors include plug-in and rotary types.

1. GN19-10 series indoor HV disconnectors

(1) Structural characteristics. For GN19-10 series plug-in indoor HV disconnector, its appearance is shown in Fig. 2-2, and the physical drawing is shown in Fig. 2-3. GN19-10 series plug-in indoor HV disconnector has the three-phase common-support frame structure, which is mainly composed of static contact terminal, base, post insulator, tie rod insulator and moving contact terminal. The conduction part of the disconnector consists of moving contact terminal, static contact terminal and contact base. The conduction part of each phase is fixed on the base through two post insulators, and the three phases are installed in parallel. The moving contact terminal of each phase is made of two grooved copper sheets, which not only increases the heat dissipation area of the moving contact terminal and contributes to reducing the temperature, but also improves the mechanical strength of the moving contact terminal and the dynamic stability of the disconnector. One end of the moving contact terminal of the disconnector is installed on the contact base through the shaft pin, and the other end is connected with the static contact terminal. Its contact pressure is maintained by the contact springs at both ends. The moving contact terminal of each phase is connected with the tie rod insulator which is connected with the main shaft installed on the base. The main shaft rotates to drive the tie rod insulator to operate the moving contact terminal so as to complete opening and closing. Both ends of the main shaft extend out of the base, and any end of the main shaft can be connected with the manual operating mechanism.

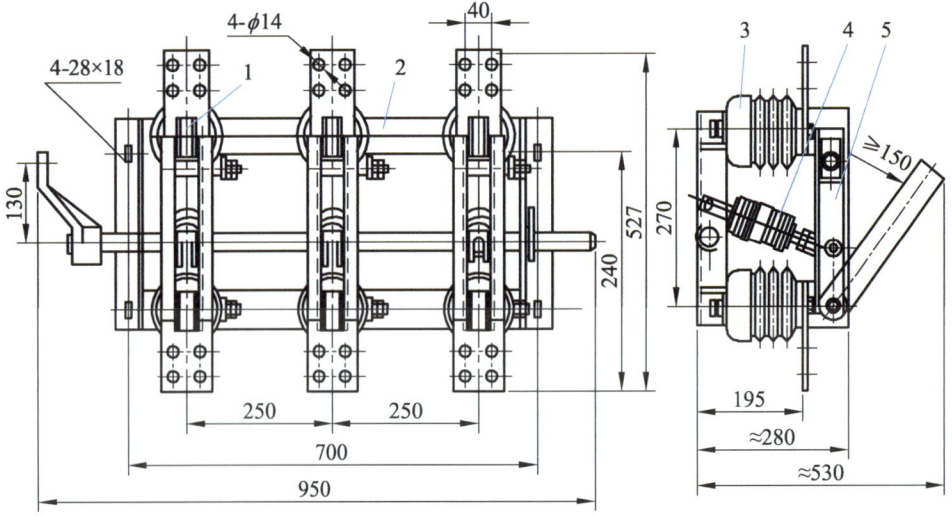

1-Static contact terminal; 2-Base; 3-Post insulator; 4-Tie rod insulator; 5-Moving contact terminal.

Fig. 2-2 GN19-10 Indoor HV Disconnector

Fig. 2-3　GN19-10 Disconnector

The GN19-10/1000 and GN19-10/1250 models are equipped with two magnetic locking plates at the contact of the moving and static contact terminals. When large short-circuit current passes, the attractive electromagnetic force generated by the magnetic locking plates increases the contact pressure of the moving and static contact terminals, thus increasing the dynamic and thermal stability of the contact terminals.

(2) Action principle. During opening, the operating mechanism drives the main shaft to rotate through the tie rod crank arm, and the operating insulator pushes the knife switch upward to separate the knife switch and the static contact terminal. Besides, the knife switch rotates around the contact base, and the moving contact terminal is also driven by the knife switch to move upward to the open position to complete the opening action.

During closing, the operating mechanism drives the main shaft to rotate through the tie rod crank arm, and the operating insulator drives the knife switch to rotate downward around the contact base. After the moving contact terminal and the static contact terminal meet, the moving contact terminal continues to move and insert into the static contact terminal to the closed position to complete the closing action.

2. GN30-10 series indoor HV disconnectors

For GN30-10 rotary indoor HV disconnector, its basic dimensions are shown in Fig. 2-4, and the physical drawing is shown in Fig. 2-5. GN30-10 disconnector is a new type of rotary contact knife disconnector, which is especially suitable for installation in the HV switch cabinet, making the HV switch cabinet compact and simple in structure, occupying less space and improving its safety and reliability. The disconnector body is fixed on the upper and lower surfaces of the disconnector support frame by two sets of insulators, and the upper and lower surfaces are completely separated by the diaphragm fixed on the disconnector support frame. The disconnector is closed and opened by rotating the contact knife. Because the static contact terminals are installed on the upper and lower surfaces of the switch cabinet respectively, the live part and the dead part are completely separated in the switch cabinet, which ensures the safety of maintainers. GN30-10D disconnectors are equipped with grounding knife switch to meet the requirement of grounding.

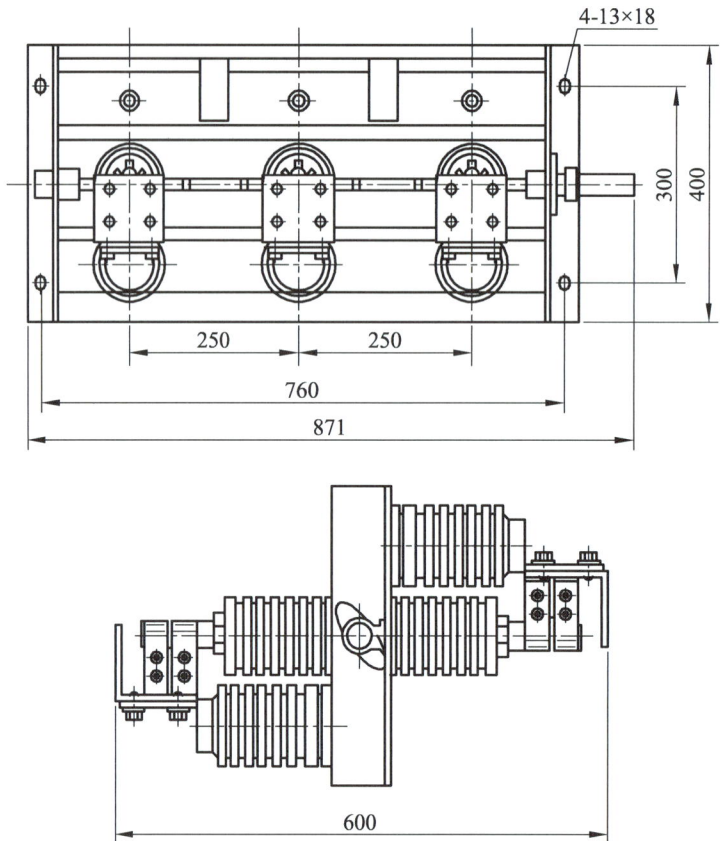

Fig. 2-4　Appearance and Installation Dimensions of GN30-10 Disconnectors

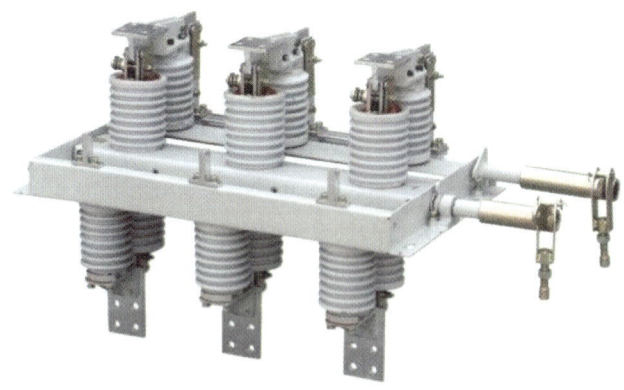

Fig. 2-5　Physical Drawing of GN30-10 Series Disconnectors

2.2.2 Structural Type of Outdoor Disconnectors

2.2.2.1 Structural form of outdoor disconnectors

According to the number of post insulators and the opening mode of conductive movable arm, there are generally four types: single-column vertical contraction type, double-column horizontal opening type, double-column horizontal telescopic type and three-column horizontal rotary type.

(1) Single-column vertical contraction type. The upper half is a folding conductive rod, and the lower half is a vertical insulating pillar. The conductive rod can move up and down and form a vertical air gap when it is disconnected. It can be arranged directly under the bus. According to different bus design schemes, the three-phase disconnector can be arranged perpendicular to the center line of the bus or staggered.

(2) Double-column horizontal rotary type. The two pillars are separated and vertically arranged. During opening, the movable rods at the top rotate 90° in the horizontal plane, forming a horizontal air gap. There are also two pillars arranged in a V shape, that is, a V-shaped disconnector.

(3) Double-column horizontal telescopic type. A static contact terminal is installed on one pillar, and a folding conductive rod is installed on the other pillar. The conductive rod can extend horizontally and form a horizontal air gap during opening.

(4) Three-column horizontal rotary type. The three pillars are separated and vertically arranged. The side pillars on both sides are fixed, and a horizontal movable conductive rod is installed on the upper part of the middle pillar, which rotates about 60° in the horizontal plane during opening, forming two horizontal air gaps.

2.2.2.2 Outdoor disconnectors of common models

1. Double column disconnector

(1) GW4-110 disconnector.

GW4-110 disconnector is a horizontal open structure with double columns and single break, which consists of the base, insulating pillar, conduction part and operating mechanism. The physical diagram is shown in Fig. 2-6. Each pole has two solid rod-type insulation pillars, which are installed on the bearing seats at both ends of the base respectively and connected by the cross connecting rod, and can rotate horizontally. The conductive knife switch is divided into two sections, which are respectively fixed at the top ends of the two insulating pillars. The contact point of the contact terminal is in the middle of the two insulating pillars, and the finger contact terminal is equipped with a protective cover to prevent rain, snow and dust. The single-pole structure of GW4-110 disconnector is shown in Fig. 2-7.

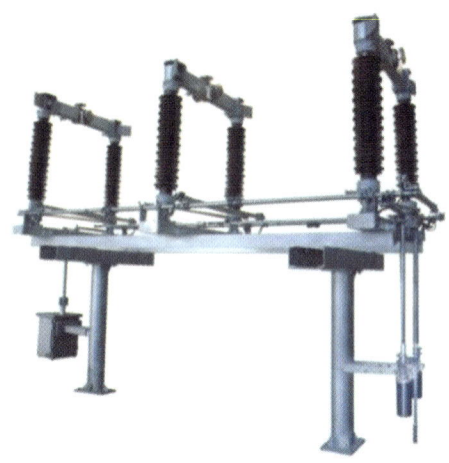

Fig. 2-6 Physical Drawing of GW4-110 Disconnector

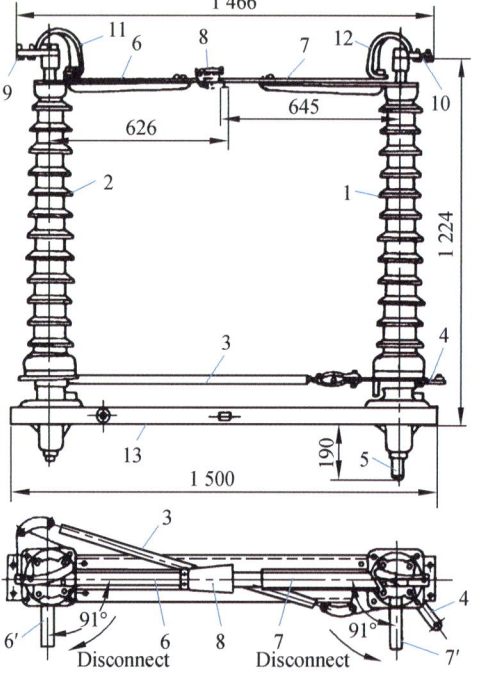

1, 2-Insulating post; 3-Connecting rod; 4-Traction rod of operating mechanism; 5-Shaft of insulating post; 6, 7-Knife switch; 8-Contact terminal; 9, 10-Terminal; 11, 12-Flexibly connected conductor; 13-Base.

Fig. 2-7 GW4-110 Double-column Disconnector

During opening and closing the disconnector, the operating mechanism rotates 180° through the vertical tie rod and the main crank arm, so that the main tie rod drives the left insulator of the primary phase to rotate 90°. In addition, the insulator on the other side reversely rotates 90° through the tie rod in the same cycle, and then the B and C phases are synchronously disconnected or closed through the linkage of the alternate horizontal tie rods. In order to prevent the outgoing line from twisting with the rotation of the pillar, a flexibly connected conductor is installed between the knife switch and the outgoing line terminal.

The GW4 disconnector can be equipped with manual, electric and pneumatic operating mechanisms as well as three-phase linkage operation, and the electric and pneumatic operation can realize remote control. It can also be equipped with the grounding switch as needed. This type of disconnector has a simple and compact structure, small size and light weight, which is widely used in 10–110 kV power distribution units. Because the knife switch rotates in the horizontal plane, a large distance between phases is required.

The action principle of GW4 body: When the operating mechanism is working, it drives the transmission shaft in the middle of the support frame to rotate by 180°, and the supporting insulator (installed on the rotating lever) on one side is driven to rotate by 90° through the horizontal tie rod. Meanwhile, the supporting insulator on the other side rotates by 90° through the tie rod in the same cycle, so that the two main knife switches and the contact terminal between them can achieve closing and opening. When the operating mechanism of the grounding knife switch is open and closed, it makes the rotating shaft of the grounding knife switch rotate by an angle with the help of the transmission shaft and the horizontal fixed rod, so that the grounding knife switch is closed or open. The fan-shaped plate on the rotating shaft of the grounding knife switch and the arc-shaped plate fastened on the flange of the porcelain knob insulator form interlocking, which can ensure the correct action sequence of main knife opening-grounding knife closing and grounding knife opening-main knife closing.

(2) GW5-110D disconnector.

the GW5-110D disconnector is improved from the double-column disconnector. The switch consists of the base, rod-type post insulator, conductive knife switch, left and right contact terminals and transmission part, etc. The two rod-type insulators of the disconnector of each phase are arranged in a V-shape with an intersection angle of 50° and fixed on a base, so it is also called V-type disconnector. The rod-type insulator is equipped with a knife switch and the movable contact terminal is connected in a wedge shape. The single-pole shape of GW5-110D disconnector is shown in Fig. 2-8. One three-phase disconnector consists of three single-poles.

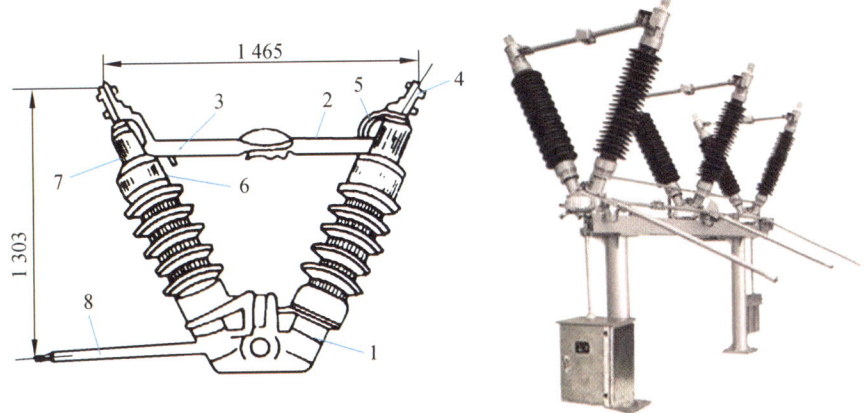

1-Base; 2, 3-Knife switch; 4-Terminal; 5-Flexibly connected conductor; 6-Rod-type insulator; 7-Support base; 8-Grounding knife switch.

Fig. 2-8 GW5-110D Disconnector

The operating mechanism can be equipped with a manual, electric or pneumatic operating mechanism. The disconnector can be equipped with a grounding knife switch as required. During operation, the two rod-type insulators rotate in opposite directions (one rotates clockwise and the other rotates counterclockwise) at the same speed, and the two halves of the knife switch rotate 90 degrees around the insulator axis at the same time to connect or disconnect the disconnector.

The GW5 disconnector is widely used in the voltage classes of 35–110 kV. It is characterized by simple structure, small size and light weight. The knife switch is divided into two halves to reduce the length of the conductive rod of the knife switch, and the knife switch moves horizontally at the same speed during operation, so that the ice layer is easily broken due to great shearing force. During closing, the supporting insulator is subject ti the bending force, so the insulator should have high strength. Because the knife switch rotates horizontally, the distance between poles is required to be large.

(3) GW11-252 disconnector.

① Structural characteristics. GW11-252 disconnector is of double-column horizontal telescopic structure, and its physical diagram is shown in Fig. 2-9. The action mode of the knife switch is horizontal telescopic type, and a single insulation break in the horizontal direction is formed after opening. The opening and closing state is clear, which is convenient for inspection. The disconnector is of the single-post structure, and three single posts form one three-phase disconnector. Its structure includes the base, insulating pillar, transmission device, conductive knife switch, static contact terminal and operating mechanism. The three-pole disconnector is operated by an electric operating mechanism, and the three-pole grounding switch is operated by a manual operating mechanism (or by an electric operating mechanism). A static contact terminal of grounding switch can be installed for grounding on the side of moving and static contact terminals of each pole of disconnector. The structure of GW11-252 disconnector is shown in Fig. 2-10.

Fig. 2-9　Physical Drawing of GW11-252 Disconnector

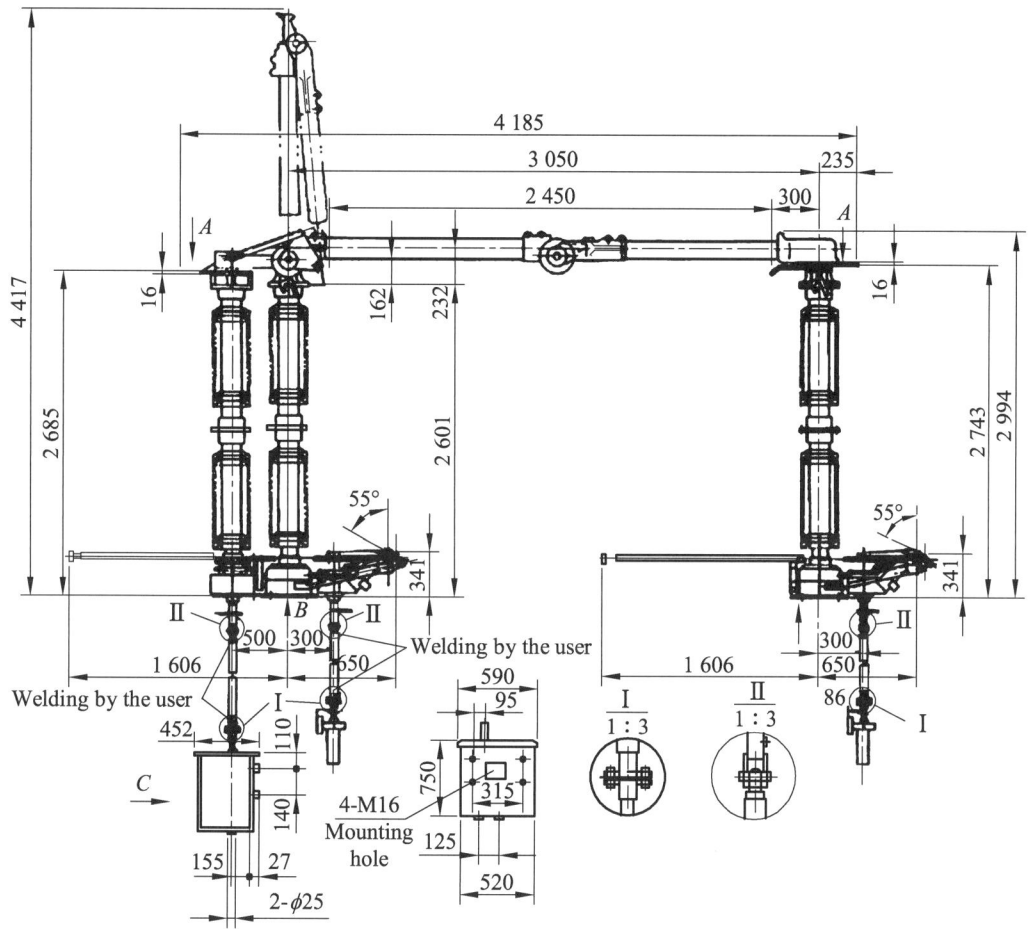

Fig. 2-10 GW11-252 Double-column Horizontal Telescopic Disconnector and Its Operating Mechanism

② Action principle. The motor operating mechanism is driven by an asynchronous motor, and the torque is transmitted to the main shaft of the mechanism through a mechanical deceleration device. Then the torque is transmitted to the operating strut of the disconnector through a connecting steel pipe. The operating strut rotates about 90° at a time, and the top of the operating strut drives the conductive knife switch through a connecting rod transmission device. In the closing process, the lower conductive rod of the conductive knife switch rotates downwards, and the upper conductive rod moves circularly around the coupling, so that the upper and lower conductive rods are connected in a straight line. Meanwhile, the moving contact terminal inserts into the static contact terminal to complete the closing action. The opening process is the opposite. The upper and lower conductive rods fold and stand at the top of the insulation pillar to ensure the safe insulation distance of the break. After the opening action is completed for the disconnector, the grounding knife switch is operated. The handle of the manual mechanism is shaken clockwise. Through worm gear and worm drive, the main shaft of the mechanism transmits the torque to the rotating shaft of the grounding switch through the connecting rod, so that the conductive rod rotates about 75° upward. When the moving and static contact terminals contact, the moving contact terminal moves upward and inserts the static contact terminal into the static contact to complete the closing action. The opening process is the opposite. The three-level linkage of the disconnector and grounding switch is realized through the tie rod between poles.

On the moving contact terminal side, the disconnector and the grounding switch achieve main knife opening-grounding knife closing and grounding knife opening-main knife closing by a mechanical interlocking device. On the static contact terminal side, an electromagnetic lock is used to ensure the correct operation sequence.

2. Three-column disconnector

(1) Structural characteristics of GW7-220 disconnector. The GW7-220 high-voltage disconnector has a three-column double-break horizontal rotary open-type structure, as shown in Fig. 2-11. It consists of the base, insulating pillar, conductive knife switch, static contact terminal, transmission device and operating mechanism. The static contact terminals are respectively arranged at the upper ends of the rod-shaped post insulators on both sides, and the middle rod-shaped post insulator is used to support the knife switch and drive the knife switch to rotate horizontally. Fig. 2-12 is the single-pole structure diagram of GW7-220 disconnector.

(2) GW7 Action principle.

① Disconnector. The electric mechanism drives the rotating shaft arranged in the main pole base to rotate 180°, and a four-connecting rod mechanism composed of the crank arm and connecting rod drives the middle porcelain knob insulator, and drives the conductive knife switch to rotate about 70° on the horizontal plane so as to complete the opening and closing actions. The main pole drives the two side poles to complete the opening and closing actions synchronously through alternate horizontal connecting pipes.

② Grounding switch. During closing, the main knife switch should be in the open position, and the main shaft of the grounding knife switch operating mechanism is rotated by 180°. The grounding switch moves upward through the crank arm and connecting rod and inserts into the static contact terminal to complete the closing process of the grounding switch. The opening process is the opposite.

Fig. 2-11 Physical Drawing of GW7-220 Disconnector

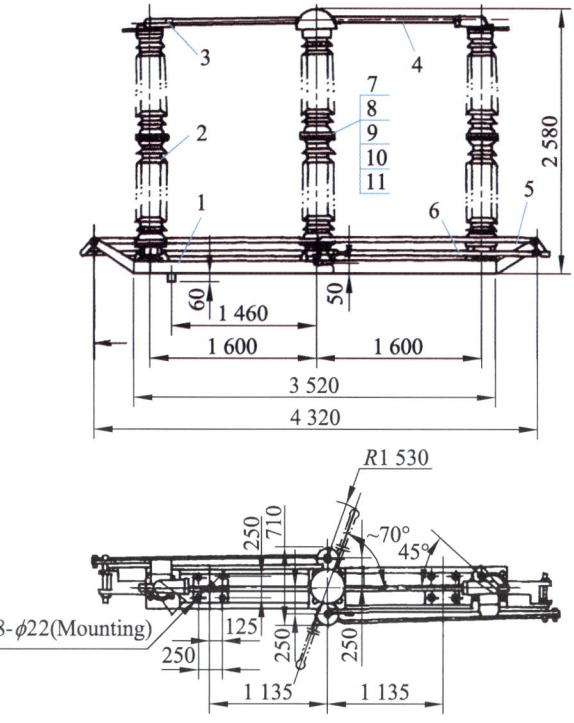

1-Base; 2-Insulating post; 3-Static contact terminal; 4-Main knife switch; 5-Grounding switch; 6-Tie rod.

Fig. 2-12 Single-pole Outline Diagram of GW7-220 Disconnectors

Type II grounding switch acts in two steps. During closing, the grounding knife switch moves upward (about 80°). After colliding with the static contact terminal, it stretches upward, and the moving contact terminal inserts into the static contact terminal. The opening process is the opposite. The grounding knife switch first retracts for a certain distance, so that the moving contact terminal is pulled out from the static contact terminal, and then falls to the horizontal position downwards. The three-pole grounding switch achieves synchronous action through the horizontal connecting pipe.

There is a mechanical interlock between the disconnector and the grounding switch to ensure that the grounding switch cannot be closed when the disconnector is closed. When the grounding switch is in the closed position, the disconnector cannot be closed. Mechanical interlocking is realized by a pair of crescent plates on the bearing seat in the middle of the main pole.

3. Single column disconnector

(1) GW6-220GD disconnector.

GW6-220GD disconnector has two porcelain knob insulators for each pole, namely, supporting porcelain knob insulator and operating porcelain knob insulator. The disconnector is of the single-column type since there is only one supporting knob. The static contact terminal is fixed on the overhead hard bus or suspended on the overhead flexible bus.

The moving contact terminal is fixed on the conductive folding frame. The operating porcelain knob insulator is rotated by the operating mechanism to drive the transmission device to operate the conductive folding frame to move up and down, so that the moving contact terminal moves vertically. The static contact terminal is clamped or released to achieve closing and opening, and form an electrical insulation break. The structure is shown in Fig. 2-13, and the dotted line in the figure is the position of the movable part in the closed position.

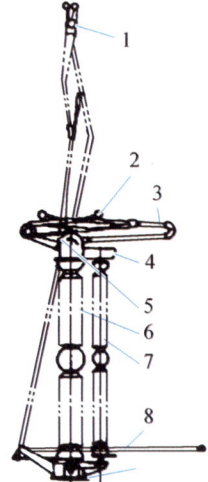

1-Stati ccontact terminal; 2-Moving contact terminal; 3-Conductive folding frame; 4-Transmission device;
5-Terminal block; 6-Supporting knob insulator; 7-Operating knob insulator; 8-Grounding switch; 9-Base.

Fig. 2-13   GW6-220GD Disconnector

(2) GW10-252 disconnector.

GW10-252 disconnector is a single-column vertical telescopic outdoor AC HV disconnector, and its unipolar structure is shown in Fig. 2-14, including the base, insulating pillar, transmission device, conductive knife switch, static contact terminal and operating mechanism. The three-phase disconnector consists of three single-poles.

The structure and action principle of GW10-252 disconnector are consistent with those of GW11-252 disconnector, and their main difference lies in the structure of conductive knife switch and static contact terminal.

The static contact terminal of the disconnector is composed of copper tube plated with silver layer and wiring splint, and the wiring splint on both sides has wire holes to connect with the bus. The static contact terminal is installed on the bus, and forms a single insulation break in the vertical direction after opening. The action mode of the knife switch is vertical expansion and contraction. The conductive knife switch is of telescopic (semi-folding frame) structure, which consists of moving contact terminal, upper conductive rod, coupling (i.e. conductive joint), gear, rack, operating rod, spring, and roller contact terminal, etc. In the open position, the upper and lower conductive rods are folded in the horizontal position through the coupling. When the mechanism drives the lower aluminum tube to move upwards, the upper aluminum tube moves in a circle around the coupling (which is installed at the top of the lower aluminum tube and moves in a circle around the base of the transmission device). In the closed position, the upper and lower conductive rods are connected in series to form a straight line, and the moving contact terminal clamps the static contact terminal. The moving contact terminal is installed at the top of the upper conductive rod, and the contact piece has a long contact surface. The contact pressure is generated by the elastic device of the transmission part (inside the upper conductive rod) and keeps a stable value.

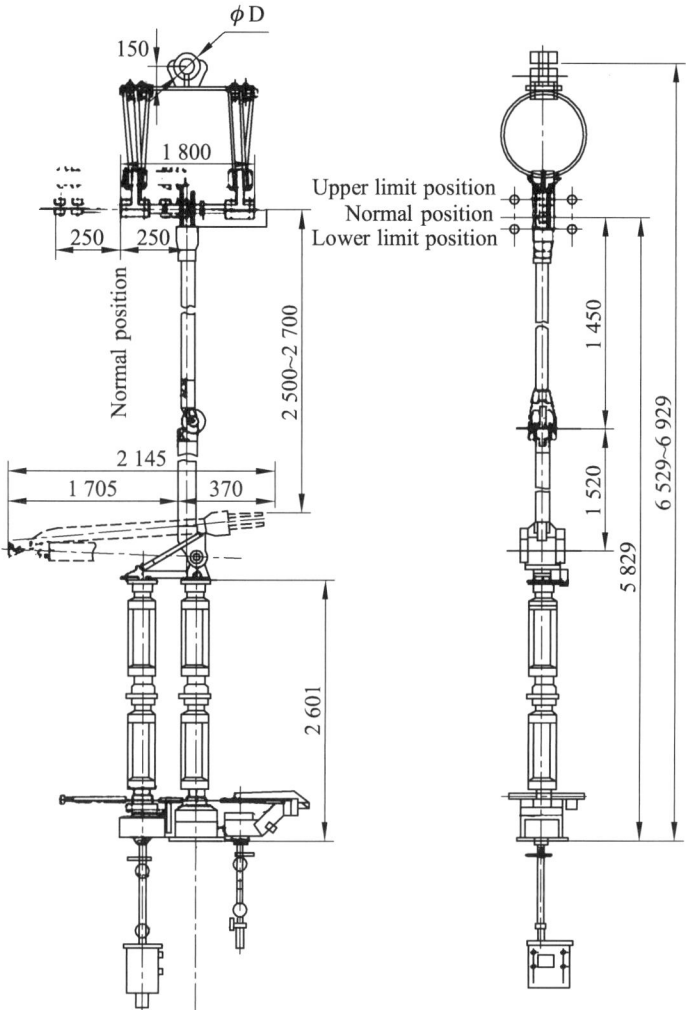

Fig. 2-14　Appearance of GW10-252 Single-column Vertical Telescopic Disconnector and Its Operating Mechanism

The disconnector of each pole is equipped with a grounding switch for grounding the lower end of the break, and the grounding switch is a single-pole step action type.

模块三 隔离开关的操动机构

一、手动操动机构

采用手动操动机构时，必须在隔离开关安装地点就地操作。手动操动机构结构简单、价格低廉、维护工作量少，而且在合闸操作后能及时检查触头的接触情况，因此被广泛应用。

手动操动机构有杠杆式和蜗轮式两种，前者一般适用于额定电流小于 3000 A 的隔离开关，后者一般适用于额定电流大于 3000 A 的隔离开关。

（1）杠杆式手动操动机构。

CS6 型手动杠杆式操动机构主要用于户内式高压隔离开关，其结构示意图如图 2-15 所示。图中实线表示隔离开关的合闸位置，虚线表示隔离开关的分闸位置，箭头表示隔离开关进行分、合闸操作时手柄的转动方向。

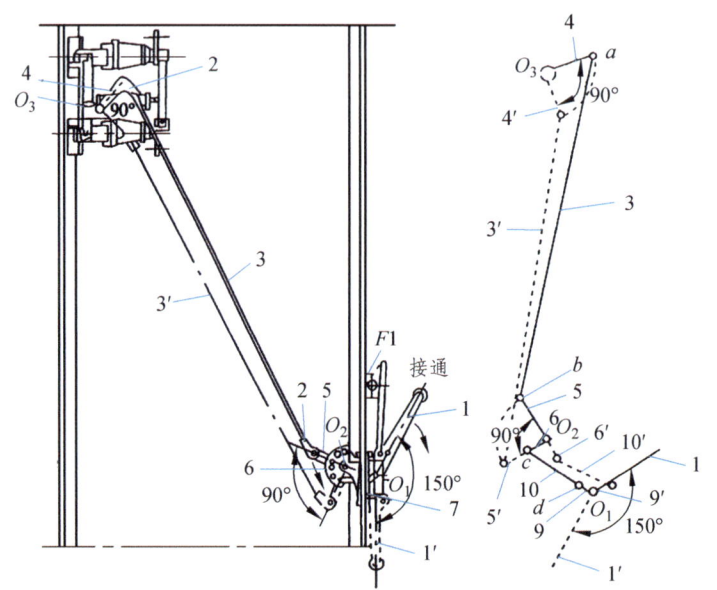

1—手柄；2—接头；3—牵引杆；4—拐臂；6—扇形杆；7—底座；5、8、9、10—连杆。

图 2-15 CS6 手动杠杆操动机构结构示意图

隔离开关在合闸位置时，连杆的绞接轴 d 处于死点位置以下，因此，可防止短路电流通过隔离开关时，刀闸因电动力作用而自行分闸。分闸操作时，拔出 O_1 轴处的销子，使手柄顺时针向下旋转 150°，则连杆随之顺时针上旋转 150°，通过连杆带动扇形杆逆时针向下旋转 90°，牵引杆被拉向下，并带动拐臂顺时针向下旋转 90°，使隔离开关分闸，O_1 轴处的销子自动弹入锁定。合闸操作顺序相反。

辅助触点盒 F 内有若干对触点，其公共小轴经杆与手柄联动。这些触点用于信号、联锁等二次回路。

（2）蜗轮式。

CS9 型手动蜗轮式操动机构安装图如图 2-16 所示。图中连杆与窄板绞接，窄板与牵引杆硬性连接。操作时摇动摇把，经蜗杆带动蜗轮转动，通过连杆系统使隔离开关分、合闸。顺时针摇动摇把,使蜗轮转过180°,隔离开关即完全合闸;逆时针摇动摇把,使蜗轮反转过180°,隔离开关即完全分闸。

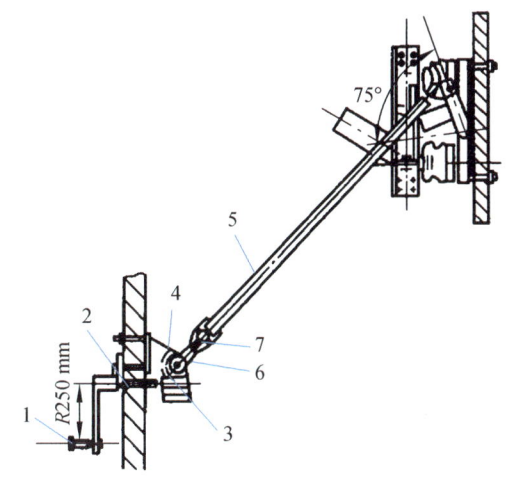

1—摇把;2—轴;3—蜗杆;4—蜗轮;5—牵引杆;6—连杆;7—窄板。

图 2-16 CS9 型手动蜗轮式操动机构安装图

二、电动操动机构

(一)电动机构的作用

(1)完成隔离开关的分、合闸操作。

电动操作机构的基本工作原理就是通过电动机向隔离开关的传动机构提供能源;电动机通过正、反转或循环转动带动隔离开关的传动机构以完成隔离开关的分、合闸操作。

(2)向监控系统提供隔离开关的辅助信号。

机构通过传动轴带动辅助开关,完成既定动作,发出辅助信号;通过手动操作电器元件,元件动作后向终端发出辅助状态信号;元件状态信息在终端动态采集。

(二)典型电动操动机构

1. CJ2-XG 型电动操动机构

该机构属于户外用动力式机构,用于 GW4、GW7 等高压隔离开关或接地开关分、合闸操作。可进行远方控制,也可就地电动控制或利用手柄进行人力操作。

CJ2-XC 型电动操动机构如图 2-17 所示,由电动机驱动齿轮及蜗轮减速装置,将力矩传递给输出轴,输出轴垂直安装,机构中设有分合闸终点限位开关及机械限位装置,使机构主轴的转角限制在准确的位置。机构设有手柄,可在现场进行手动分、合闸操作。

机构箱内设有刀开关和保护熔丝及机前电控操作的分、合闸按钮,也可用摇把进行人力分、合闸操作。机构内装有六动合、六动断的辅助开关,由转轴带动辅助开关切换,在隔离开关处于合闸或分闸位置时,发出相应的信号。为便于安装维修,机构箱为三面开门结构,用专门钥匙打开前门,从箱内两侧拧开蝶形螺母后,可打开两侧门。

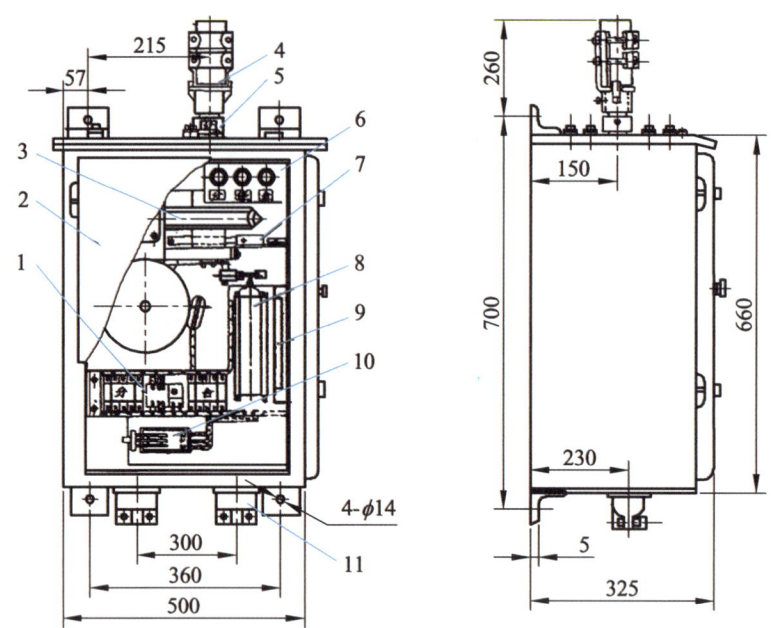

1—接触器及热继电器；2—机构箱；3—减速装置机构；4—连接器；5—分合位置指示器；
6—操动按钮；7—限位开关；8—辅助开关；9、10—接线板；11—出线盒。

图 2-17 CJ2–XG 型电动操动机构

CJ2-XC 型电动操动机构的工作原理如图 2-18 所示。

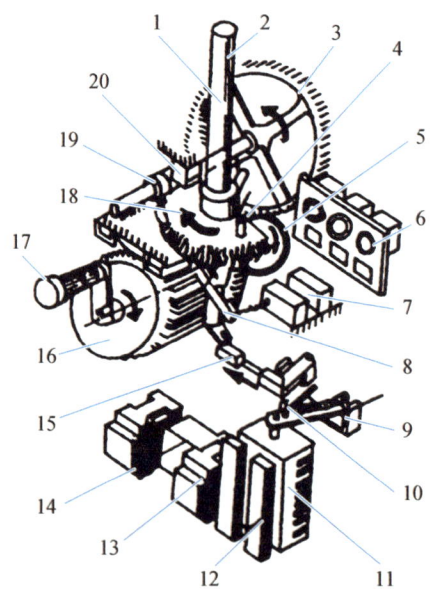

1—主轴；2—键；3—大齿轮；4—挡钉；5—小齿轮；6—按钮；7—限位开关；8—弹簧压片；
9—限位板；10—弹簧；11—辅助开关；12—接线板；13—接触器；14—热继电器；
15—连杆；16—电动机；17—摇把；18—蜗轮；19—蜗杆；20—限位块。

图 2-18 CJ2-XG 型电动操动机构传动原理图

电动分闸时，按下分闸按钮，分闸接触器的控制线圈接通，接触器触头闭合，使电动机线路接通，电动机驱动齿轮与蜗轮减速装置，带动与主轴相连的隔离开关或接地开关实现分闸。当主轴接近分闸终点位置时，装在蜗轮上的弹性压片使终点限位开关分开，切断分闸接触器的控制线圈的电流，接触器触头打开，切断电动机电源，机械限位装置使机构限制在分闸准确位置。

在分闸过程中，需要中途停止时，可按下停止按钮切断控制电源。

电动合闸时，按下合闸按钮，合闸接触器的控制线圈接通，接触器触头闭合，使电动机线路接通，主轴按分闸相反方向旋转，使隔离开关合闸。当主轴接近合闸终点位置时，终点限位开关分开，合闸接触器断电，触头断开电机电源，机构限位装置使机构限制在合闸准确位置。

对不配电磁锁的机构，可用手柄直接操作电机轴，进行分合闸操作。对装设有电磁锁的机构，先按一下电磁锁上的按钮，若指示灯亮，表示允许开锁，可以进行人力操作，这时将电磁锁上的拉板向右拉动，手动操作轴挡板被拉开。将手柄插入蜗杆轴上进行操作，操作完毕，将手柄取出电磁锁锁栓复位，使电动操作轴挡板返回原位。按下按钮后，若指示灯不亮，表明不允许人力操作。紧急情况下人力操作时，须经批准，先将机构电源开关拉开，取来应急钥匙，插入电磁锁钥匙孔中，按顺时针方向转90°后，即可将电磁锁拉开，插入手柄，进行人力操作。

2. CJ6 型电动操动机构

CJ6 型电动操动机构可用于 GW4 隔离开关的操作，其结构如图 2-19 所示，实物如图 2-20 所示，由电动机、机械减速传动系统、电气控制系统及箱壳组成。

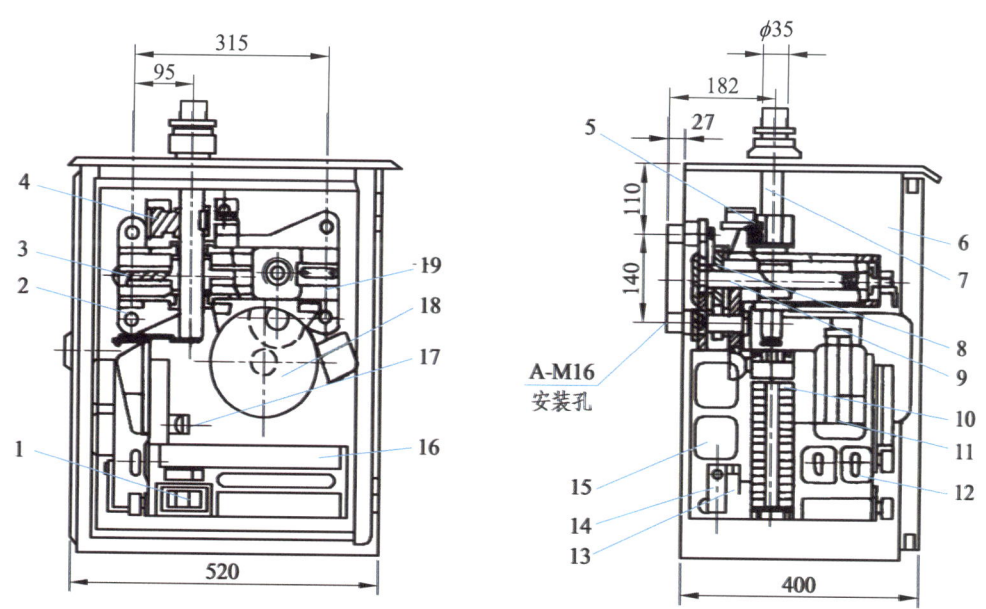

1—按钮；2—框架；3—蜗轮；4—定位件；5—行程开关；6—箱；7—主轴；8—齿轮；9—蜗杆；
10—辅助开关；11—刀开关；12—组合开关；13—加热器；14—热继电器；15—接触器；
16—接线端子；17—照明灯座；18—电动机；19—手动闭锁开关。

图 2-19 CJ6、CJ6-I 型电动操动机构示意图

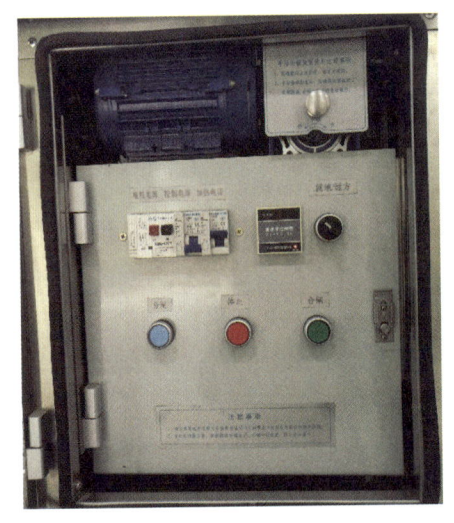 

（a）操作面板　　　　　　　　　　（b）操作面板背面及机构箱内部

图 2-20　电操机构实物图

电动机为三相交流异步电动机；机械减速传动系统包括齿轮、蜗杆、蜗轮及输出转轴。输出转轴用钢管连接，使隔离开关主开关或接地开关分、合闸；蜗杆端部为方轴，供手动摇柄进行手动操作。

电气控制部分包括电源转换开关、控制按钮（分、合、停各一个）、交流接触器、行程开关、热继电器及辅助开关等。

箱壳由钢板制成，起支撑及保护作用，在正面及侧面各有一门。

电气控制系统控制电动机，电动机经两对齿轮传递给蜗杆—蜗轮，带动输出主轴。减速系统三级减速，第一、二级为齿轮减速，第三级为蜗杆蜗轮减速。齿轮减速使用规格不同的齿轮可组成两种传动比，因此使总的传动比也有两种：第一种使电动机构分闸或合闸一次的动作时间为 7.5 s，第二种使电动机构分闸或合闸一次的动作时间为 3 s。

隔离开关电操机构的原理实际上是一个带行程开关的电动机正反转控制电路，如图 2-21 所示。操作操动机构时，先将电源转换开关接通电源，分闸时，按下分闸按钮（或远方控制），将分闸用交流接触器的控制线圈接通，分闸接触器触头闭合，使三相交流电接通，电动机向分闸方向旋转，通过二级齿轮变速，再经蜗杆、蜗轮减速后将力矩传送给机构主轴，使主轴旋转 180°。当主轴至分闸终点位置时，装在主轴上的定位件使微动开关动作，切断分闸接触器的控制线圈电流，触头分开，随之电动机三相电源也被切断。装在盖板上的橡皮缓冲定位装置，使机构主轴转动角度准确限制为 180°。

合闸时，按下合闸按钮，合闸接触器触头闭合，主轴按分闸相反方向旋转使隔离开关合闸，其程序原理与分闸相同。

除分、合闸按钮外，还设有停止按钮以满足异常情况下使用，当发生异常情况，可立即按"停"，机构停止转动。

机构主轴下装有六常开、六常闭或八常开、八常闭的辅助开关，供电器联锁及信号指示之用。为了避免当电动机过载，机械卡死或发生其他意外情况而烧坏电动机，箱内控制板上装有热继电器，电流整定使电动机短路过载时 20～25 s 动作。

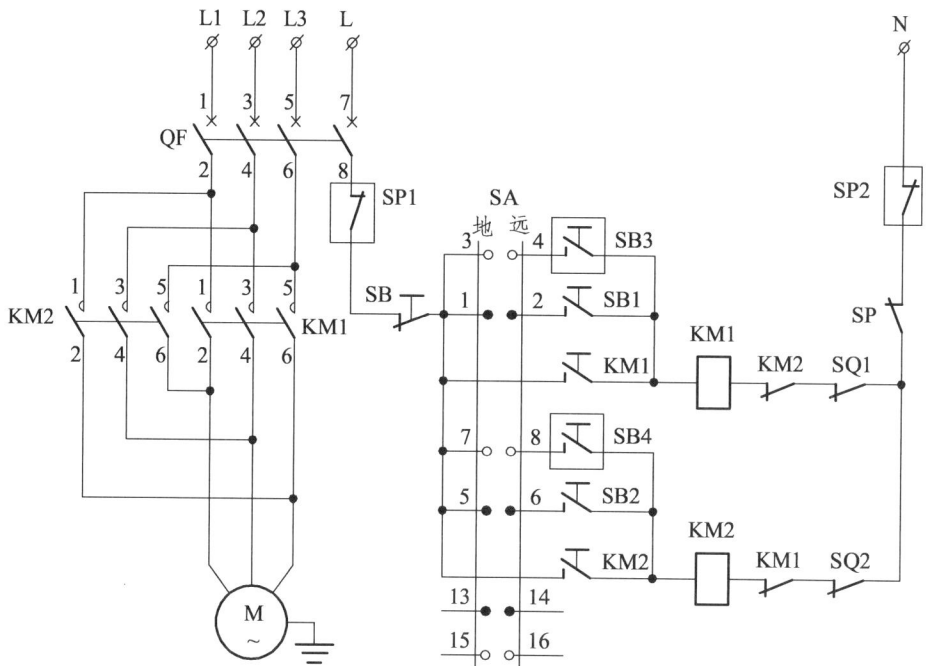

图 2-21 隔离开关电动机构原理图

Module 3 Operating Mechanism of Disconnector

2.3.1 Manual Operating Mechanism

When the manual operating mechanism is adopted, it must be operated locally on the installation site of disconnector. The manual operating mechanism is widely used because of its simple structure, low price, less maintenance workload and timely inspection of contact of the contact terminal after closing.

There are two kinds of manual operating mechanisms: lever type and worm gear type. The former is generally applicable to the disconnectors with rated current less than 3000 A, while the latter is generally suitable for the disconnectors with rated current greater than 3000 A.

(1) Lever-type manual operating mechanism.

CS6 manual lever-type operating mechanism is mainly used for indoor HV disconnectors, and its structural schematic diagram is shown in Fig. 2-15. In the figure, the solid line indicates the closed position of the disconnector; the dotted line indicates the open position of the disconnector; the arrow indicates the rotating direction of the handle when the disconnector is opened and closed.

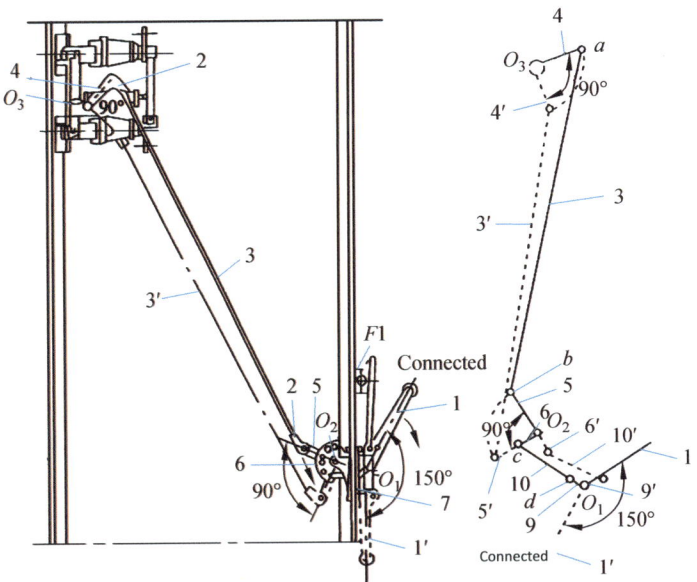

1-Handle; 2-Joint; 3-Drawbar; 4-Crank arm; 6-Fan-shaped rod; 7-Base; 5, 8, 9, 10-Connecting rod.

Fig. 2-15 Schematic Diagram of CS6 Manual Lever Operating Mechanism

When the disconnector is in the closed position, the hinge axis d of the connecting rods and is below the dead point position, so as to prevent the knife switch from opening by itself due to the action of electrodynamic force when the short-circuit current passes through the disconnector. During the opening operation, the pin at the O_1 axis is pulled out, so that the handle rotates clockwise downward by 150°. Then, the connecting rod rotates clockwise upward by 150°, and the fan-shaped rod rotates counterclockwise downward by 90° through the connecting rod. The traction rod is pulled downward, driving the crank arm to rotate clockwise downward by 90°, so that the disconnector is opened and the pin at the O_1 axis is automatically locked. The closing operation sequence is the opposite.

There are several pairs of contacts in the auxiliary contact box F, and their common small shafts are linked with the handle through rods and. These contacts are used for secondary circuits such as signal and interlock.

(2) Worm gear type.

The installation diagram of CS9 manual worm gear operating mechanism is shown in Fig. 2-16. In the figure, the connecting rod is hinged with the narrow plate, and the narrow plate is rigidly connected with the traction rod. During operation, shake the crank, drive the worm wheel to rotate by the worm gear, and make the disconnector open and closed by the connecting rod system. Shake the crank clockwise to make the worm gear rotate 180°, and the disconnector will be completely closed; shake the crank handle anticlockwise to make the worm gear reversely rotate180°, and the disconnector will be completely opened.

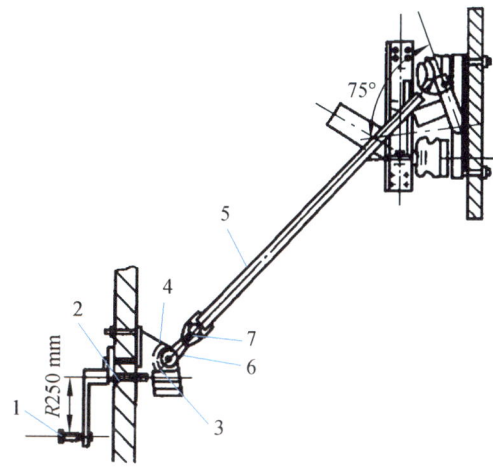

1-Cranking bar; 2-Axle; 3-Worm; 4-Worm gear; 5-Drawbar; 6-Connecting rod; 7-Narrow plate.

Fig. 2-16 Installation Diagram of CS9 Manual Worm Gear Operating Mechanism

2.3.2 Electric Operating Mechanism

1. Function of electric mechanism

(1) Complete the opening and closing of the disconnector.

The basic working principle of electric operating mechanism is to provide energy for the transmission mechanism of the disconnector through the motor. The motor drives the transmission mechanism of the disconnector through forward, reverse or cyclic rotation to complete the opening and closing of the disconnector.

(2) Provide auxiliary signals of the disconnector to the monitoring system.

The mechanism drives the auxiliary switch through the transmission shaft to complete the set action and give the auxiliary signal. By manually operating the electrical component, the component sends an auxiliary status signal to the terminal after the component acts. The status information of the component is dynamically collected at the terminal.

2. Typical electric operating mechanism

(1) CJ2-XG electric operating mechanism.

The mechanism is an outdoor dynamic mechanism, which is used for opening and closing of high-voltage disconnectors or grounding switches such as GW4 and GW7. It can be remotely controlled, electrically controlled locally, or manually operated with a handle.

CJ2-XC electric operating mechanism is shown in Fig. 2-17. The motor drives the gear and worm gear decelerator to transmit the torque to the output shaft which is installed vertically. The mechanism is equipped with opening/closing end limit switches and mechanical limit devices to limit the rotation angle of the main shaft of the mechanism to an accurate position. The mechanism is equipped with a handle for manually opening and closing on site.

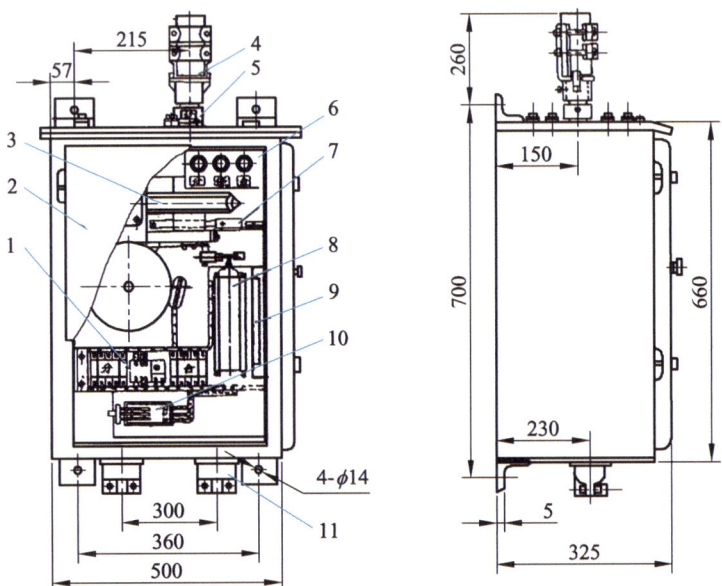

1-Contactor and thermal relay; 2-Mechanism box; 3-Decelerator mechanism; 4-Connector; 5-Opening and closing position indicator; 6-Operation button; 7-Limit switch; 8-Auxiliary switch; 9, 10-Terminal block; 11-Outgoing line.

Fig. 2-17 CJ2-XG Electric Operating Mechanism

There is a knife switch and a protective fuse in the mechanism box as well as a opening/closing button operated by electric control in front of the machine. Besides, a crank can also be used for manual opening and closing operation. The mechanism is equipped with six-switch-opening and six-switch-closing auxiliary switches, and the rotating shaft drives the auxiliary switch. When the disconnector is in the closed or open position, a corresponding signal will be sent. In order to facilitate installation and maintenance, the mechanism box has a three-side door structure. The front door can be opened with a special key, and the two-side doors can be opened after unscrewing the butterfly nut from both sides in the box.

The working principle of CJ2-XC electric operating mechanism is shown in Fig. 2-18.

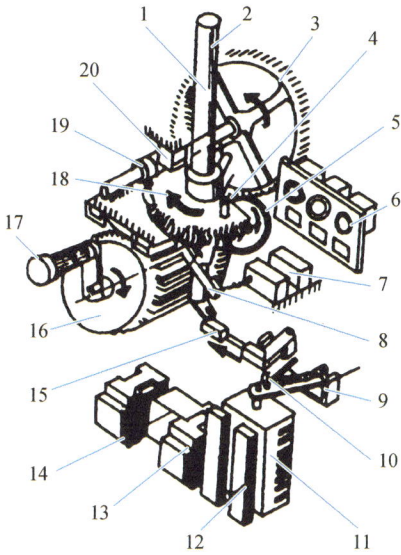

1-Main shaft; 2-Key; 3-Large gear; 4-Stop nail; 5-Small gear; 6-Button; 7-Limits witch; 8-Spring tablet; 9-Limit plate; 10-Spring; 11-Auxiliary switch; 12-Terminal block; 13-Contactor; 14-Thermal relay; 15-Connecting rod; 16-Motor; 17-Cranking bar; 18-Worm gear; 19-Worm; 20-Stop block.

Fig. 2-18 Transmission Schematic Diagram of CJ2-XG Electric Operating Mechanism

In case of electric opening, after pressing the opening button, the control coil of the opening contactor is on, and the contact terminal of the contactor is closed, so that the motor circuit is connected, and the motor drives the gear and the worm gear decelerator to drive the disconnector or the grounding switch connected with the main shaft to realize opening. When the main shaft is close to the end position of opening, the elastic tablet installed on the worm gear separates the end limit switch, and cuts off the current of the control coil of the opening contactor. The contact terminal of the contactor is opened, and the power supply of the motor is cut off. The mechanical limit device restricts the mechanism to the accurate open position.

In the process of opening, when it is necessary to stop halfway, the stop button can be pressed to cut off the control power supply.

During electric closing, after pressing the closing button, the control coil of the closing contactor is on, the contact terminal of the contactor is closed, so that the motor circuit is connected. The main shaft rotates in the opposite direction of opening to make the disconnector is closed. When the main shaft is close to the end position of closing, the end limit switch is separated and the closing contactor is power-off. The contact terminal is power-off, and the mechanism limit device limits the mechanism to the accurate closed position.

For the mechanism without the electromagnetic lock, the motor shaft can be directly operated by the handle for opening and closing operation. For the mechanism equipped with the electromagnetic lock, first press the button on the electromagnetic lock. If the indicator light is on, it means that unlocking is allowed and manual operation can be carried out. At this time, the pull plate on the electromagnetic lock is pulled to the right, and the baffle plate of the manual operation shaft is pulled open. Insert the handle into the worm shaft for operation. When the operation is completed, take out the handle and the electromagnetic lock bolt is reset, so that the baffle plate of the electric operation shaft returns to its original position. After pressing the button, if the indicator light is not on, it means that manual operation is not allowed. The emergency manual operation should be approved. First, pull open the power switch of the mechanism, take the emergency key, and insert the key into the keyhole of the electromagnetic lock. After rotating 90° clockwise, the electromagnetic lock can be pulled open and inserted into the handle for manual operation.

(2) CJ6 electric operating mechanism.

CJ6 electric operating mechanism can be used to operate GW4 disconnector. Its structure is shown in Fig. 2-19, and its physical object is shown in Fig. 2-20. It consists of motor, mechanical deceleration transmission system, electrical control system and box shell.

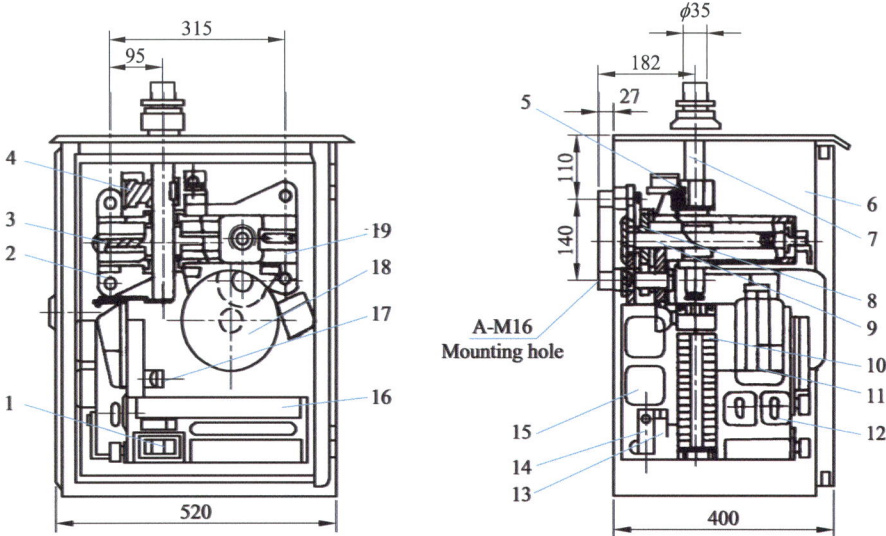

1-Button; 2-Frame; 3-Worm gear; 4-Locating piece; 5-Travel switch; 6-Box; 7-Main shaft; 8-Gear; 9-Worm; 10-Auxiliary switch; 11-Knife switch; 12-Combined switch; 13-Heater; 14-Thermal relay; 15-Contactor; 16-Wiring terminal; 17-Lighting lampholder; 18-Motor; 19-Manual locking switch.

Fig. 2-19   Schematic Diagram of CS6 and CJ6-I Electric Operating Mechanism

(a) Operation Panel

(b) Back of Operation Panel and Interior of Mechanism Box

Fig. 2-20   Physical Diagram of Electric Operating Mechanism

The motor is a three-phase AC asynchronous motor. The mechanical deceleration transmission system includes gear, worm, worm gear and output shaft. The output rotating shaft is connected with the steel pipe, so that the main switch of the disconnector or grounding switch is open and closed. A square shaft is at the end of the worm, which is used for manual operation by manually shaking the handle.

The electrical control part includes power supply switch, control buttons (one for opening, closing and stop), AC contactor, travel switch, thermal relay and auxiliary switch.

The box shell is made of steel plates, which plays the role of support and protection, and there is a door on the front and side respectively.

The electric control system controls the motor, and the motor is transmitted to the worm-worm gear through two pairs of gears to drive the output main shaft. The deceleration system has three-stage deceleration. The first and second stages are gear deceleration, and the third stage is worm and worm gear deceleration. According to the gears with different specifications, gear deceleration can form two transmission ratios, so there are two total transmission ratios: in the first one, it takes 7.5 s for the electric mechanism to open or close, and in the second one, it takes 3 s for the electric mechanism to open or close.

The principle of the electric operating mechanism for disconnectors is essentially a forward and reverse control circuit of an electric motor with limit switches, as shown in Fig. 2-21. When operating the operating mechanism, the power supply switch is first turned on. During the opening operation, the opening (or remote control) button is pressed, energizing the control coil of the AC contactor used for opening. The contact terminals of the opening contactor close, allowing three-phase AC power to flow and the electric motor rotates in the opening direction. The torque is transmitted to the main shaft of the mechanism through secondary gear reduction, and then further reduced through a worm stem and worm gear, causing the main shaft to rotate 180°. When the main shaft reaches the end position for opening, the positioning device mounted on the main shaft actuates a microswitch, cutting off the control coil current of the opening contactor, and the contact terminals open, disconnecting the three-phase power supply to the electric motor. The rubber cushioning positioning device mounted on the cover plate ensures that the rotation angle of the main shaft is accurately limited to 180°.

To close the disconnector, the closing button is pressed to close the contact terminals of the closing contactor. The main shaft rotates in the opposite direction to perform the closing operation, following the same principle as the opening operation.

In addition to the opening and closing buttons, a stop button is provided for use in case of abnormal situations. When an abnormal situation occurs, pressing the "stop" button immediately stops the mechanism from rotating.

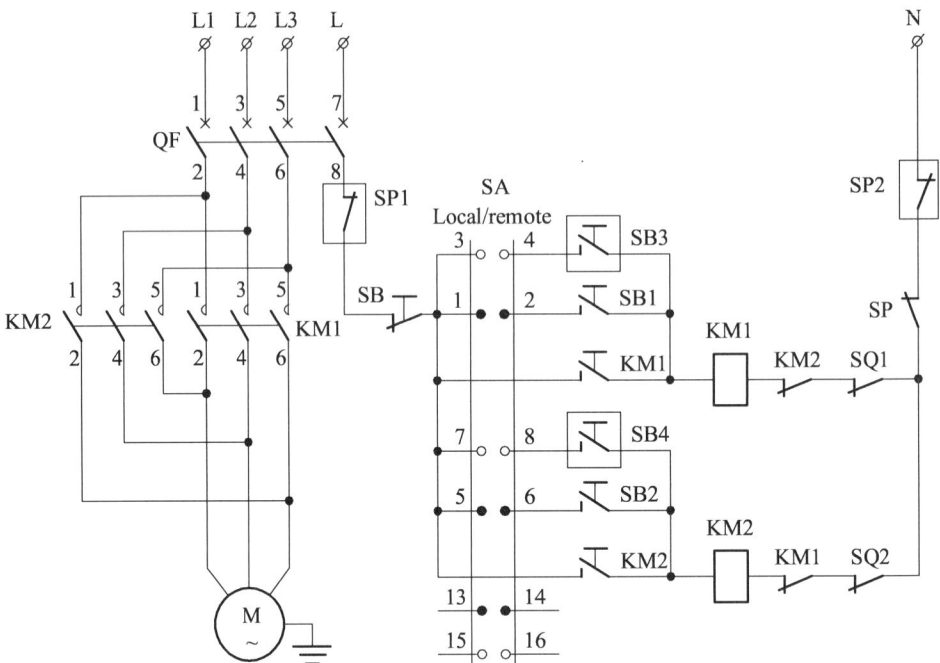

Fig. 2-21　Schematic Diagram of Electric Mechanism of Disconnector

The main shaft of the mechanism is equipped with auxiliary switches, either six normally open and six normally closed or eight normally open and eight normally closed, for electrical interlocking and signal indication purposes. To prevent the electric motor from burning out due to overload, mechanical jamming, or other unexpected situations, a thermal relay is installed on the control board inside the cabinet. The current setting of the thermal relay allows the electric motor to trip in 20 to 25 seconds in case of short-circuit overload.

模块四　隔离开关的运行与维护

一、高压隔离开关运行规定

隔离开关运行与维护工作必须遵守已颁布的标准及技术规程，同时结合各变电站（所）地理环境等实际情况编制的现场运行措施及制度执行。

（一）一般规定

（1）隔离开关应满足装设地点的运行工况，在正常运行和检修或发生短路情况下应满足安全要求。

（2）隔离开关和接地开关所有部件和箱体上，尤其是传动连接部件和运动部位不得有积水出现。

（3）隔离开关应有完整的铭牌、规范的运行编号和名称，相序标志明显，分合指示、旋转方向指示清晰正确，其金属支架、底座应可靠接地。

（二）导电部分

（1）隔离开关导电回路长期工作温度不宜超过 80 ℃。

（2）隔离开关在合闸位置时，触头应接触良好，合闸角度应符合产品技术要求。

（3）隔离开关在分闸位置时，触头间的距离或打开角度应符合产品技术要求。

（三）绝缘子

（1）绝缘子爬电比距应满足所处地区的污秽等级，不满足污秽等级要求的应采取防污闪措施。

（2）定期检查隔离开关绝缘子金属法兰与瓷件的胶装部位防水密封胶的完好性，必要时联系检修人员处理。

（3）未涂防污闪涂料的瓷质绝缘子应坚持"逢停必扫"，已涂防污闪涂料的绝缘子应监督涂料有效期限，在其失效前复涂。

（四）操动机构和传动部分

（1）隔离开关与其所配装的接地开关间有可靠的机械闭锁，机械闭锁应有足够的强度，电动操作回路的电气联锁功能应满足要求。

（2）接地开关可动部件与其底座之间的铜质软连接的截面面积应不小于 50 mm²。

（3）隔离开关电动操动机构操作电压应在额定电压的 85%～110%之间。

（4）隔离开关辅助接点应切换可靠，操动机构、测控、保护、监控系统的分合闸位置指示应与实际位置一致。

（5）同一间隔内的多台隔离开关的电机电源，在端子箱内应分别设置独立的开断设备。

（6）操动机构箱内交直流空开不得混用，且与上级空开满足级差配置的要求。

（7）电动操动机构的隔离开关手动操作时，应断开其控制电源和电机电源。

（8）电动操作时，隔离开关分合到位后电动机应自动停止。
（9）接地开关的传动连杆及导电臂(管)上应按规定设置接地标识。

（五）其他

（1）机构箱应设置可自动投切的驱潮加热装置，定期检查驱潮加热装置运行正常、投退正确。
（2）应结合设备停电对机构箱二次设备进行清扫。

（六）紧急停运规定

发现下列情况，应立即向值班调控人员申请停运处理：
（1）线夹有裂纹、接头处导线断股散股严重。
（2）导电回路严重发热达到危急缺陷，且无法倒换运行方式或转移负荷。
（3）绝缘子严重破损且伴有放电声或严重电晕。
（4）绝缘子发生严重放电、闪络现象。
（5）绝缘子有裂纹。
（6）其他根据现场实际认为应紧急停运的情况。

二、高压隔离开关的运行

1. 高压隔离开关的例行巡视

投入电网运行和处于备用状态的高压隔离开关，正常例行巡视检查项目及标准如表 2-2 所示。

表2-2　隔离开关例行巡视检查项目及标准

序号	巡视项目	巡视标准
1	导电部分	（1）合闸状态的隔离开关触头接触良好，合闸角度符合要求；分闸状态的隔离开关触头间的距离或打开角度符合要求，操动机构的分、合闸指示与本体实际分、合闸位置相符； （2）触头、触指（包括滑动触指）、压紧弹簧无损伤、变色、锈蚀、变形，导电臂（管）无损伤、变形现象； （3）引线弧垂满足要求，无散股、断股，两端线夹无松动、裂纹、变色等现象； （4）导电底座无变形、裂纹，连接螺栓无锈蚀、脱落现象； （5）均压环安装牢固，表面光滑，无锈蚀、损伤、变形现象
2	绝缘子	（1）绝缘子外观清洁，无倾斜、破损、裂纹、放电痕迹或放电异声； （2）金属法兰与瓷件的胶装部位完好，防水胶无开裂、起皮、脱落现象； （3）金属法兰无裂痕，连接螺栓无锈蚀、松动、脱落现象
3	传动部分	（1）传动连杆、拐臂、万向节无锈蚀、松动、变形现象； （2）轴销无锈蚀、脱落现象，开口销齐全，螺栓无松动、移位现象； （3）接地开关平衡弹簧无锈蚀、断裂现象，平衡锤牢固可靠；接地开关可动部件与其底座之间的软连接完好、牢固

续表

序号	巡视项目	巡视标准
4	基座、机械闭锁及限位部分	（1）基座无裂纹、破损，连接螺栓无锈蚀、松动、脱落现象，其金属支架焊接牢固，无变形现象； （2）机械闭锁位置正确，机械闭锁盘、闭锁板、闭锁销无锈蚀、变形、开裂现象，闭锁间隙符合要求； （3）限位装置完好可靠
5	操动机构	（1）隔离开关操动机构机械指示与隔离开关实际位置一致； （2）各部件无锈蚀、松动、脱落现象，连接轴销齐全
6	其他	（1）名称、编号、铭牌齐全清晰，相序标识明显； （2）超 B 类接地开关辅助灭弧装置分合闸指示正确、外绝缘完好无裂纹、SF_6 气体压力正常； （3）机构箱无锈蚀、变形现象，机构箱锁具完好，接地连接线完好； （4）基础无破损、开裂、倾斜、下沉，架构无锈蚀、松动、变形现象，无鸟巢、蜂窝等异物； （5）接地引下线标志无脱落，接地引下线可见部分连接完整可靠，接地螺栓紧固，无放电痕迹，无锈蚀、变形现象； （6）五防锁具无锈蚀、变形现象，锁具芯片无脱落损坏现象； （7）原存在的设备缺陷是否有发展

2. 高压隔离开关的全面巡视

全面巡视在例行巡视的基础上增加以下项目：

（1）隔离开关"远方/就地"切换把手、"电动/手动"切换把手位置正确。

（2）辅助开关外观完好，与传动杆连接可靠。

（3）空气开关、电动机、接触器、继电器、限位开关等元件外观完好。二次元件标识、电缆标牌齐全清晰。

（4）端子排无锈蚀、裂纹、放电痕迹；二次接线无松动、脱落，绝缘无破损、老化现象；备用芯绝缘护套完备；电缆孔洞封堵完好。

（5）照明、驱潮加热装置工作正常，加热器线缆的隔热护套完好，附近线缆无烧损现象。

（6）机构箱透气口滤网无破损，箱内清洁无异物，无凝露、积水现象。

（7）箱门开启灵活，关闭严密，密封条无脱落、老化现象，接地连接线完好。

（8）五防锁具无锈蚀、变形现象，锁具芯片无脱落损坏现象。

3. 高压隔离开关的熄灯巡视

重点检查隔离开关触头、引线、接头、线夹有无发热，绝缘子表面有无放电现象。

4. 高压隔离开关的特殊巡视

（1）新安装或 A、B 类检修后投运的隔离开关应增加巡视次数，巡视项目按照全面巡视执行。

（2）异常天气时的巡视：

① 大风天气时，检查引线摆动情况，有无断股、散股，均压环及绝缘子是否倾斜、断裂，

各部件上有无搭挂杂物。

② 雷雨天气后，检查绝缘子表面有无放电现象或放电痕迹，检查接地装置有无放电痕迹。

③ 大雨、连阴雨天气时，检查机构箱、端子箱有无进水，驱潮加热装置工作是否正常。

④ 冰雪天气时，检查导电部分是否有冰雪立即熔化现象，大雪时还应检查设备积雪情况，及时处理过多的积雪和悬挂的冰柱。

⑤ 覆冰天气时，观察外绝缘的覆冰厚度及冰凌桥接程度，覆冰厚度不超过 10 mm，冰凌桥接长度不宜超过干弧距离的 1/3，爬电不超过第二伞裙，不出现中部伞裙爬电现象。

⑥ 冰雹天气后，检查引线有无断股、散股，绝缘子表面有无破损现象。

⑦ 大雾、重度雾霾天气时，检查绝缘子有无放电现象，重点检查污秽部分。

⑧ 高温天气时，检查触头、引线、线夹有无过热现象。

（3）高峰负荷期间，增加巡视次数，重点检查触头、引线、线夹有无过热现象。

（4）故障跳闸后，检查隔离开关各部件有无变形，触头、引线、线夹有无过热、松动，绝缘子有无裂纹或放电痕迹。

三、高压隔离开关的维护

高压隔离开关的维护工作应根据运行记录、缺陷情况，制定相应的维护措施，并尽可能配合停电机会进行，对负荷特别重的隔离开关根据运行情况，制定应急处理方案。

（1）对各导电部分及引线加以紧固，保证接触良好。

（2）清扫绝缘子表面，检查法兰及铁瓷结合部位；对 110 kV 及以上隔离开关支柱绝缘子按规定进行绝缘子探伤检查。

（3）清除传动机构各部分锈蚀，检查传动杆件、拐臂连接是否可靠，并对传动机构转动点加注润滑脂。

（4）检查操动机构内各元器件应完好且安装牢固，二次回路接线正确，接触良好；清除机械活动部分锈蚀，按规定加注润滑脂。

（5）电动、手动操作灵活，动作准确，分合闸位置正确。

（6）按规定完成隔离开关预防性试验项目要求的各项内容，试验结果应符合规程要求。

四、高压隔离开关的验收与投运

（一）高压隔离开关的验收

高压隔离开关的交接验收应按有关标准、规程的要求进行。

（1）隔离开关安装的基本要求：

① 隔离开关、接地开关导电管应合理设置排水孔，确保在分、合闸位置内部均不积水。垂直传动连杆应有防止积水的措施，水平传动连杆端部应密封。

② 传动连杆应采用装配式结构，不应在施工现场进行切焊配装。连杆应选用满足强度和刚度要求的热镀锌无缝钢管，无扭曲、变形、开裂。

③ 检查传动摩擦部位磨损情况，补充适合当地条件的润滑脂。

④ 单柱垂直伸缩式在合闸位置时，驱动拐臂应过死点。

⑤ 定位螺钉应按产品的技术要求进行调整，并加以固定。

⑥ 均压环无变形，安装方向正确，与本体连接良好，安装应牢固、平正，不得影响接线板的接线；安装在环境温度零度及以下地区或 500 kV 以上的均压环，应在均压环最低处打排水孔，排水孔位置、孔径应合理。

⑦ 检查破冰装置应完好。

⑧ 设备出厂铭牌齐全、运行编号、相序标志清晰可识别。

（2）隔离开关的安装资料包括：① 订货技术协议或技术规范；② 出厂试验报告；③ 使用说明书；④ 交接试验报告；⑤ 安装报告；⑥ 施工图纸。

（3）隔离开关的竣工验收外观检查标准：

① 隔离开关安装牢固，外表清洁，油漆完整，相序色标志正确，按规定接地。

② 隔离开关引线连接可靠，整齐美观。

③ 触头接触良好，位置正确，导电固定接触面涂有电力脂，导电活动接触面涂有中性凡士林。

④ 瓷瓶完好无裂纹、损伤，表面清洁；瓷、铁浇装处粘接牢固有防水措施。

⑤ 操动机构、传动装置、辅助开关、闭锁装置安装牢固，动作灵活，位置指示正确，各转动部分涂有润滑脂。

⑥ 隔离开关防误装置达到"五防"要求。

⑦ 操动机构箱门关闭良好，封堵严密，照明、加热、除湿装置工作正常。

⑧ 隔离开关分合闸位置符合技术条件要求，相间距离、带电部分对地距离满足有关规定。

⑨ 新安装或检修后的调试符合技术要求，安装、检修资料，产品的备品备件、专用工具按规定移交。

（二）隔离开关投运

（1）全部缺陷消除，运行单位组织人员对设备验收合格并办理移交手续。

（2）完善设备的调度名称编号，相应的标志应醒目齐全。

（3）技术手册及运行规程齐全，并根据系统运行方式，编制反事故预案。

（4）操作所需的专用工具、安全工器具、常用备品备件齐全、完整。

上述工作全部完结，投运手续按规定齐全完备，由设备所属主管部门按预先准备的投运方案组织投运。

Module 4 Operation and Maintenance of Disconnectors

2.4.1 Operating Regulations for High-voltage Disconnectors

The operation and maintenance of disconnectors must comply with the published standards and technical specifications, and should also be in line with the on-site operating measures and regulations developed based on the actual conditions of each substation, such as the geographical environments.

2.4.1.1 General regulations

(1) Disconnectors shall meet the operational conditions of the installation site and fulfill safety requirements during normal operation, maintenance, or short-circuit situations.

(2) There shall be no water accumulation on any components or enclosure of the disconnectors and grounding switches, especially on the transmission connection parts and motion parts.

(3) Disconnectors shall have complete nameplates, clear and standardized operation numbering and labels, obvious phase sequence marks, and clear and correct indications for opening and closing directions and rotation directions. The metal brackets and bases shall be reliably grounded.

2.4.1.2 Conduction parts

(1) The working temperature of the conduction circuit of disconnectors should not exceed 80°C in the long term.

(2) When the disconnector is in the closed position, the contact terminals shall have good contact and the closing angle should meet the technical requirements of the product.

(3) When the disconnector is in the open position, the distance or opening angle between the contact terminals shall meet the technical requirements of the product.

2.4.1.3 Insulators

(1) The creepage distance of insulators shall meet the pollution level of the area. For those that do not meet the pollution level requirements, anti-pollution measures shall be taken.

(2) The integrity of the water-sealing adhesive for the metal flanges and porcelain parts of disconnector insulators shall be regularly checked, and the maintenance personnel shall be contacted for treatment if necessary.

(3) Porcelain insulators without anti-pollution flashover coatings shall be cleaned on every outage. For those with anti-pollution flashover coatings, their effective period shall be monitored, and reapplication shall be performed before expiration.

2.4.1.4　Operating mechanism and transmission parts

(1) There shall be reliable mechanical interlocking between the disconnector and the grounding switch it is equipped with, and the mechanical interlocking shall have sufficient strength. The electrical interlocking function of the electric operation circuit shall meet the requirements.

(2) The cross-sectional area of the copper flexible connection between the movable components of the grounding switch and its base shall not be less than 50 mm^2.

(3) The operating voltage of the electric operating mechanism of the disconnector shall be between 85% and 110% of the rated voltage.

(4) The auxiliary contacts of the disconnector shall switch reliably, and the indications for the open and closed positions of the operating mechanism, measuring and control systems, protection, and monitoring systems shall be consistent with the actual positions.

(5) For the motor power supplies of multiple disconnectors in the same bay, independent switching devices shall be provided in the terminal box.

(6) AC and DC air switches shall not be mixed in the operating mechanism box, and they shall satisfy the requirements of graded configuration with the upstream air switch.

(7) When manually operating the disconnector of the electric operating mechanism, its control power and motor power shall be disconnected.

(8) During electric operation, the electric motor shall automatically stop when the disconnector reaches the final opening or closing position.

(9) Grounding marks shall be set according to regulations on the transmission connecting rod and conductive arm (tube) of the grounding switch.

2.4.1.5　Other regulations

(1) The mechanism box shall be equipped with automatically switched dehumidifying and heating devices, and regular checks shall be conducted to ensure the proper operation and correct switching of the dehumidifying and heating devices.

(2) The secondary equipment of the mechanism box shall be cleaned during equipment power outages.

2.4.1.6　Emergency shutdown regulations

In the following situations, immediate application for shutdown processing shall be made to the duty control personnel:

(1) There are cracks in line clamps, severe strand breaking, or loose strands at the joints of conductors.

(2) Severe heat is generated in the conduction circuit, which reaches a critical defect level, and in which condition the operation mode can't be alternated or the load transferred.

(3) Severe damage occurs to insulators accompanied by discharging noise or severe corona discharge.

(4) Serious discharging or flashover occurs to insulators.

(5) There are cracks in insulators.

(6) There are other situations that are considered urgent for shutdown based on the actual site conditions.

2.4.2 Operation of High-voltage Disconnector

1. Routine inspection of high-voltage disconnectors

Before being put into operation within the power grid or when in standby, high-voltage disconnectors shall undergo routine inspections as per the items and criteria shown in Table 2-2.

Table 2-2　Routine Inspection Items and Criteria for Disconnectors

S/N	Routine inspection items	Routine inspection criteria
1	Conduction part	(1) For disconnectors in the closed status, the contact terminals shall have good contact, and the closing angle shall meet the requirements. For disconnectors in the open status, the distance between contact terminals or the opening angle shall meet the requirements. The closing and opening indications of the operating mechanism shall correspond to the actual positions of the disconnector. (2) The contact terminals, contact fingers (including sliding contact fingers), and compression springs shall be free from damage, discoloration, corrosion, or deformation. The conductive arms (tubes) shall be free from damage or deformation. (3) The lead shall have sag meeting the requirements and be free from any loose or broken strands. The lead clamps on both ends shall be free from loosening, cracks, discoloration, or other abnormalities. (4) The conduction base shall be free from deformation or cracks, and the connecting bolts shall be free from corrosion or detachment. (5) The grading rings shall be securely installed, with a smooth surface and without any signs of rust, damage, or deformation
2	Insulator	(1) The appearance of insulators shall be clean and without any inclination, damage, cracks, discharge marks, or unusual sounds during discharge. (2) The bonding of metal flanges and ceramic parts shall be intact, and the waterproof adhesive shall be free from cracking, peeling, or detachment. (3) The metal flanges shall be free from cracks, and the connecting bolts shall be free from corrosion, loosening, or detachment
3	Transmission components	(1) Transmission connecting rods, crank arms, and universal joints shall be free from rust, loosening, or deformation. (2) The shaft pins shall be free from rust or detachment, and the open-end pins shall be complete. The bolts shall be free from loosening or displacement. (3) The balance spring of the grounding switch shall be free from rust or fractures, and the balancing hammer shall be firm and reliable. The flexible connection between the movable components of the grounding switch and its base shall be intact and secure
4	Base, mechanical interlocks, and limitation components	(1) The base shall be free from cracks or damage, and the connecting bolts shall be free from rust, loosening, or detachment. The metal support shall be securely welded without any deformation. (2) The mechanical interlock shall be in the correct position, and the mechanical interlock plate, interlock board, and interlock pin shall be free from rust, deformation, or cracks. The interlock clearance shall meet the requirements. (3) The limitation device shall be intact and reliable

Continued

S/N	Routine inspection items	Routine inspection criteria
5	Operating mechanism	(1) The mechanical indication of the disconnector's operating mechanism shall correspond to the actual positions of the disconnector. (2) All components shall be free from rust, loosening, or detachment, and the connecting shaft pins shall be complete
6	Others	(1) The names, numbers, and nameplates shall be complete and clear, with obvious phase sequence marking. (2) For Super-B Class grounding switches, the auxiliary arc extinguishing device shall indicate the correct closing and opening positions, and the external insulation shall be intact without cracks. The SF_6 gas pressure shall be normal. (3) The mechanism box shall be free from rust or deformation, and the locking device shall be intact. The grounding connection line shall be intact. (4) The foundation shall be free from damage, cracks, inclination, or sinking, and the structure shall be free from rust, loosening, or deformation. There shall be no foreign materials such as nests or honeycombs. (5) The grounding down lead markers shall be intact without detachment, and the visible part of the grounding down lead shall have a complete and reliable connection. The grounding screws shall be tightened, without any discharge marks, rust, or deformation. (6) The five-proof locks shall be free from rust or deformation, and the lock cylinder shall not be detached or damaged. (7) Whether any existing equipment defects are developing

2. Comprehensive inspection of high-voltage disconnectors

Comprehensive inspection includes the following additional items on top of routine inspection:

(1) The correct positioning of the "remote/local" switch and "electric/manual" switch handles of the disconnector shall be verified.

(2) It shall be ensured that the auxiliary switch is in good condition and reliably connected to the transmission connecting rod.

(3) The appearance of components such as air switches, motors, contactors, relays, limit switches, etc shall be intact. The secondary components shall have complete and clear identification labels and cable tags.

(4) The terminal blocks shall be free from any rust, cracks, or discharge marks; the secondary wiring shall be secure and without any looseness or detachment, and the insulation shall be free from any sing of damage or aging. The insulation sheath of spare core shall be complete and intact; and the cable duct shall be sealed.

(5) The illumination and dehumidification and heating devices shall function normally. The heat-insulating sheath of heater cables shall be intact, and there shall be no signs of burning in nearby cables.

(6) The vent screen of the mechanism box shall be inspected for any damage. There shall be no foreign objects inside the box and no condensation or water accumulation.

(7) The door of the box shall open and close smoothly, with a tight seal, and the sealing strip shall not be loose or aged. The integrity of the grounding connection wire shall be confirmed.

(8) The five-proof locks shall be free from rust or deformation, and the lock cylinder shall not be detached or damaged.

3. Dark inspection of high-voltage disconnectors

The focus shall be on checking the contact terminals, leads, connectors, and clamps of the disconnector for signs of overheating. Additionally, the insulator surface shall be inspected for any discharge phenomena.

4. Special inspection of high-voltage disconnectors

(1) For disconnectors newly installed or put into operation after Class A/B maintenance, the inspection frequency shall be increased and the inspection items shall follow the comprehensive inspection.

(2) Inspection during abnormal weather conditions:

① During strong wind weather, the swing of leads shall be examined for any broken or loose strands, and the grading ring and insulators shall be examined for inclination or fracture. Also, various components shall be checked for any debris attached.

② After thunderstorms, the insulator surface shall be inspected for any signs or traces of discharges, and the grounding device shall be examined for any discharge signs.

③ During heavy rain or continuous rainy weather, the mechanism box and terminal box shall be checked for water ingress, and the dehumidification and heating device shall be checked for proper functioning.

④ During icy or snowy weather, conducting parts shall be checked for immediate melting of ice or snow. In heavy snow, the equipment shall also be inspected for snow accumulation, and excessive snow and hanging icicles shall be promptly removed.

⑤ During icing weather, the ice thickness on the external insulators and the degree of icicle bridging shall be observed. The ice thickness shall not exceed 10 mm, and the icicle bridging length shall not exceed 1/3 of the dry arcing distance. The creepage distance shall not exceed the second skirt, and there shall be no creepage on the middle skirt.

⑥ After hailstorms, leads shall be inspected for any broken or loose strands, and the insulator surface shall be checked for any damage.

⑦ During heavy fog or severe haze weather, the insulators shall be inspected for any signs of discharge, with a focus on checking polluted areas.

⑧ During high-temperature weather, contact terminals, leads, and clamps shall be checked for any sign of overheating.

(3) During peak load periods, the inspection frequency shall be increased, with a focus on checking for overheating on contact terminals, leads, and clamps.

(4) After a fault trip, the disconnector components shall be inspected for any deformation, and contact terminals, leads, and clamps shall be checked for overheating or looseness. Additionally, insulators shall be inspected for any cracks or signs of discharge.

2.4.3　Maintenance of High-voltage Disconnectors

Maintenance work for high-voltage disconnectors shall be based on operational records and defect conditions. Corresponding maintenance measures shall be formulated, and whenever possible, maintenance should be coordinated with power outage opportunities. Emergency handling plans shall be developed for heavily loaded disconnectors based on their operational status.

(1) Conduction parts and leads shall be tightened to ensure good contact.

(2) The surface of insulators shall be cleaned, and the flanges and porcelain-metal joints shall be inspected. Post insulators for disconnectors rated at 110 kV and above shall undergo insulation testing as per regulations.

(3) Rust on all parts of the transmission mechanism shall be removed, and the transmission connecting rods and crank arm connections shall be inspected for reliability. Lubricating grease shall be applied to the rotating points of the transmission mechanism.

(4) Components inside the operating mechanism shall be checked for completeness and secure installation. Correct secondary circuit wiring and good contact shall be ensured. Rust on mechanical moving parts shall be removed, and lubricating grease shall be applied as specified.

(5) Electric and manual operations shall be flexible, and precise, and reach the correct open and closed positions.

(6) The required preventive test items for disconnectors shall be completed according to regulations, and the test results shall comply with the specifications.

2.4.4　Acceptance and Commissioning of High-voltage Disconnectors

1. Acceptance of high-voltage disconnectors

The handover acceptance of high-voltage disconnectors shall be conducted according to relevant standards and regulations.

(1) Basic installation requirements for disconnectors:

① The conduction tubes of disconnectors and grounding switches shall be reasonably designed with drainage holes to prevent internal water accumulation in both open and closed positions. Measures to prevent water accumulation shall be taken for vertical transmission connecting rods, and the ends of the horizontal transmission connecting rods shall be sealed.

② Transmission connecting rods shall adopt assembled structures and shall not be cut or welded at the construction site. The connecting rods shall be made of hot-dip galvanized seamless steel pipes that meet the strength and rigidity requirements without distortion, deformation, or cracking.

③ Frictional parts of the transmission shall be checked for wear, and appropriate lubricating grease suitable for local conditions shall be supplemented.

④ In single-column pantograph disconnectors, the drive crank arm shall pass the dead point when in the closed position.

⑤ Positioning screws shall be adjusted according to technical requirements for the product and be fixed.

⑥ Grading rings shall be free from deformation, correctly installed, securely connected to the main body, and installed firmly and levelly without affecting the wiring of the power strip. For grading rings installed in environments with temperatures below zero or rated above 500 kV, drainage holes shall be created at the lowest point with reasonable positions and hole sizes.

⑦ The ice-breaking devices shall be checked for completeness.

⑧ The equipment's factory nameplate, operating number, and phase sequence marking shall be clear and recognizable.

(2) Installation documents of disconnectors shall include: ① Technical agreement or specifications for orders. ② Factory test report. ③ User manual. ④ Handover test report. ⑤ Installation report. ⑥ Construction drawings.

(3) Appearance inspection criteria for the completion acceptance of disconnectors:

① The disconnector shall be installed firmly, with a clean appearance, intact paint, correct color markings of phase sequence, and grounded as specified.

② The connections of the disconnector leads shall be reliable and neat.

③ The contact terminals of the contacts shall be in good condition, positioned correctly, and coated with electrical grease on the fixed contact surface and neutral Vaseline on the movable contact surface.

④ The porcelain insulators shall be intact without cracks or damages, with clean surfaces. The joints between porcelain and iron castings shall be firmly bonded with waterproof measures.

⑤ The operating mechanism, transmission device, auxiliary switches, and interlocking devices shall be installed firmly, with flexible operation, correct position indication, and lubricating grease applied to all rotating parts.

⑥ The misoperation prevention devices of the disconnector shall meet the five-proof requirements.

⑦ The doors of the operating mechanism box shall be closed smoothly, sealed tightly, and the lighting, heating, and dehumidification devices shall work properly.

⑧ The open and closed positions of the disconnector shall comply with technical requirements. The distances between phases and between live parts and the ground shall meet relevant regulations.

⑨ Commissioning after new installation or maintenance shall meet technical requirements. Installation and maintenance data, spare parts, and special tools for the product shall be handed over as required.

2. Commissioning of high-voltage disconnectors

(1) All defects shall be eliminated, and the operating unit shall organize personnel to carry out equipment acceptance and complete the handover procedures.

(2) The equipment's dispatching name and number shall be well-established, and corresponding signs shall be prominent and complete.

(3) The technical manuals and operating procedures shall be complete, and emergency plans

shall be prepared based on system operating methods.

(4) Special tools, safety appliances, and common spare parts required for operation shall be complete and intact.

After all the above tasks are completed and commissioning procedures are fully prepared, the responsible department of the equipment shall organize the commissioning as planned.

模块五 隔离开关检修的基本要求

一、隔离开关检修的分类

依据企业相关规范,隔离开关检修工作分为四类:A 类检修、B 类检修、C 类检修、D 类检修,具体检修定义及检修项目参见表 2-3。

表 2-3 隔离开关检修分类表

检修分类	检修范围	检修项目	检修周期
A 类检修	整体性检修	包含整体更换、解体检修	按照设备状态评价决策进行,应符合厂家说明书要求
B 类检修	局部性检修	包含部件的解体检查、维修及更换	按照设备状态评价决策进行,应符合厂家说明书要求
C 类检修	例行检查及试验	包含本体及外观检查维护、操动机构检查维护及整体调试	基准周期 35 kV 及以下 4 年、11(066)kV 及以上 3 年,可依据设备状态、地域环境、电网结构等特点,在基准周期的基础上酌情延长或缩短检修周期,具体参见相关规程
D 类检修	在不停电状态下进行的检修	包含专业巡视、辅助二次元器件更换、金属部件防腐处理、传动部件润滑处理、箱体维护等不停电工作	依据设备运行工况,及时安排,保证设备正常功能

二、隔离开关检修作业的基本要求

(一)检修作业的安全要求

检修作业人员必须严格执行《国家电网公司电力安全工作规程》及相关规程规定,明确停电范围、工作内容、停电时间。

(1)施工用电设施安装完毕后,应有专业班组或指定专人负责运行及维护。

(2)现场如需进行电、气焊工作,要办理动火手续,由专业人员操作。检查电机电源和控制电源确已断开,二次电源隔离措施符合现场实际条件。

(3)隔离开关检修前必须对检修作业危险点进行分析,结合现场实际条件适时装设个人保安线。施工现场的大型机具及电动机具金属外壳接地良好、可靠。每次检修作业前,应针对被检修隔离开关的具体情况,对危险点进行详细分析,做好充分的预防措施,并组织所有检修人员共同学习。

(4)在隔离开关转动前,要进行认真检查;隔离开关转动时,应密切注视设备的动作情况,防止绝缘子断裂等造成人身伤害和设备损坏。工作中禁止将安全带系在支柱绝缘子及均压环上。

（二）检修作业的技术要求

隔离开关各检修部位的检修项目及技术要求见表 2-4。

表 2-4　隔离开关的检修项目和技术要求

检修部位	检修项目	技术要求
导电部分	1. 主触头的检修。 2. 触头弹簧的检修。 3. 导电臂的检修。 4. 接线座的检修	1. 主触头接触面无过热、烧伤痕迹，镀银层无脱落现象。 2. 触头弹簧无锈蚀、分流现象。 3. 导电臂无锈蚀、起层现象。 4. 接线座无腐蚀，转动灵活，接触可靠。 5. 接线板应无变形、无开裂，镀层应完好
机构和传动部分	1. 轴承座的检修新换轴销应采用防腐材料。 2. 轴套、轴销的检修。 3. 传动部件的检修平连杆端部应密封，内部无积水。 4. 机构箱检查。 5. 辅助开关及二次元件检查机构箱门无变形。 6. 机构输出轴的检查。 7. 主刀闸和接地刀闸的联锁的检修	1. 轴承座应采用全密封结构，加优质二硫化钼锂基润滑脂。 2. 轴套应具有自润滑措施，应转动灵活，无锈蚀。 3. 传动部件应无变形、无锈蚀、无严重磨损，水辅助开关与传动杆的连接可靠。 4. 机构箱应达到防雨、防潮、防小动物等要求。 5. 二次元件及辅助开关接线无松动，端子排无锈蚀。 6. 机构输出轴与传动轴的连接紧密，挡销无松动。 7. 主刀闸与接地刀闸的机械联锁可靠，具有足够的机械强度，电气闭锁动作可靠
绝缘子	绝缘子检查	1. 绝缘子完好、清洁，无掉瓷现象，上下节绝缘子同心度良好。 2. 法兰无开裂，无锈蚀，油漆完好。法兰与绝缘子的结合部位应涂防水胶。 3. 超声波探伤无异常

三、隔离开关检修作业的主要工作流程

（一）准备工作

（1）接受任务后进行现场勘察，收集技术资料，并熟悉图纸和安装检修工艺。

（2）编制作业指导书或"三措"（安全措施、技术措施和组织措施），危险点分析及预控措施，并审批。

（3）准备工器具，编制材料计划，并领取。

（4）场地准备。

（5）在开工前召开班前会，学习作业指导书，进行安全和技术交底，落实危险点分析及预控措施。

（二）开工作业

（1）办理工作票。

（2）安装或检修前的设备检查。

（3）作业中的工艺流程及质量标准应符合技术规范要求。

（4）作业中应严格执行安全措施要求。

（三）收尾工作

（1）工作结束后应进行班组自检并会同验收人员验收对各项检修、试验项目进行验收。

（2）按相关规定，关闭检修和试验电源。

（3）清理工作现场，将工器具全部收拢并清点，废弃物按相关规定处理，材料及备品备件回收清点。

（4）会同验收人员对现场安全措施及检修设备的状态进行检查，要求恢复至工作许可时状态。

（5）工作人员全部撤离工作现场。

（6）填写记录报告，并提交技术文件资料，办理工作票终结手续。

（7）班会总结，验收资料整理，并存档保管。

Module 5 Basic Maintenance Requirments of Disconnectors

2.5.1 Classification of Disconnector Maintenance

In accordance with relevant enterprise standards, disconnector maintenance is divided into four class: Class A maintenance, Class B maintenance, Class C maintenance, and Class D maintenance. For specific definitions and maintenance items, please refer to Table 2-3.

Table 2-3 Disconnector Maintenance Classification

Maintenance category	Maintenance scope	Maintenance items	Maintenance cycle
Class A maintenance	Comprehensive maintenance	The maintenance shall include complete replacement and disassembly maintenance	The maintenance shall be carried out based on the evaluation and decisions of equipment status, and must comply with the manufacturer's specifications
Class B maintenance	Partial maintenance	The maintenance shall include the inspection, repair, and replacement of disassembled components	The maintenance shall be carried out based on the evaluation and decisions of equipment status, and must comply with the manufacturer's specifications
Class C maintenance	Routine inspection and testing	The maintenance shall include body and external appearance inspection and maintenance, operating mechanism inspection and maintenance, and comprehensive testing	The reference cycle shall be 4 years for 35 kV and below and 3 years for 110 (66) kV and above. The maintenance cycle can be extended or shortened based on equipment status, regional environment, power grid structure, etc. Refer to relevant regulations for details
Class D maintenance	Maintenance under live condition	The maintenance shall include specialized inspection, replacement of auxiliary secondary components, corrosion protection of metal components, lubrication of transmission components, and enclosure maintenance work	The maintenance shall be arranged promptly based on equipment operating conditions to ensure normal functionality

2.5.2 Basic Requirements for Disconnector Maintenance Operations

1. Safety requirements for maintenance operations

Maintenance personnel must strictly comply with the *Electric Power Safety Working Regulations of State Grid Corporation of China* and related regulations, and clearly define the power-off range, work content, and power-off time.

(1) After the installation of construction power facilities, a professional team or designated personnel shall be responsible for operation and maintenance.

(2) If electric or gas welding is required on-site, hot work procedures must be carried out by professional personnel. It shall be checked that the motor power supply and control power supply

has been disconnected and that secondary power isolation measures meet on-site conditions.

(3) Before disconnector maintenance, an analysis of the hazards at the work site must be conducted, and personal safety lines should be installed as needed in conjunction with actual site conditions. The large-scale machinery and electric tools on the construction site shall have a well-grounded and reliable metal casing. Prior to each maintenance operation, a detailed analysis of the hazards associated with the specific disconnector to be maintained shall be performed, and sufficient preventive measures shall be taken, with all maintainers participating in the process.

(4) Before turning the disconnector, a careful inspection must be carried out. During the rotation of the disconnector, the movement of the equipment shall be closely monitored to prevent personal injury or equipment damage caused by insulator fracture, etc. The safety belt shall not be attached to the insulator or grading ring during work.

2. Technical requirements for maintenance operations

For maintenance items and technical requirements for each part of the disconnector, refer to Table 2-4.

Table 2-4   Disconnector Maintenance Items and Technical Requirements

Maintenance part	Maintenance items	Technical requirements
Conduction part	1. Maintenance of main contact terminals. 2. Maintenance of contact terminal spring. 3. Maintenance of conductive arms. 4. Maintenance of wire holder	1. The contact surface of the main contact terminal shall be free from overheating and burn marks, and the silver plating shall not peel off. 2. The contact terminal springs shall be free from rust and bypass signs. 3. The conductive arms shall be free from rust and peeling. 4. The wire holder shall be free from corrosion, flexible in rotation, and have reliable contacts. 5. The power strip shall be free from deformation and cracks, and the plating shall be intact
Mechansim and tranmission parts	1. Maintenance of bearing seats. 2. Maintenance of sleeve and shaft pin. 3. Maintenance of transmission components. 4. Inspection of mechanism box. 5. Inspection of auxiliary switches and secondary components. 6. Inspection of mechanism output shaft. 7. Maintenance of main knife switch and grounding knife switch interlock	1. Bearing seats shall be of fully sealed structure and filled with high-quality molybdenum disulfide lithium-based grease. 2. The sleeves shall have self-lubricating measures, rotate flexibly, and be free from rust. Newly replaced shaft pins shall be made of anti-corrosion materials. 3. Transmission components shall be free from deformation, rust, and serious wear. The end of the horizontal connecting rod shall be sealed, with no water accumulation inside. 4. The mechanism box shall meet requirements for rain, moisture, and vermin protection, and the door of the mechanism box shall be free from deformation. 5. Secondary components and auxiliary switch connections shall be secure and without looseness, and terminal blocks shall be free from rust. The connection between the auxiliary switch and the transmission connecting rod shall be reliable. 6. The connection between the mechanism output shaft and the transmission shaft shall be tight, and the locking pin should not be loose. 7. The mechanical interlock of the main knife switch and grounding knife switch shall be reliable, with sufficient mechanical strength and reliable electrical interlocking operation

Continued

Maintenance part	Maintenance items	Technical requirements
Insulator	Inspection of insulators	1. Insulators shall be intact, clean, without chipped porcelain, and with good concentricity for the upper and lower sections of the insulator. 2. Flanges shall be free from cracks and rust and with intact paint. Waterproof glue shall be applied at the junction of the flange and the insulator. 3. Ultrasonic flaw detection shall show no anomalies

2.5.3 Main Workflow of Disconnector Maintenance Operations

1. Preparatory work

(1) Upon receiving the assignment, a site survey shall be conducted to gather technical information and familiarize oneself with drawings and installation and maintenance procedures.

(2) Work instructions or "three measures" (safety measures, technical measures, and organizational measures), hazard analysis, and preventive and control measures shall be prepared and approved.

(3) Tools and equipment shall be prepared, material plans shall be developed, and materials shall be collected.

(4) Site preparation shall be carried out.

(5) A pre-shift meeting shall be held before commencing work, during which the work instructions shall be reviewed, safety and technical briefings shall be conducted, and hazard analysis and preventive and control measures shall be confirmed.

2. Commencement of work

(1) Work ticket shall be obtained.

(2) Equipment checks shall be conducted prior to installation or maintenance.

(3) The process flow and quality standards during operations shall meet the requirements of technical specifications.

(4) Safety measures shall be strictly observed during operations.

3. Close-out activities

(1) After completing the work, a self-inspection by the shift team shall be conducted, and together with acceptance personnel, the inspection and acceptance of maintenance and test items shall be carried out.

(2) According to relevant regulations, maintenance and test power sources shall be cut off.

(3) Clean up the work site, collect and count all tools and instruments, dispose the waste according to relevant regulations, and recycle and count materials and spare parts.

(4) Together with acceptance personnel, a condition inspection of the site's safety measures and maintenance equipment shall be conducted, and it shall be ensured that everything is restored to the state of work permit issuance.

(5) All personnel shall exit the work site.

(6) Records and reports shall be completed, technical documentation shall be submitted, and the completion procedures for the work ticket shall be processed.

(7) A team debriefing shall be held. The acceptance documentation shall be organized, archived, and stored.

模块六 隔离开关常见故障分析及处理

一、隔离开关常见故障类型

隔离开关故障从整体结构分类可分为四种：导电回路故障、支柱式绝缘子故障、传动部分故障、操动机构故障。

（一）导电回路故障

1. 触头过热

（1）触指与触头接触不良，引起触头过热。
（2）触指、触头烧损严重，接触不良引起过热。
（3）触指弹簧失效，压力不够引起过热。
（4）各连接部分松动引起过热。

2. 接线座过热

（1）导电管与接线座接触不良引起过热。
（2）接线座内导电带两端接触面接触不良引起过热。
（3）出线端子与接线板接触不良引起过热。

（二）支柱式绝缘子故障

（1）支柱式绝缘子外绝缘闪络。
（2）支柱式绝缘子断裂。

（三）传动部分故障

（1）传动连杆轴销生锈卡死。
（2）转动轴承生锈损坏卡死。
（3）主刀闸与地刀闸闭锁板卡死。
（4）伞形齿轮脱齿。
（5）垂直连杆进水冬天冻冰，严重时使操作机构变形，无法操作。

（四）操作机构故障

（1）电动机主回路故障。
（2）控制回路公用部分故障。
（3）控制回路分闸部分故障。
① 分闸回路不通。
② 分闸回路通，但保持不住。
（4）控制回路合闸部分故障。
① 合闸回路不通。

② 合闸回路通但保持不住。
（5）分闸终了时电动机不停止或分闸不到位。
（6）合闸终了时电动机不停止或合闸不到位。

二、常见故障处理

（一）接触部分过热处理

（1）应停电处理，处理时应认真执行导电回路检修工艺及质量标准。

（2）解体检修时，严禁使用有缺陷的劣质线夹、螺栓等零部件，用压接式设备线夹替换螺栓式设备线夹，接头接触面要清洗干净并及时涂抹导电脂，螺栓使用正确、紧固力度适中。

（3）对过热频率较高的母线侧隔离开关，要保证检修到位、保证检修质量。对接线座部位，要重点检查导电带两端的连接情况，保证两端面清洁、平整、涂抹导电脂、压接紧密。对触头部位，要保证触头的光洁度，并涂抹中性凡士林，检查触头的烧伤情况，必要时要更换触头、触指，左触头的触指座要打磨干净，有过热、锈蚀现象的弹簧应更换。要保证三相分合闸同期，右触头的插入深度符合要求和两侧触指压力均匀。为检验检修质量，还应测量回路接触电阻，保证各接触面接触良好。

（4）对老型号的 GW4、GW5 型隔离开关左触头处过热，应采取加装分流带的处理方法，即在每个触指和触指座相应的地方，各钻一个 6 mm 螺孔，然后用螺钉将叠起的铜质软连接片固定在触指与触指座之间。

（5）对老型号的 GW4、GW5 型隔离开关左触头更换为新式触头，新式触头弹簧中间有绝缘块，消除了弹簧分流的可能性，使弹簧不易退火变形，弹性减弱。

（6）涂在隔离开关动触头及静触杆上导电膏的量不易掌握，致使开关发热。处理方法是针对这种活动导电接触面，应严格控制导电膏的涂抹量。首先将活动接触面使用无水酒精清洗干净，在导电面上抹一层均匀少量的导电膏，马上用布擦干净，使导电面上只留下微量的薄层导电膏。

（二）支柱式绝缘子断裂和闪络放电处理

（1）应停电处理，处理时应认真执行支柱式绝缘子检修工艺及质量标准。

（2）新支柱式绝缘子采用是高强度瓷柱，使用超声波无损探伤仪对瓷柱进行检测，测试合格后方可使用。

（3）对运行中的支柱式绝缘子加强维护工作，在探伤诊断良好的基础上，在瓷柱所在水泥结合面处涂敷绝缘子专用防护胶。

（4）更换新的瓷柱，增加爬电距离和瓷柱高度、提高整体绝缘水平。采取带电清扫，加强清扫力度，给隔离开关绝缘子增加硅橡胶伞裙以增大爬距和利用 RTV 涂料的憎水性喷涂 RTV。

（三）拒绝拉、合闸处理

1. 传动机构及传动系统造成的拒分拒合

（1）原因。机构箱进水，各部轴销、连杆、拐臂、底架甚至底座轴承锈蚀卡死，造成拒分拒合。

（2）处理方法。对传动机构及锈蚀部件进行解体检修，更换不合格元件。加强防锈措施，涂润滑脂，加装防雨罩。传动机构问题严重或有先天性缺陷时应更换。

2. 电气问题造成的拒分拒合

（1）原因。三相电源开关未合上、控制电源断线、电源熔丝熔断、热继电器误动切断电源、二次元件老化损坏使电气回路异常而拒动、电动机故障等原因都会造成电动机构分、合闸时，电动机不启动，隔离开关拒动。

（2）处理方法。电气二次回路串联的控制保护元器件较多，包括小型断路器、转换开关、交流接触器、限位开关及连锁开关、热继电器等。任一元件故障，就会导致隔离开关拒动。当按分合闸按钮不启动时，要首先检查操作电源是否完好，然后检查各相关元件。发现元件损坏时应更换，并查明原因。二次回路的关键是各个元件的可靠性，必须选择质量可靠的二次元件。

（四）分、合闸不到位

1. 机构及传动系统造成的分、合闸不到位

（1）原因。机构箱进水，各部轴销、连杆、拐臂、底架甚至底座轴承锈蚀，造成分合不到位。连杆、传动连接部位、闸刀触头架支撑件等强度不足断裂，造成分合闸不到位。

（2）处理方法。对机构及锈蚀部件进行解体检修，更换不合格元件。加强防锈措施，采用二硫化钼锂。更换带注油孔的传动底座。

2. 隔离开关分、合闸不到位或三相不同期

（1）原因。分、合闸定位螺钉调整不当。辅助开关及限位开关行程调整不当。连杆弯曲变形使其长度改变，造成传动不到位等。

（2）处理方法。检查定位螺钉和辅助开关等元件，发现异常进行调整，对有变形的连杆，应查明原因及时消除。此外，在操作现场，当出现隔离开关合不到位或三相不同期时，应拉开重合，反复合几次，操作时应符合要求，用力适当。如果还未完全合到位，不能达到三相完全同期，应安排计划停电检修。

（五）电动操动机构不动作

（1）机构问题主要表现为操作失灵，如拒动或分合闸不到位，往往发生在倒闸操作时，影响系统的安全运行。由于机构箱密封不好或进水造成机构锈蚀严重，润滑干涸，操作阻力增大，在操作困难的同时，还会发生零部件损坏，如变速齿轮断裂，连杆扭弯等。

（2）二次回路的可靠性将直接影响高压隔离开关的动作可靠性，辅助开关和行程开关切换不到位或者触点接触不良均会造成隔离开关拒动。接线端子接触不良、接触器不吸合、电动机烧坏、二次线绝缘破坏等会造成远方操作失灵。二次回路的关键是各个元件的可靠性，必须选用质量可靠的二次元件。

（3）应停电处理，处理时应认真执行操动机构检修工艺及质量标准。

Module 6　Common Fault Analysis and Handling of Disconnectors

2.6.1　Common Types of Disconnector Faults

Disconnector faults can be categorized into four types based on the overall structure: conduction circuit faults, post insulator faults, transmission component faults, and operating mechanism faults.

1. Conduction circuit faults

(1) Overheating of contact terminals.

① Poor contact between the contact finger and the contact terminal leads to overheating.

② Severe burning and damage of the contact finger and the contact terminal lead to poor contact and overheating.

③ Failure of the contact finger spring results in insufficient pressure and overheating.

④ Loose connections lead to overheating.

(2) Overheating of wire holder.

① Poor contact between the conduction pipe and the wire holder causes overheating.

② Poor contact at the contact surfaces on both ends of the conduction strip inside the wire holder causes overheating.

③ Poor contact between the outgoing terminal and power strip leads to overheating.

2. Post insulator faults

(1) Flashover occurs to the external insulation of post insulators.

(2) The post insulators break.

3. Transmission component faults

(1) The shaft pin of the transmission connecting rod is rusted or seized.

(2) The rotation shaft is damaged or seized due to rust.

(3) The locking plates of the main knife switch and grounding knife switch are seized.

(4) The bevel gear is out of gear.

(5) Water infiltrates into the vertical connecting rod and freezes in Winter, causing severe deformation of the operating mechanism and rendering it inoperable.

4. Operating mechanism faults

(1) The main circuit of the motor fails.

(2) The common part of the control circuit fails.

(3) The opening part of the control circuit fails.

① The opening circuit isn't connected.

② The opening circuit can't be constantly connected.

(4) The closing part of the control circuit is partially faulty.

① The closing circuit isn't connected.

② The closing circuit can't be constantly connected.

(5) The electric motor doesn't stop after the opening operation is completed or the opening position isn't reached.

(6) The electric motor doesn't stop after the closing operation is completed or the closing position isn't reached.

2.6.2　Handling of Common Faults

1. Handling of overheating in contact area

(1) The power shall be cut off for handling. During handling, the maintenance procedures and quality standards for conduction circuits shall be rigorously followed.

(2) During disassembly maintenance, the use of defective or inferior components such as poor-quality cable clamps and bolts is strictly prohibited. Bolt-type cable clamps shall be replaced with compression-type ones. The contact surface of the joint shall be cleaned and promptly coated with conductive grease, and bolts shall be used correctly and tightened to an appropriate torque.

(3) For disconnectors with frequent overheating on the bus side, the maintenance must be thorough and of high quality. For the wire holder section, special attention shall be given to the connection of the conduction strips at both ends to ensure that the end faces are clean, even, coated with conductive grease, and tightly compressed. For the contact terminal section, the cleanliness of the contact terminals shall be ensured, and neutral vaseline shall be applied. The burning of contact terminals shall be checked, and contact terminals or contact fingers shall be replaced if necessary. The finger base of the left contact finger shall be polished, and springs showing signs of overheating or corrosion shall be replaced. The synchrony of three-phase opening and closing operations shall be ensured. The insertion depth of the right contact terminal shall meet the requirements. The pressure on both sides of the contact finger shall be uniform. In order to verify the quality of maintenance, the contact resistance of the circuit shall also be measured to ensure good contact of all contact surfaces.

(4) For older GW4 and GW5 disconnectors with overheating on the left contact terminal, an additional shunt strip shall be installed. This involves drilling a 6 mm screw hole at the corresponding position of each contact finger and finger base and then fixing a stacked copper soft connection strip between the contact finger and finger base with a screw.

(5) For older GW4 and GW5 disconnectors, the left contact terminals shall be replaced with new models. The new model of contact terminal has an insulating block in the middle of the spring, which eliminates the possibility of spring shunting, makes the spring less susceptible to annealing and deformation, and reduces the elasticity.

(6) Improper application of conductive paste on the moving contact terminals and fixed contact levers of the disconnector leads to overheating. The solution involves strictly regulating the amount

of conductive paste applied to these active conductive contact surfaces. Firstly, the active contact surface shall be thoroughly cleaned with anhydrous alcohol. Next, a thin, even layer of conductive paste shall be applied on the conductive surface and then immediately wiped with a cloth to leave only a minimal thin layer of conductive paste.

2. Handling of fractures and flashovers of post insulators

(1) The power shall be cut off for handling. During handling, the maintenance procedures and quality standards for post insulators shall be rigorously followed.

(2) New post insulators, which utilize high-strength porcelain columns, shall undergo testing using ultrasonic non-destructive testing equipment. Only those passing the test shall be used.

(3) Enhanced maintenance shall be performed on operational post insulators. Based on sound diagnostics from non-destructive testing, a specialized protective coating for the insulator shall be applied to the cement-bonded surface where the porcelain column is located.

(4) The porcelain columns shall be replaced with new ones to increase the creepage distance and column height and improve the overall insulation levels. The creepage distance shall be increased by carrying out live cleaning, strengthening cleaning intensity, adding silicone rubber skirts to disconnector insulators, and spraying hydrophobic RTV coating.

3. Handling of refusal to trip and close

(1) Refusal to open and close caused by transmission mechanism and transmission system.

① Cause: Water ingress into the mechanism box, rust and seizure of various shafts, pins, connecting rods, crank arms, base frames, and even base bearings lead to refusal to open and close.

② Solution: The transmission mechanism and rusted components shall be disassembled for disassembly maintenance, and any non-compliant parts shall be replaced. Strengthening of anti-corrosion measures, application of lithium molybdenum disulfide, and installation of louvered rain guard shall be undertaken. The transmission mechanism with serious or inherent defects shall be replaced.

(2) Refusal to open and close caused by electrical problems.

① Cause: Failure to close the three-phase power supply switch, control power supply disconnection, blown power fuse, malfunctioning thermal relay causing power interruption, anomalies in the electrical circuit due to aging or damage of secondary components resulting in operation failure, motor malfunction, and other issues can lead to the motor mechanism not starting and the disconnector failing to operate during opening and closing operations of the motor mechanism.

② Solution: The secondary electrical circuit contains numerous series connected control and protection components, including miniature circuit breakers, transfer switches, AC contactors, limit switches, interlock switches, thermal relays, etc. Any failure of these components will lead to the disconnector failing to operate. When the opening or closing button doesn't initiate operations, it shall be checked firstly if the operating power supply is intact and then various relevant components. If any components are found damaged, they shall be replaced, and the cause identified. The reliability of secondary circuits depends on the reliability of individual components; therefore, only reliable secondary components shall be selected.

4. Incomplete opening and closing

(1) Incomplete opening and closing caused by the mechanism and transmission system

① Cause: Water ingress into the mechanism box, and rust of various shafts, pins, connecting rods, crank arms, base frames, and even base bearings lead to incomplete opening and closing. Fractures resulting from inadequate strength at connecting rods, connecting points of transmission, and support components of knife switches and contact terminals, cause incomplete opening and closing.

② Solution: The mechanism and rusted components shall be disassembled for disassembly maintenance, and any non-compliant parts shall be replaced. The anti-corrosion measures shall be strengthened and the lithium molybdenum disulfide shall be applied. The replacement of the transmission base with an oil injection hole shall be undertaken.

(2) Incomplete opening and closing or three-phase asynchrony of disconnector.

① Cause: The opening and closing positioning screws are adjusted improperly. The travel of the auxiliary switches and limit switches is adjusted improperly. The bending and deformation of connecting rods lead to changes in length, incomplete transmission, etc.

② Solution: The components such as positioning screws and auxiliary switches shall be inspected. Adjustments shall be made if any anomalies are found. If connecting rods are deformed, prompt identification and elimination of the cause shall be undertaken. Additionally, when the disconnector fails to close or when there's a three-phase asynchrony in the operational environment, the disconnector shall be opened and closed several times with proper force in accordance with the operation requirements. If complete closing hasn't yet been achieved, and complete synchronization of the three phases has not been realized, a interruption maintenance shall be scheduled.

5. Inoperability of electric operating mechanism

(1) Mechanism-related problems manifest as operational failures, such as refusal to operate or incomplete opening and closing. These often occur during reclosing operations, which compromises the system's safety. Poor sealing of the mechanism box or water ingress causes severe corrosion, along with dried lubrication, and increased operating resistance, resulting in difficulties in operation. Concurrently, component damage may occur, such as gear fractures and twisted connecting rods.

(2) The reliability of secondary circuits will directly impact the reliability of high-voltage disconnector operation. Insufficient switching or poor contact of auxiliary switches and travel switches can lead to the refusal to operate of the disconnector. Poor contact with terminal connections, failure to engage contactors, motor burnout, destruction of secondary wire insulation, etc., can cause the failure of remote operation. The reliability of secondary circuits depends on the reliability of individual components; therefore, only reliable secondary components shall be selected.

(3) The power shall be cut off for handling. During handling, the maintenance procedures and quality standards for operating mechanism shall be rigorously followed.

任务一　GW4-126型隔离开关的整体调试

一、工作任务

对GW4-126型隔离开关进行整体调整。

二、引用标准

(1)《电气装置安装工程高压电器施工及验收规范》(GB 50147—2010)。
(2)《国家电网公司变电检修管理规定(试行)第4分册隔离开关检修细则》。
(3)《国家电网公司变电运维管理规定》。
(4)《国家电网公司变电检修管理规定》。
(5)《国家电网公司电力安全工作规程》(变电部分)。

三、工作要求

(1)在6级及以上的大风以及暴雨、雷电、冰雹、大雾、沙尘暴等恶劣天气下,应停止露天高处作业。
(2)被检修设备与周围设备均应使用围栏隔离,面向通道处设置唯一出入口。
(3)隔离开关两侧引线均应停电并且各挂装一组三相短路接地线。
(4)作业人员精神状态良好,熟悉工作中安全措施、技术措施以及现场工作危险点。

四、工作准备

(一)危险点及预控措施

1. 高压触电

危险点:围栏外所有电气设备均视为带电,工作时跨出围栏可能导致高压触电。

预控措施:用围栏将被检修间隔与相邻带电设备(间隔)隔离,并且向作业现场内悬挂"止步,高压危险"标示牌,在通道处设置唯一出入口,悬挂"从此进出"标示牌;被检修隔离开关两侧引线均应停电并且各挂装一组三相短路接地线;工作时至少需要两人,一人监护一人操作,听工作负责指挥。

2. 低压触电

危险点:控制回路、照明回路空气开关进线侧(上端)均带低压电;合上空开后,隔离开关二次回路均视为带电状态,直接触摸可能造成低压触电。

预控措施:断开控制回路、照明回路电源空气开关,并上悬挂"禁止合闸,有人工作"标示牌,只允许手动操作隔离开关;工作时至少需要两人,一人监护一人操作,听工作负责指挥。

3. 机械伤害与物件伤人

危险点:操作人员在分合隔离开关、接地开关时,未通知其他工作班成员,可能会对工作班成员造成机械伤害；工作班上下抛掷工具，或检修人员掉落工具，可能会造成落物伤人。

预控措施：隔离刀闸分合、接地开关分合时，需要大声呼唱，得到工作班所有人员大声回应之后，方可操作；禁止将工具及材料上下投掷，应用绳索拴牢传递，以免打伤下方工作人员或击毁脚手架；高处作业应一律使用工具袋，较大的工具应用绳拴在牢固的构件上，工件、边角余料应放置在牢靠的地方或用铁丝扣牢并有防止坠落的措施，不准随便乱放，以防止从高空坠落发生事故；工作时至少需要两人，一人监护一人操作，听工作负责指挥。

4. 高处坠落

危险点：工作班成员爬上隔离开关底座时，使用爬梯无人扶持，也未固定，爬梯滑落造成坠落受伤；工作班成员在隔离开关底座检修时，可能坠落受伤。

预控措施：在工作人员上下铁架或梯子上，应悬挂"从此上下！"的标示牌；梯子应坚固完整，有防滑措施。梯子的支柱应能承受作业人员及所携带的工具、材料攀登时的总重量；工作人员在上下爬梯时，应有人扶持爬梯，或将爬梯固定在牢固构架上；在没有脚手架或者在没有栏杆的脚手架上工作，高度超过 1.5 m 时，应使用安全带；安全带的挂钩或绳子应挂在结实牢固的构件上，或专为挂安全带用的钢丝绳上，并应采用高挂低用的方式；工作时至少需要两人，一人监护一人操作，听工作负责指挥。

5. 设备损坏

危险点：操作隔离开关或接地开关时发生闭锁卡死，野蛮操作，将损坏设备；操作隔离开关时，工具遗落在机构箱内，导致操作时损坏机构。

预控措施：禁止野蛮操作；操作过程中，禁止遗留任何工具于柜内，若出现不能操作的现象，应当立即停止，不得使用蛮力，立即汇报负责人；工作时至少需要两人，一人监护一人操作，听工作负责指挥。

（二）工器具及材料选择

GW4-126 型隔离开关调整工器具及材料见表 2-5。

表 2-5　GW4-126 型隔离开关调整工器具及材料

类别	名称	规格型号	数量	备注
专用工具	隔离开关操作把手		1 把	
通用工具	组合套筒扳手		1 套	
	开口呆扳手	17-19 16-18 22-24	各 1 把	
	卷尺	5 m	1 把	
	钢板尺	300 mm	1 把	
	活动扳手	300 mm	1 把	
	木榔头		1 把	
	塞尺	（0.02～1）mm×150 mm	1 套	

续表

类别	名称	规格型号	数量	备注
通用工具	安全带	全身式	2根	
	线垂	3 m	1个	
	工具包		1个	
材料中性	凡士林	白	1瓶	

（三）作业人员分工

GW4-126型隔离开关调整人员分工见表2-6。

表2-6 GW4-126型隔离开关调整人员分工

序号	工作岗位	数量	职责
1	工作负责人	1	负责本次工作的人员分工、现场查勘、作业方案制定、召开班前会、作业过程中安全监督、工作中突发状况的处理、工作质量的监督、班后会总结
2	操作人员	1	负责隔离开关调整操作
3	辅助操作人员（可无）	1	辅助操作人员进行隔离开关分合位置的检查

五、作业程序

GW4-126型隔离开关调整作业流程见表2-7。

表2-7 GW4-126型隔离开关调整作业流程

序号	作业内容	作业步骤及标准	安全措施及注意事项	责任人
1	工作票及查勘阶段	（1）规范填写和签发工作票，正确履行工作票手续； （2）现场查勘，必须由2人进行，核对现场将停电的停电范围，核对待检修设备双重名称正确无误，检查工作现场安全净距足够	（1）查勘时正确穿戴安全帽、工作服、工作鞋； （2）查勘时禁止操作设备； （3）查勘时严禁无关人员进入现场	工作负责人或工作票签发人
2	工作前准备工作	（1）检查工器具是否齐全，检查工器具外观和试验合格； （2）工作负责人同工作许可人巡视待检修设备，确认工作票所列安全措施已经正确执行，安全措施是否完备，现场是否具备开工条件，必要时进行补充； （3）执行工作许可手续； （4）对工作班成员召开班前会； （5）抄写设备铭牌参数	（1）工器具无损伤、变形、失灵现象，需要试验的工器具合格证在有效期内； （2）巡视现场时禁止无关人员进入现场； （3）班前会应包含工作地点双重名称；工作时间与内容；工作分工；工作危险点及预控措施；停电范围及工作现场安全措施；	工作负责人与工作许可人

续表

序号	作业内容	作业步骤及标准	安全措施及注意事项	责任人
2	工作前准备工作		（4）全体工作成员应当正确穿戴安全帽、工作服、工作鞋、劳保手套等劳动保护用品	
3	本体外观检查	（1）目测检查各支柱瓷瓶无明显倾斜、缺损； （2）目测检查基础无沉降，各部件、连接及接地明显无弯曲、锈蚀、裂纹、破损； （3）手动操作隔离开关，目测检查主刀和地刀的断口、闭锁、连杆连接情况，并记录； （4）检查瓷瓶底座螺栓有无缺失、松动	（1）瓷瓶目测有倾斜时，需要用线锤确认； （2）分合隔离开关时注意引流线上所挂三相短路接地线不能松脱	工作班
4	主刀调整	（1）取下机构输出主拉杆和水平拉杆； （2）松开各个机械限位，确保首相行程满足需要； （3）将首相和电机置于合闸位（触头臂触指臂在一条直线；圆柱形触头与两排触指同时接触缝隙<0.02 mm；触头臂触指臂上下差不大于5 mm）； （4）连接主拉杆； （5）手动操作检查首相分闸是否合规（角度 90°±1°；平行距离≥1200 mm；前后偏差≤10 mm）； （6）首相与另外两项置于合闸位置，连接水平拉杆； （7）手动操作检查各分闸是否合规； （8）置于同期位置，检查三相同期并调整同期拉杆（同期≤10 mm）； （9）锁定各并帽螺栓，恢复各个限位	（1）调整拉杆长度时，只能转动拉杆本体调整，不能转动端部螺栓； （2）若因同期拉杆原因，导致某相不能合闸，则应先调整期同期拉杆使其能正常合闸，不得野蛮操作强行至于合位	工作班

续表

序号	作业内容	作业步骤及标准	安全措施及注意事项	责任人
5	调整地刀	（1）检查连杆抱箍、触头螺栓、导电臂连接均无松动； （2）手动将接地刀闸由分位缓慢推向合位，观察触指与触头中心是否重合，各相是否同期并且调整（合闸同期≤10 mm）； （3）手动快速合上接地刀闸，观察触指与触头是否有效解除，且未达到限位，并调整； （4）手动快速分开接地刀闸，观察分闸间距；（分闸间距≤900 mm）； （5）检查并调整首相分闸位置； （6）将首相置合闸位置，依次连接中相和尾相； （7）手动操作，复查三相分合闸位置； （8）检查相间同期位置； （9）调整地刀—主刀机械闭锁； （10）润滑，紧固并帽、开口销，手动操作复查分合		工作班
6	现场恢复至初始状态	（1）分开所有隔离开关、接地开关； （2）将"禁止合闸，有人工作"标示牌挂回机构箱		工作班
7	工作终结	（1）清理作业现场，做到"工完料尽场地清"； （2）召开班后会； （3）终结工作票	严禁在清理场地的同时负责人前去终结工作票	工作负责人

Task 1 Integral Commissioning of GW4-126 Disconnector

1.1 Work Tasks

Integral commissioning of the GW4-126 disconnector.

1.2 References

(1) *Code for Construction and Acceptance of High-voltage Electrical Apparatus of Electric Equipment Installation Engineering* (GB 50147-2010).

(2) *Substation Maintenance Management Regulations of State Grid Corporation of China (Trial Implementation), Volume 4—Rules for Maintenance of Disconnectors*.

(3) *Regulations of State Grid Corporation of China on Management of Substation Operation and Maintenance*.

(4) *Substation Maintenance Management Regulations of State Grid Corporation of China*.

(5) *Electric Power Safety Working Regulations (Power Transformation) of State Grid Corporation*.

1.3 Work Requirements

(1) During severe weather conditions such as wind level 6 or higher, heavy rain, thunderstorms, hail, dense fog, sandstorms, etc., outdoor work at elevated locations shall be halted.

(2) Equipment under maintenance and surrounding devices shall be enclosed by barriers, with a single entrance arranged facing the passage.

(3) Leads on both sides of the disconnector shall be de-energized, and each shall be equipped with a set of three-phase short-circuit grounding wires.

(4) Operators are to be in good mental state, and are aware of safety measures, technical measures and hazards to site operation during operation.

1.4 Preparation for Work

1.4.1 Hazards and preventive and control measures

1. High-voltage electric shock

Hazard: all electrical equipment outside the barrier is considered live. Stepping out of the barrier during work may result in high-voltage electric shock.

Preventive and control measures: barriers shall be used to isolate the equipment under maintenance from adjacent live equipment (bay). The "Stop! High Voltage, Danger!" sign shall be hung within the work area. A single entrance shall be set up and the "Entrance/Exit" sign board shall be hung at the passage. Leads on both sides of the disconnector under maintenance shall be de-energized, and each shall be equipped with a set of three-phase short-circuit grounding wires. At least two persons are required during work, one for supervision and the other for operation under instructions of person in charge of work.

2. Low-voltage electric shock

Hazard: the incoming side (upper end) of the air switches of the control circuit and lighting circuit carries low voltage. After closing the air switch, the secondary circuit of the disconnector is considered energized. Direct contact may result in low-voltage electric shock.

Preventive and control measures: the power supply air switches of control and lighting circuits shall be disconnected and the "No Closing, Work in Progress" sign board shall be hung. Only manual operation of the disconnector shall be permitted. At least two persons are required during work, one for supervision and the other for operation under instructions of person in charge of work.

3. Mechanical injuries and object hazards

Hazard: failure to notify other work crew members when operating the disconnector or grounding switch may cause mechanical injuries to them. Throwing tools within the work crew or dropping tools during maintenance may cause injuries from falling objects.

Preventive and control measures: when opening and closing the isolating knife switch and grounding switch, It's necessary to shout loudly and wait for a loud response from all work crew members before proceeding. To prevent injuries to personnel below or scaffold damage, it shall be prohibited to throw tools and materials, and they shall be passed using ropes. Tools shall be carried in tool bags for elevated work. Larger tools shall be securely fastened with ropes to stable components. Workpieces and leftover materials shall be placed securely or fastened with iron wire and equipped with anti-falling measures. Arbitrary placement shall be prohibited to prevent accidents from falling objects. At least two persons are required during work, one for supervision and the other for operation under instructions of person in charge of work.

4. Falls from heights

Hazard: work crew members may suffer injury from falling when climbing the base of the disconnector without assistance or proper fixation of the ladder. The work crew members may also suffer injury from falling when carrying out maintenance on the base of the disconnector.

Preventive and control measures: the sign board reading "Climb Up And Down Here!" shall be hung on iron frames or ladders used by personnel. Ladders shall be sturdy, complete, and equipped with anti-slip measures. Ladder supports shall be able to withstand the total weight of personnel and tools and materials carried along during climbing. Personnel shall be assisted or ladders shall be secured to stable structures when climbing up or down. When working without scaffolding or on scaffolding without railings and higher than 1.5 meters, safety belts shall be used. Safety belt hooks or ropes shall be attached to robust and secure members or to steel ropes specifically designed for safety belt attachment in a position higher than the user. At least two persons are required during work, one for supervision and the other for operation under instructions of person in charge of work.

5. Equipment damage

Hazard: locking or seizure and forcible operation during the operation of the disconnector or grounding switch may result in equipment damage. Leaving tools inside the mechanism box while operating the disconnector may damage the mechanism during operation.

Preventive and control measures: rough handling is prohibited. It is forbidden to leave any tool in the cabinet during operation. If there is a phenomenon that operation cannot be given during the operation, the operation shall be stopped immediately, no brute force shall be used, and report shall be sent to the person in charge immediately; At least two persons are required during work, one for supervision and the other for operation under instructions of person in charge of work.

1.4.2 Tools, equipment, and material selection

For the adjusting tools, equipment, and material of GW4-126 disconnector, refer to Table 2-5.

Table 2-5　Adjusting Tools, Equipment, and Materials of GW4-126 Disconnector

Category	Name	Specification and model	Quantity	Remarks
Specialized tools	Operating handle of disconnector		1	
General tools	Combination socket wrench		1 set	
	Single-ended open jaw spanner	17-19	1 pcs of each type	
		16-18		
		22-24		
	Band tape	5 m	1	
	Steel ruler	300 mm	1	
	Monkey wrench	300 mm	1	
	Wood hammer		1	
	Feeler gauge	(0.02−1) mm ×150 mm	1 set	
	Safety belt	Full body type	2	
	Plumb	3 m	1	
	Tool kit		1 drum	
Material	Neutral petroleum jelly	White	1 bottle	

1.4.3 Division of workforce

For the adjustment of GW4-126 type disconnector, refer to Table 2-6 for the division of workforce.

Table 2-6　Division of Workforce for Adjustment of GW4-126 Type Disconnector

S/N	Job	Quantity	Responsibilities
1	Person in charge of work	1	Be responsible for work division of working staffs, site survey, stipulation of operation scheme, pre-shift meeting, safety supervision during operation, handling of emergencies during work, supervision of work quality and summary of post-shift meeting
2	Operator	1	Take charge of the adjustment operation of disconnector
3	Assistant operator (optional)	1	Assist in inspecting the opening and closing positions of disconnector

1.5 Operation Procedures

Refer to Table 2-7 for the workflow of the GW4-126 type disconnector adjustment.

Table 2-7 Workflow of the GW4-126 Type Disconnector Adjustment

S/N	Scope of work	Operational steps and standards	Safety measures and precautions	Person-in-charge
1	Work ticket and survey phase	(1) The work tickets shall be properly filled out and issued and the work ticket procedures shall be followed; (2) the site survey shall be conducted by 2 persons, including verifying the power-off range, checking the correct and accurate identification of the equipment to be maintained, and ensuring adequate safe clearance at the work site	(1) Safety helmets, work clothes, and work boots shall be worn during the survey; (2) the equipment shall not be operated during the survey; (3) no unrelated personnel shall be allowed on site during the survey	Person in charge of work or work ticket issuer
2	Preparations before work	(1) Check if all instruments and tools are complete, and if their appearance and test are acceptable; (2) The person in charge of work and work permitter shall inspect the equipment to be maintained, confirm the implementation of safety measures listed in the work ticket, check if safety measures are complete, verify if the work site is ready for operation, and provide supplement when necessary; (3) Implement work permit procedures; (4) Call in toolbox meeting participated by members of work team; (5) Record parameters on the equipment nameplate	(1) Instruments and tools are free of damage, deformation and malfunction. Qualification certificates for instruments and tools to be tested are within the term of validity; (2) Prevent other persons from entering the site during tour inspection; (3) Toolbox meeting shall cover dual designations of work place; Working hours and contents; Work division; Working hazards as well as preventive and control measures; Power-cut scope and safety measures on work site; (4) All work members shall properly wear such labor protection appliances as safety helmet, working clothes, working shoes and protective gloves	Person in charge of work and work permitter
3	Appearance inspection of main body	(1) No obvious tilt or damage shall be observed in each post and porcelain during visual inspection; (2) Upon visual inspection, the foundation shall exhibit no settling, and components, connections, and grounding shall be visibly free from apparent bending, rust, cracks, or damage; (3) The disconnector shall be manually operated, and the main knife switch and grounding knife switch shall be visually inspected for break, locking, and connecting conditions of connecting rod, with a record made;	(1) Tilting of the porcelain insulator shall be confirmed using a plumb line; (2) When opening and closing the disconnector, it shall be ensured that the three-phase short-circuit grounding wire hung on the leading wire doesn't become detached	Work shift

Continued

S/N	Scope of work	Operational steps and standards	Safety measures and precautions	Person-in-charge
3	Appearance inspection of main body	(4) It shall be checked whether there are any missing or loose bolts on the porcelain insulator base		
4	Main knife switch adjustment	(1) The mechanism's output main tie rod and horizontal tie rod shall be removed; (2) All mechanical limiters shall be released to ensure that the first-phase travel meets requirements; (3) The first phase and motor shall be switched to the closing position; (The contact terminal arm and contact finger shall be aligned in a straight line; the simultaneous contact gap of the cylindrical contact terminal and contact fingers from two rows shall be smaller than 0.02 mm; the contact terminal arm and contact finger arm shall have a difference of no more than 5 mm in upper and lower positions); (4) The main tie rod shall be connected; (5) It shall be checked through manual operation if the first-phase opening meets specifications (angle of 90°±1°; Parallel distance ≥ 1200 mm; Front-rear deviation ≤ 10 mm; (6) Place the first phase and the other two phases at closed position and connect horizontal tie rod; (7) Manually check the opening operations for compliance; (8) Place it at synchronization position, check three-phase synchronization and adjust synchronization tie rod (synchronization ≤ 10 mm); (9) Lock the two-cap bolts and reset the limit switches	(1) In adjusting tie rod length, turn tie rod body only, and do not turn end bolts; (2) In the case of failure to close a phase due to synchronous tie rod, adjust the synchronous tie rod to allow proper closing, and avoid rough handling and forced closing	Work shift
5	Adjust grounding knife switch	(1) Make sure that connecting rod clamp, contact terminal bolt and conductive arm connection are tight; (2) Slowly push grounding knife switch to closing position from open position by hand, make sure that contact finger and contact terminal center coincide and the phases are synchronous, and make adjustment as required (closing synchronization ≤ 10 mm); (3) Quickly close the grounding knife switch by hand, make sure that the contact finger and the contact terminal are effectively released and have not reached the limit, and make adjustment as required;		Work shift

Continued

S/N	Scope of work	Operational steps and standards	Safety measures and precautions	Person-in-charge
5	Adjust grounding knife switch	(4) Quickly open the grounding knife switch by hand and observe opening spacing; (Opening spacing ≤ 900 mm); (5) Check and adjust open position of the first phase; (6) Place the first phase at closed position and connect the mid phase and the end phase in turn; (7) Manually check closed positions of three phases; (8) Check interphase synchronous position; (9) Adjusting ground knife switch-main knife switch mechanical locking; (10) Lubricate and tighten two caps and split pins, and manually recheck opening and closing		
6	Restore the site to its initial state	(1) Open all disconnector and grounding switches; (2) Replace "No Closing, Work in Progress" sign board to the mechanism box		Work shift
7	End of work	(1) Clean up the work site, make sure that the elements and materials are fully utilized and the site is cleaned; (2) Convene a post-shift meeting; (3) Terminate the work ticket	It is strictly prohibited for the person in charge to terminate the work ticket while cleaning the site	Person in charge of work

任务二　GW7-126 型隔离开关触头过热故障处理

一、工作任务

对 GW7-126 型隔离开关 B 相触头过热故障进行处理。

二、引用标准

（1）《隔离开关和接地开关设备评价导则》（Q/GDW 450—2010）。
（2）《国家电网公司变电检修管理规定（试行）第 4 分册隔离开关检修细则》。
（3）《国家电网公司变电运维管理规定》。
（4）《国家电网公司变电检修管理规定》。
（5）《国家电网公司电力安全工作规程》（变电部分）。

三、工作要求

（1）隔离开关触头过热处理应在良好、干燥天气下进行，在测量过程中，遇到 6 级以上大风以及雷暴雨、冰雹、大雾、沙尘暴等恶劣天气时应停止工作。
（2）测量仪器、安全防护工具必须有校验合格证。
（3）高处作业时必须使用安全带，严禁低挂高用。
（4）操作时，隔离开关停电，设备在断开位置，且被检修刀闸的两个各挂一组接地线。
（5）进行回路电阻测试时，要确保被测试设备和仪器可靠接地。

四、工作准备

（一）现场勘察的基本要求及条件

（1）查阅故障隔离开关运行情况、了解作业场地条件。查阅该隔离开关历年试验报告、运行记录和缺陷情况记录。
（2）收集故障设备的状况，包括故障相、故障位置、故障严重程度等。
（3）查勘现场设备停电范围及安全措施。
（4）查勘场地检修电源箱，核实电源容量是否满足要求。
（5）现场工器具、作业车辆、机具、材料等定置摆放位置。

（二）工器具及材料选择

GW7-126 型隔离开关触头过热故障处理工器具及材料见表 2-8。

表 2-8　GW7-126 型隔离开关触头过热故障处理工器具及材料清单

序号	名称	规格	单位	数量
1	组合套扳手		套	
2	钢板尺	100 mm	块	1
3	塞尺	50 丝	个	1

续表

序号	名称	规格	单位	数量
4	平口钳	6寸	把	1
5	尖嘴钳	6寸	把	1
6	美工刀		双	1
7	一字改刀	4寸	把	2
8	细齿扁锉		把	1
9	松动剂		瓶	1
10	砂布	00号	张	1
11	中性凡士林		瓶	1
12	导电膏		瓶	1
14	尼龙刷子		把	1
15	25%~28%的氨水		公升	若干
16	白布		米	若干
18	台虎钳		把	1
19	白棕绳	3米	根	1
20	回路电阻测试仪		台	1

（三）危险点及预防措施

1. 高压触电

危险点：误入带电间隔可能导致高压触电。

预控措施：用围栏将被检修间隔与相邻带电设备（间隔）隔离，并且向作业现场内悬挂"止步，高压危险"标示牌，在靠近道路侧设置唯一出入口，悬挂"从此进出"标示牌；被测试断路器的两侧，各挂装一组三相短路接地线；工作时至少需要两人，一人监护一人操作，听工作负责指挥。

2. 低压触电

危险点：隔离开关机构箱内控制回路、照明回路空气开关进线侧（上端）均低压电；合上空开后，二次回路均视为带电状态，触摸可能造成低压触电。外搭接电源及测试过程中，触碰到带电部分可能会造成低压触电。

预控措施：若运行人员在空气开关上悬挂了"禁止合闸，有人工作"标示牌时，工作负责人需要向运行人员申请变更安全措施,得到许可后，方能合上空开；工作负责人反复对工作班成员强调开关空气开关时，切勿将手指伸入空气开关上端头；外搭接电源时，工作时至少需要两人，一人监护一人操作，听工作负责指挥。

3. 机械伤害

危险点：隔离开关在进行分、合闸调试操作时，可能会对工作班成员造成机械夹持伤害。

预控措施：隔离开关在分、合闸时，需要大声呼唱，得到工作班所有人员大声回应之后，方可操作；工作时至少需要两人，一人监护一人操作，听工作负责指挥。

4. 设备、仪器损坏

危险点：未按照要求进行测试、野蛮操作，对仪器和被测设备损坏。作业过程中工器具敲打瓷瓶等部件。

预控措施：禁止野蛮操作，操作过程中，按步骤和要求进行测试，防止仪器损坏。作业时注意工器具不要碰伤瓷瓶。若出现不能操作的现象，应当立即停止，不得使用蛮力，立即汇报负责人；工作时至少需要两人，一人监护一人操作，听工作负责指挥。

5. 高处坠落

危险点：工作班成员爬上隔离开关时，使用爬梯无人扶持，也未固定，爬梯滑落造成坠落受伤；工作班成员在隔离开关上操作时，未正确使用安全带可能坠落受伤。

预控措施：在工作人员上下的梯子上，应悬挂"从此上下！"的标示牌；梯子应坚固完整，有防滑措施。梯子的支柱应能承受作业人员及所携带的工具、材料攀登时的总重量；工作人员在上下爬梯时，应有人扶持爬梯，或将爬梯用绳索固定在刀闸构架上；在没有脚手架或者在没有栏杆的脚手架上工作，高度超过 1.5 米时，应使用安全带；安全带的挂钩或绳子应挂在结实牢固的构件上，或专为挂安全带用刀闸检修架上，并应采用高挂低用的方式；工作时至少需要两人，一人监护一人操作，听工作负责指挥。

（四）作业人员分工

GW7-126 型隔离开关触头过热故障处理人员分工如 2-9 所示。

表 2-9　GW7-126 型隔离开关触头过热故障处理人员分工

序号	工作岗位	数量/人	工作性质
1	操作人员	1	专门负责操作
2	监护人员	1	专职监护
3	辅助人员	1	辅助接线、工器具传递、记录等工作

五、作业程序

本任务操作流程如表 2-10 所示。

表 2-10　GW7-126 型隔离开关触头过热故障处理操作流程

序号	作业内容	作业标准	安全注意事项	责任人
1	前期工作准备	（1）履行工作票手续； （2）核对现场的设备位置和情况； （3）装设安全围栏、悬挂标示牌； （4）按规程要求，正确使用劳动防护用品，工作服穿戴整齐	（1）按照安规要求办理工作票的相关手续； （2）进入作业现场要正确戴好安全帽，穿工作服，着软底鞋	

续表

序号	作业内容	作业标准	安全注意事项	责任人
2	工作环境确认	（1）检查隔离开关初始状态，控制电源开关在断开位置。记录设备铭牌参数。询问过往检修记录、试验记录； （2）核查围栏、标示牌； （3）检查需要使用的工器具、材料、机具与仪表，检查方式正确； （4）工作负责人与辅工召开班前会，按照"四清楚"原则，向辅助工进行工前交底，明确工作任务、安全措施、危险点、工作流程并履行确认手续	（1）设备初始状态应该是分闸状态； （2）围栏和标示牌设置正确； （3）安全用具、工器具外观检查合格、无损伤、变形、破损等情况； （4）开工前，对辅助工进行安全交底，正确做到"四清楚"，正确履行确认手续	
3	拆除	（1）作业人员穿戴好安全带，准备好作业工器具，登上刀闸检修架上进行操作； （2）首先是拆除 B 相两侧引流线，并将拆下的引流线固定在本相瓷瓶上； （3）拆除左右触指座，解除限位螺栓的闭锁，使触头座可以左右偏转，便于取下固定螺栓； （4）拆除后的触指座用放置在平整的位置； （5）拆除 B 相导电杆，分别在导电杆上作标记以便于装复，拆除导电杆的连接螺栓，取下导电杆及夹板； （6）将单相导电回路放于工作平台上	（1）作业人员通过扶梯登上隔离开关时，要有人扶梯； （2）登上刀闸后，首先是将安全带挂在在刀闸检修架上； （3）高处作业时，工器具的传递切勿抛掷，要用专门的工器具包或绳索系牢上下传递； （4）正确将引流线固定在本相瓷瓶上； （5）拆卸导电杆过程中，注意不要损伤导电管	
4	分解左、右触指座	（1）首先明确拆卸后的所有零部件按照由远及近的摆放顺序整齐地放置到油盘中； （2）拆除固定防雨罩螺帽，取下防雨罩及垫圈，放置到油盘中； （3）拆除固定触指两侧的垫片及固定销，放置到油盘中； （4）拆除固定触指部分的长、短圆柱销，放置到油盘中； （5）取下触指部分，分解连接触指部分的外侧垫片、弹簧、销钉，并整齐放置到油盘中；	（1）注意拆卸的零部件按顺序摆放，便于后期回装； （2）拆卸过程中防止零部件掉落	

续表

序号	作业内容	作业标准	安全注意事项	责任人
4	分解左、右触指座	（6）取下固定触指架上的开口销和垫片，取下触指架； （7）取出触指底座中的复位弹簧，并将触指底座和复位弹簧放置到油盘中，完成整个触指座的拆除工作。（左右触头座的结构一致，此处仅说明右触头座的分解情况）		
5	检修左、右触指座	（1）将所有拆卸的部件都用酒精进行清洗，去除脏污、杂质并晾干； （2）检查防雨罩是否完好，有无掉漆，如锈蚀严重或开裂应更换。固定防雨罩的螺栓、垫圈有无锈蚀、堵塞。少量锈蚀可用砂纸轻微打磨，堵塞应进行清理。不适宜再次使用的零部件应更换； （3）检查两侧固定垫片及销钉有无锈蚀、孔洞有无堵塞，垫片是否平整，有无形变，垫片与直尺想靠，无间隙则表示平直，有明显间隙，则表示垫片有形变； （4）检查圆柱销、开口销有无形变、锈蚀，用砂纸除去锈蚀，如严重锈蚀或变形应予更换； （5）检查连续触指部分的垫片、销钉、弹簧有无形变、锈蚀，弹簧间隙是否均匀，若形变严重应更换零部件。弹簧是否形变的检查也同样用直尺测量弹簧的长度，若长度变化，则表示出现形变，应更换。检查触指有无明显烧灼痕迹，用手触摸触指导电部分有无毛刺、镀银层脱落，如镀银层脱落或烧伤严重应更换。轻微毛刺可用小细扁锉局部轻微修理。检查后将拆卸下来的触指放到油盘中，倒入浓度为25%~28%的氨水，浸泡15分钟；用尼龙刷来回轻拭触指表面，以刷除表面氧化物质，并用白布擦拭晾干备用；	（1）切记用砂纸使劲在触指等导电部位打磨； （2）检查过程中防止零部件掉落； （3）要清楚用清洁剂对零部件进行清洗； （4）检查有明显变形、破损、锈蚀的部件应更换	

续表

序号	作业内容	作业标准	安全注意事项	责任人
5	检修左、右触指座	（6）检查触指架有无裂纹、锈蚀、孔洞有无堵塞； （7）检查触指底座导电接触部分有无明显烧伤痕迹、焊接部分有无裂纹、孔洞有无堵塞； （8）复位弹簧间隙均匀、无变形、锈蚀等情况		
6	检查导电杆	（1）检查导电杆有无损伤，如有轻微变形应校正，检查导电杆导电接触面有无过热情况，用砂纸清除无镀银层接触面氧化层； （2）检查导电杆的连接螺栓及夹板是否有锈蚀或破损，若轻微锈蚀可用细刀锉修理，若锈蚀或破损则应更换； （3）从触头和导电杆连接部分若有有锈蚀、开裂、破损等严重现象，应予以更换	（1）导电杆应用白布包裹后夹持在台虎钳上； （2）检查过程中防止零部件掉落； （3）检查有明显变形、破损、锈蚀的部件应更换	
7	按分解时的相反顺序装复左、右触指座	（1）装复前，用清洗剂清洗各零部件，待干后，在导电接触面涂凡士林，螺纹孔洞涂润滑脂； （2）更换锈蚀的连接、紧固件，潮湿或腐蚀较严重的地区应使用不锈钢螺栓； （3）复装时与拆卸顺序相反：依次是在触指底座上装复位弹簧、触指架、触指、限位圆柱销、外侧垫片、防雨罩，每个环节均按照顺序操作，切勿随意改变回装顺序。着重要指出地是在进行触指组装时，注意其触指是否在同一平面上。安装限位圆柱销时，要注意长圆柱销的位置； （4）装复后，检查所有连接件紧固良好、可靠	（1）在涂抹凡士林、润滑脂等物质时，应厚薄均匀涂抹； （2）触指在安装时，同侧触指安装在同一平面； （3）严重锈蚀、变形、损坏的部件应进行更换； （4）各部件连接处都应紧固牢靠	

续表

序号	作业内容	作业标准	安全注意事项	责任人
8	回装	（1）安装触指座，先将限位圆柱销顶起，再将固定触指座的螺栓进行紧固，完成螺栓紧固后，将限位圆柱销松开，回到适宜位置； （2）安装导电杆，确保导电杆安装平直，同时注意调节同期在合格范围内	（1）高处作业时，正确使用安全带，切记低挂高用； （2）设备安装后，应按规定要求调试合格，并手动分、合闸三次	
9	测试导电回路	（1）将隔离开关合闸，按顺序接好导电回路测试仪的接地线、测试线、电源线； （2）接通电源，打开测试仪开关，选择"100 A，10 s"模式，点击开始测试，读取读数； （3）回路电阻回路电阻值不超过厂家规定值的1.2倍，若不符合要求，重新检查各导电接触面，直至符合要求； （4）关闭测试仪，断开电源，拆除接线，整理仪器	（1）测试前被测试设备和仪器都可靠接地，接地时要注意先接地端，后接导体端；拆除顺序与之相反； （2）测量的电流不小于100 A； （3）测试数据要至少测三次； （4）测试过程中出现掉线等情况，一定要先断电源； （5）通电前后均要进行呼唱	

Task 2   Response to GW7-126 Disconnector Contact Terminal Overheating

2.1   Work Tasks

Take actions against GW7-126 disconnector phase-B contact terminal overheating.

2.2   References

(1) *Guide for Condition Evaluation of Disconnector and Earthing Switch* (Q/GDW 450-2010).

(2) *Substation Maintenance Management Regulations of State Grid Corporation of China (Trial Implementation), Volume 4−Rules for Maintenance of Disconnectors*.

(3) *Regulations of State Grid Corporation of China on Management of Substation Operation and Maintenance*.

(4) *Substation Maintenance Management Regulations of State Grid Corporation of China*.

(5) *Electric Power Safety Working Regulations (Power Transformation) of State Grid Corporation*.

2.3   Work Requirements

(1) Disconnector contact terminal overheat shall be handled in fine and dry weather. During measurement, work shall be suspended when there is bad weather such as force 6 and higher wind and thunderstorm, hail, dense fog and sandstorm.

(2) Measuring instruments and safety protection tools must have calibration certificates.

(3) For work at height, safety belt must be used, and hanging safety belts at lower part is strictly forbidden.

(4) During operation, the disconnector shall be powered off, the equipment shall be at open position, and a grounding wire is provided for each of the two knife switches for maintenance.

(5) In circuit resistance testing, make sure that the equipment and instruments to be tested are reliably grounded.

2.4   Preparation for Work

1. Basic requirements and conditions for on-site investigation

(1) Check status of faulty disconnector operation and know well work site conditions. Check disconnector test reports, operation records and defect records over the years.

(2) Collect status of faulty equipment, including faulty phase, fault locations, fault severity, etc.

(3) Check range of on-site equipment power cut and safety measures.

(4) Check maintenance power box on the site, and make sure that power capacity meet requirements.

(5) Place tools and appliances, operating vehicles, machines, materials, etc. at designated positions.

2. Tools, equipment, and material selection

Tools and appliances for handling GW7-126 disconnector contact terminal overheating and list of materials, refer to Table 2-8.

Table 2-8　Tools and Appliances for Handling GW7-126 Disconnector Contact Terminal Overheating and List of Materials

S/N	Name	Specifications	Unit	Quantity
1	Combination wrench		set	1
2	Steel ruler	100 mm	pieces	1
3	feeler gauge	50 threads	nos	1
4	Flat-nose pliers	6 in.	pcs	1
5	Long-nose pliers	6 in.	pcs	1
6	Art knife		pair	1
7	Slotted screwdriver	4 in.	pcs	2
8	Fine toothed flat file		pcs	1
9	Loosening agent		bottle	1
10	Emery cloth	00	sheet	1
11	Neutral petroleum jelly		bottle	1
12	Conductive paste		bottle	1
14	Nylon brush		pcs	1
15	25-28% ammonia		L	Numerous
16	White cloth		m	Numerous
18	Bench vice		set	1
19	Manila rope	3 m	Nr.	1
20	Circuit resistance measuring instrument		set	1

3. Hazards and preventive measures

(1) High-voltage electric shock.

Hazards: entering charged bay by mistake may cause high-voltage electric shock.

Preventive and control measures: provide fences to isolate the maintenance bay from adjacent live equipment (bay), and provide "Stop! High Voltage, Danger!" sign board on the work site, provide the only access near road side, and provide "Entrance/Exit" sign boards; Provide a set of three-phase short-circuit grounding wires on both sides of circuit breaker to be tested; At least two persons are required during work, one for supervision and the other for operation under instructions of person in charge of work.

(2) Low-voltage electric shock.

Hazards: low-voltage power at incoming line side (upper end) of the control circuit and lighting circuit air switch in disconnector mechanism box; after closing and opening air switch, secondary circuits shall considered as live circuits, and touching the circuits may cause low-voltage electric shock. Touching live parts during external power connection and testing may cause low-voltage electric shock.

Preventive and control measures: if an operator provides a "No Closing, Work in Progress" sign board on an air switch, the person in charge of work must request for safety precautions change from the operator before closing the air switch; When the person in charge of work repeatedly emphasizes the switching of air switch to the members of the work group, it is forbidden to stretch the fingers into the upper end of the air switch; In external power source connection, at least two persons are required, one for supervision and the other for operation as instructed by the work leader.

(3) Mechanical injury.

Hazards: during disconnector opening and closing tests, workers may suffer mechanical crushing;

Preventive and control measures: during disconnector opening and closing, loud shouting is required, and work can be started only after receiving a reply from all members of the work shift. At least two persons are required during work, one for supervision and the other for operation under instructions of person in charge of work.

(4) Equipment and instrument damage.

Hazards: failure to carry out testing as required and rough handling giving rise to instruments and tested equipment damage. Striking insulators and other components in the process of work.

Preventive and control measures: rough handling is forbidden, and tests shall be carried out by following the steps and requirements to prevent instrument damage. Take care to avoid damage to insulators during work. If there is a situation disallowing operation, work shall be suspended without delay, rough handling is not allowed, and the person in charge shall be informed without delay; At least two persons are required during work, one for supervision and the other for operation under instructions of person in charge of work.

(5) High falling.

Hazards: a worker who climbs up a disconnector through an unfixed ladder stand without support by someone may suffer injury from falling. An operator working on a disconnector without using safety belt may suffer injury from falling.

Preventive and control measures: on the ladders through which the workers climb up and down, "Climb Up And Down Here!" sign board shall be provided; Ladders shall be sturdy, complete, and equipped with anti-slip measures. Ladder supports shall be able to withstand the total weight of personnel and tools and materials carried along during climbing. When a worker is climbing up/down through a ladder stand, the ladder stand shall be support by a person or fixed on knife switch frame using a rope; When working without scaffolding or on scaffolding without railings and higher than 1.5 meters, safety belts shall be used. Hook or rope of safety belt shall be hung on a sturdy and firm member or on a knife switch maintenance frame specifically designed for the safety belt, and safety belt shall be attached high above the user; At least two persons are required during work, one for supervision and the other for operation under instructions of person in charge of work.

4. Division of work among operators

Division of work for handling GW7-126 disconnector contact terminal overheating is shown in Table 2-9.

Table 2-9 Division of Work for Handling GW7-126 Disconnector Contact Terminal Overheating

S/N	Job	Number of persons	Nature of the job
1	Operator	1	Exclusively for operation
2	Supervisor	1	Full-time supervision
3	Supporting staff	1	Provide assistance in wiring, tool transmission, recording, etc.

2.5 Operation Procedures

Operation processes for the task is shown in Table 2-10.

Table 2-10 Process of GW7-126 Disconnector Contact Terminal Overheating Handling

S/N	Scope of work	Operation standards	Precautions for safety	Person-in-charge
1	Early preparations	(1) Apply for work ticket; (2) Check equipment positions and conditions on the site; (3) Provide safety fences and sign boards; (4) Correctly use labor protection appliances and wear work clothes neatly as required in the regulations	(1) Go through procedures for work ticket application as required in the safety regulations; (2) Before entering the work site, correctly wear safety helmet, work clothes and soft soled shoes	
2	Validation of working environment	(1) Check initial state of the disconnector and keep power switch at OFF position. Record parameters on equipment nameplate. Check maintenance records and test records; (2) Check fences and sign boards; (3) Correctly check necessary tools, materials, machines and instruments; (4) The person in charge of work and auxiliary workers shall convene a pre-shift meeting to provide pre-shift briefing, ascertain work tasks, safety measures, hazards, work processes and go through procedures for confirmation by following the "Four-understanding" principle	(1) Initial equipment status shall be open status; (2) Fences and sign boards are provided correctly; (3) Make sure that safety equipment and tools are qualified and free of damage, deformation or breakage; (4) Before start of work, provide safety instructions to auxiliary workers to enable them to know well work site, work task, risk points and control measures; go through procedures for confirmation	
3	Removal	(1) Workers shall wear safety belts, prepare work tools and get on knife switch maintenance frame for work; (2) Firstly, leading wires on both sides of phase B shall be removed, and the leading wires shall be fixed on insulator of this phase; (3) Remove left and right contact finger bases, release locking by the limit bolt so that the contact terminal base can be deflected left and right to allow easy removal of the fixing bolts;	(1) When a worker gets up disconnector through a staircase, the staircase must held up by someone; (2) Hang the safety belt on knife switch maintenance frame after getting up the knife switch; (3) During work at heights, do not throw tools and appliances, and transfer tools and appliances by placing them in special tool bags or firmly fixing them on ropes;	

Continued

S/N	Scope of work	Operation standards	Precautions for safety	Person-in-charge
3	Removal	(4) The removed contact finger base shall be placed in a flat position; (5) Remove B-phase conductive rod, provide marking on it to allow easy reassembly, remove connecting bolts of the conductive rod, and remove the conductive rod and clamp; (6) Place single-phase conductive rod on the working platform	(4) Correctly fix leading wire on insulator of the very phase; (5) Take care to avoid damage to conduit during conductive rod removal	
4	Disassemble left and right finger bases	(1) Make sure that all components disassembled are neatly placed in oil pan from far to near; (2) Remove rain cover fixing nut, remove rain cover and washer and place them in an oil pan; (3) Remove gaskets and fixing pins at both sides of the fixed contact finger and place them in an oil pan; (4) Remove the long and short cylindrical pins fixing the contact finger and place them in an oil pan; (5) Remove contact fingers, disassemble outer gaskets, springs and pins connecting the contact fingers, and place them neatly in an oil pan; (6) Remove split pins and gaskets fixing the contact finger stand and remove the contact finger stand; (7) Remove the return spring from the contact finger base and place the contact finger base and return spring in an oil pan and complete removal of the entire contact finger base (Left and right contact terminal bases have consistent structures, and only disassembly of right contact base is explained here)	(1) Disassembled components shall be placed in order to allow easy reassembly; (2) Prevent parts from falling during removal	
5	Check and repair left and right finger bases	(1) Clean all disassembled components with alcohol to remove dirt and impurities and air dry the components; (2) Make sure that rain cover is intact and has intact coating, and heavily rusted or cracked rain cover must be replaced. Check rain cover fixing bolts and washers for rust and blocking. Slight rust can be gently polished with sandpaper, and blocking shall be cleared. Parts and components that are not available for reuse shall be replaced;	(1) Remember to vigorously polish conductive parts such as contact fingers using sandpaper; (2) Prevent the parts from falling during inspection; (3) Clean parts and components with detergent; (4) Replace component with obvious deformation, breakage or rust	

Continued

S/N	Scope of work	Operation standards	Precautions for safety	Person-in-charge
5	Check and repair left and right finger bases	(3) Check that fixing gaskets and pins at both sides are free of rust, the holes are unblocked and the gaskets are flat and free of deformation; absence of clearance when the gaskets abut against a ruler indicates that they are straight; presence of obvious clearance indicates that the gaskets have deformed; (4) Check cylindrical pins and split pins for deformation and rust, remove the rust using sand papers, replace heavily rusted or deformed pins; (5) Check gaskets, pins and springs of continuous contact finger for deformation or rust, check spring clearance for evenness, and replace heavily deformed parts and components. For checking spring deformation, spring length can also be measured with a ruler, variation of length indicates presence of deformation and replacement is required. Check if there are obvious burn marks on the contact finger, touch conductive portion of the contact finger with hand to make sure if there are burrs or silver plating falling off, and replacement is required if silver plating falls off or there is severe burn. Small burrs can be removed with a small flat file. After inspection, place removed contact fingers in an oil pan, fill 25%–28% ammonia water and immerse the contact fingers for 15 min; Gently wipe touch finger surface back and forth to remove surface oxides, and wipe it with white cloth and air dry it for later use; (6) Check contact finger frame for crack and rust and check the holes for clogging; (7) Check conductive part of the contact finger base for visible burn marks, and check the welded portions for crack and check holes for clogging; (8) Return spring clearance shall be uniform and free of deformation, rust, etc		

Continued

S/N	Scope of work	Operation standards	Precautions for safety	Person-in-charge
6	Check conductive rod	(1) Check conductive rods for damage, rectify slightly deformed ones; check conductive contact surface of conductive rod for overheat, and remove oxide layer on the contact surface without silver plating; (2) Check conductive rod connecting bolts and clamps for rust and damage, remove slight rust (if any) with finishing file, and replace the heavily rusted or damaged ones; (3) Replace the contact terminals and conductive rods if there is heavy rust, crack and breakage at the connection between them	(1) Conductive rods shall be wrapped in white cloth and clamped on a vise; (2) Prevent the parts from falling during inspection; (3) Replace component with obvious deformation, breakage or rust	
7	Reassemble left and right touch finger bases in a reverse order.	(1) Before reassembly, clean all parts with cleaning agent; after drying, apply vaseline on the conductive contact surface and fill grease in threaded holes; (2) Replace rust eaten connectors and fasteners, and use stainless steel bolts in damp areas or areas exposed to heavy corrosion; (3) Reassembly in reverse order of disassembly: reinstall return spring, finger frame, finger, cylindrical limit pin, outer gasket and rain cover into finger base, complete every step in sequence, never change reassembly sequence arbitrarily. It must be particularly noted that during finger assembly, make sure that contact fingers are on a same plane. In installing cylindrical limit pin, pay attention to position of cylindrical limit pin; (4) After reassembly, make sure that all connectors are well and firmly fastened	(1) Uniformly apply Vaseline, grease and other substances; (2) In contact finger installation, contact finger on a same side must be installed in a same plane; (3) Replace components that are heavily rusted, deformed or damaged; (4) Make sure that all component connections are tight and firm	
8	Reassembly	(1) For installing a finger base, jack up the cylindrical limit pin, tighten finger base fixing bolts, after tightening the bolts, loosen the cylindrical limit pin and allow it to return to appropriate position; (2) In conductive rod installation, make sure that it is installed flat and straight and adjust synchronism into acceptable range	(1) For work at height, correctly use safety belts, and remember to fasten safety belt at lower part; (2) After equipment installation, commissioning is required to make sure that it is qualified, and three cycles of manual opening and closing are required	

Continued

S/N	Scope of work	Operation standards	Precautions for safety	Person-in-charge
9	Test conductive loop	(1) Close the disconnector and connect grounding wires, test wires and power wire of the conductive circuit tester in sequence; (2) Turn on power source, turn on the tester switch, select "100 A, 10 s" mode, click to start testing, and read the values; (3) Circuit resistance shall be no more than 1.2 time of the value specified by the manufacturer; if it does not meet the requirement, recheck the conductive contact surfaces until the requirement is met; (4) Turn off the tester, disconnect power source, remove the wires and sort out the instruments	(1) Before the test, the equipment and instruments to be tested shall be reliably grounded. In grounding, grounding terminal must be connected and then conductive terminal can be connected; Sequence of removal is the opposite; (2) Current measured shall be no less than 100 A; (3) The test data should be measured for at least three times; (4) Cut off power supply if there is disconnection during the test; (5) Loudly shouting is required before and after energization	

任务三 CJ6 型电动机操动机构检修

一、工作任务

对指定型号隔离开关的 CJ6 型电动操动机构进行分解检修,本次主要进行二次元件的检修。

二、引用标准

(1)《国家电网公司变电检修管理规定(试行)第 4 分册隔离开关检修细则》。
(2)《国家电网公司电力安全工作规程》(变电部分)。
(3)《高压交流隔离开关和接地开关》(GB 1985—2014)。

三、工作要求

本次作业为室外作业,要求天气晴好,无降雨,湿度不大于 85%。

四、准备工作

(一)现场勘察的基本要求及准备工作

(1)检修前应认真查阅电动操动机构安装记录、大修记录、设备运行记录、故障情况记录和缺陷情况记录。对所查阅的结果进行详细、全面的调查分析,以判定传动系统及电动操动机构的综合状况,为现场制定检修方案打好基础。

(2)准备好电动操动机构使用说明书、记录本、表格、检修报告等。

(3)编制标准化作业指导书。

(4)拟订检修方案,确定检修项目,编排工期进度。

(5)场地准备。在检修现场四周设一留有通道口的封闭式遮栏,并在周围背向带电设备的遮栏上挂适当数量的"止步,高压危险"标示牌,在通道入口处挂"从此进出"标示牌;在作业现场按定置图摆放检修工具、量具、材料、备品备件和测试仪器及垃圾箱。

(二)工器具及材料选择

准备工具、机具、材料、备品备件、试验仪器和仪表等,并运至检修现场。仪器仪表、工器具应检验合格,满足本次施工的要求,材料应齐全,图纸及资料应符合现场实际情况。

工器具见表 2-11,材料见表 2-12,备品备件见表 2-13。

表 2-11 工器具

序号	名称	规格型号	单位	数量
1	手锤		把	2
2	活动扳手		把	3
3	梅花扳手		套	3
4	套筒扳手	10~32	套	1

续表

序号	名称	规格型号	单位	数量
5	一字改刀		把	1
6	十字改刀		把	1
7	万用表		块	1
8	兆欧表	1000 V、2500 V	块	1
9	卷尺	5 m	把	1
10	电源盘及电源线		套	1

表 2-12 材料

序号	名称	规格型号	单位	数量
1	清洗剂		千克	10
2	低温润滑脂		千克	2
3	凡士林		千克	0.3
4	砂布		张	10
5	抹布		千克	1.5
6	钢丝刷		把	3
7	毛刷	40 mm	把	3

表 2-13 备品备件

序号	名称	规格型号	单位	数量
1	行程开关		个	1
2	交流接触器		个	1
3	热继电器		个	1
4	辅助开关		个	1
5	按钮		个	2

（三）危险点及预防措施

作业中危险点分析及控制措施见表 2-14。

表 2-14 作业中危险点分析及控制措施

序号	危险点	控制措施
1	触电	（1）工作人员之间做好相互配合，拉、合电源开关时发出相应口令。 （2）使用完整合格的安全开关，装合适的熔丝。 （3）接、拆试验电源必须在电源开关拉开的情况下进行。 （4）要正确操作绝缘电阻表，防止感电伤人
2	误入、误登带电间隔	（1）工作前向作业人员交代清楚临近带电设备，并加强监护。 （2）工作人员应走指定通道，在遮栏内工作，严禁擅自移动和跨越遮栏。 （3）严禁攀登运行设备构架

续表

序号	危险点	控制措施
3	机械伤害	严格执行一般工具的使用规定，使用前严格检查，不合格的工具禁止使用
4	作业空间窄小，碰伤头部和手脚	（1）工作中必须戴好安全帽。 （2）统一指挥，注意作业配合和动作呼应

（四）作业中安全注意事项

（1）参加作业人员必须经过危险点及标准化作业指导书、工作票的学习并签字后方可参加工作。

（2）开工前，工作负责人应核对设备双重名称、编号与工作票所列检修内容相同，方可带领工作班成员进入现场。

（3）工作过程中，工作班成员因故离开现场返回时，应核对一次设备名称、编号。

（4）接取临时电源时，应从现场动力箱中接取，按规定接入漏电保护器，严禁在开关动力箱中取临时电源。

（5）梯子使用前应认真检查，不合格的梯子不准使用，使用中应有专人扶梯或采取其他防止梯子倾倒的措施。

五、作业程序

1. CJ6电动操动机构检修前检查项目及标准

（1）电动机、传动齿轮、蜗轮、蜗杆、转轴、辅助开关及电动机控制附件等操作是否灵活、可靠。

（2）行程开关、按钮、交流接触器是否能正常动作，辅助开关是否正常转换，电动机是否有异常响声，各转动部位有无松动，电动、手动是否相互闭锁等。

2. CJ6电动操动机构分解检修流程

（1）CJ6电动操动机构检修前准备工作。

① 断开电动机电源及控制电源。

② 拆除二次元件装配接线端子与进线电缆导线的固定螺钉，拆下与电动机相连的电源线，松开电缆线夹，从机构箱中抽出进线电缆。在拆下进线电缆前，应做好相应记录。

③ 拆除机构箱与基础相连的4个螺栓，将机构箱拆下并放置在检修平台上。

④ 拆除接线端子板与辅助开关相连的二次接线的螺钉，抽出二次接线。

⑤ 拆除电动机接线盒上2个固定螺栓，取下罩。拧下其与接线端子板间二次接线固定螺钉，抽出电缆线。

⑥ 拧下分、合闸接触器上与行程开关相连接的二次接线螺钉，拆下二次接线。

（2）CJ6电动操动机构二次元件分解。

① 拆除接线端子板与分、合闸按钮，急停按钮，分、合闸接触器，组合开关，刀开关（空气开关）相连接的二次接线螺钉，拆除二次接线。

②从L形接线板上分别拆下接线端子板，分、合闸按钮，急停按钮，分、合闸接触器，组合开关，刀开关（空气开关）。

③拆除辅助开关上的二次接线固定螺钉，拆下二次接线，拆卸前应做好记录。

④拆除辅助开关与减速器箱底座下部间的固定螺栓，拆下辅助开关传动板，取出辅助开关，拧下辅助开关转动盘分、合闸切换块的2个螺钉，取出分、合闸切换块。

（3）CJ6电动操动机构二次元件检修工艺要求。

①检查行程开关，分、合闸按钮，急停按钮等动作是否灵活、正确，触点是否烧伤，如有烧伤痕迹，可用00号砂布处理。如破损应更换。

②检查二次线接线端子是否紧固，绝缘是否良好。

③检查接线端子板端子排编号，缺的应补齐。端子排如有破损、裂纹应更换，压线螺钉锈蚀应更换。

④检查分、合闸接触器的外观有无破损，如破损严重应更换；检查其动作情况，调整好触头开距和超行程后用万用表测试接触器通、断是否可靠，同时检查线圈有无烧伤，必要时更换。

⑤检查接触器触点是否烧伤痕迹，必要时更换。

⑥检查行程开关，分、合闸按钮，急停按钮、交流接触器、热继电器、辅助开关等弹簧及弹片，用手轻压弹簧及弹片，检查复位情况，如永久疲劳应更换。

⑦检查热继电器，如破损应更换，并用清洗剂清洗热继电器外表面。

⑧检查热继电器整定设置是否正确。

⑨检查加热器是否良好，自动控制装置动作是否准确可靠，用1000 V绝缘电阻表测量其绝缘电阻应符合要求。

⑩检查L形接线板，除去锈蚀，校正变形及作防锈处理，锈蚀严重者应更换。

⑪用清洗剂清洗所有零部件，待干后，在所有元件导电接触面涂少量的中性凡士林油。

（4）CJ6电动操动机构二次元件检修后装复。分、合闸接触器，行程开关，分、合闸按钮，组合开关，空气断路器及其接线端子板的装复，按分解相反的顺序进行。装复时应注意以下几点：

①更换锈蚀的紧固件及弹簧。

②用万用表检查行程开关，分、合闸，急停按钮，接触器触点通、断情况，并检查切换是否可靠，通、断位置是否正确。

③装复后，核对二次接线是否正确。

（5）CJ6电动操动机构二次元件检修质量标准。

①行程开关，分、合闸，急停按钮，接触器等触点分合闸位置切换应正确、灵活、无卡涩。触点接触良好，弹簧及弹片的弹性良好。

②拆下的二次回路端子线及电缆线头应有标记。

③端子排编号清晰、完整，端子排无破损。

④用1000 V绝缘电阻表测量二次元件绝缘电阻应大于2 MΩ。

（6）CJ6电动操动机构检修后装复。按分解时的相反顺序装复，并注意以下几点：

①装复前，用清洗剂清洗各零部件，待干后，在转动件上涂二硫化钼锂。

②装复时，注意辅助开关转动盘与分、合闸切换块的相对位置。

③ 检查各连接、固定螺栓（钉）是否紧固。

④ 检查一、二级齿轮啮合位置是否正确。

⑤ 用手柄转动机构，检查传动系统动作是否灵活，蜗杆及中间轴有无轴向窜动。

⑥ 复核二次接线是否正确。

⑦ 用手柄操作机构，检查机构分、合闸位置与辅助开关切换位置是否对应，接触是否可靠。

Task 3 Maintenance of CJ6 Motor Operating Mechanism

3.1 Work Tasks

CJ6 electric operating mechanism of designated disconnector shall be disassembled for maintenance, and secondary elements maintenance shall be carried out.

3.2 References

(1) *Substation Maintenance Management Regulations of State Grid Corporation of China (Trial Implementation)*, *Volume 4–Rules for Maintenance of Disconnectors*.
(2) *Electric Power Safety Working Regulations (Power Transformation) of State Grid Corporation*.
(3) *High-voltage Alternating-current Disconnectors and Earthing Switches* (GB 1985-2014).

3.3 Work Requirements

The work is outdoor work and must be completed in sunny days without rain and with humidity no more than 85%.

3.4 Preparations

1. Basic requirements and preparations for on-site investigation

(1) Before maintenance, electric operating mechanism installation records, major repair records, equipment operation records, fault records and defect records shall be carefully checked. The results shall be reviewed and analyzed in an all-round manner to determine overall condition of transmission systems and electric operating mechanism and provide a solid foundation for preparation of maintenance plan for the work site.

(2) Electric operating mechanism manuals, notebooks, forms, maintenance report, etc. shall be prepared.

(3) Prepare standard operating instructions.

(4) Prepare maintenance plan, determine maintenance items and schedule construction period.

(5) Site preparation shall be carried out. An enclosed barrier with access shall be provided around the maintenance site, and appropriate number of "Stop! High Voltage, Danger!" sign boards shall be provided on the barrier back facing the live equipment, and "Entrance/Exit" sign boards shall be provided at passageway entrances; Maintenance tools, measuring tools, materials, spare parts, testing instruments and trash cans shall be provided on the job site according to the designated layout.

2. Tools, equipment, and material selection

Tools, machines, materials, spare parts, testers and instruments shall be ready and delivered to the maintenance site. Instruments, meters and tools shall be inspected and qualified to allow construction. Materials shall be complete, and drawings and data shall comply with actual situations on site.

See Table 2-11 for tools and instruments, Table 2-12 for materials and Table 2-13 for spare parts.

Table 2-11   Tools and Instruments

S/N	Name	Specifications and models	Unit	Quantity
1	Hand hammer		pcs.	2
2	Monkey wrench		pcs.	3
3	Spline end wrench		set	3
4	Socket spanner	10~32	set	1
5	Slotted screwdriver		pcs.	1
6	Cross-head screwdriver		pcs.	1
7	Multimeter		pieces	1
8	Megger	1000 V, 2500 V	pieces	1
9	Band tape	5 m	pcs.	1
10	Power panel and power cord		set	1

Table 2-12   Materials

S/N	Name	Specifications and models	Unit	Quantity
1	Cleaning agent		kg	10
2	Low-temperature grease		kg	2
3	Vaseline		kg	0.3
4	Emery cloth		sheet	10
5	Rag		kg	1.5
6	Wire brush		pcs.	3
7	Shed	40 mm	pcs.	3

Table 2-13   Spare parts

S/N	Name	Specifications and models	Unit	Quantity
1	Travel switch		nos.	1
2	AC contactor		nos.	1
3	thermal relay		nos.	1
4	Auxiliary switch		nos.	1
5	Shed		nos.	2

3. Hazards and preventive measures

Hazards analysis and control measures during work refers to Table 2-14.

Table 2-14 Hazards Analysis and Control Measures during Work

S/N	Hazard	Control measures
1	Electric shock	(1) The staff shall cooperate with one another and issue corresponding commands when opening and closing power switch. (2) Use complete and qualified safety switches and install suitable fuses. (3) Connect and disconnect test power sources when power switch is opened. (4) Correctly operate insulation resistance meter and prevent induced electricity from causing injury
2	Entering or climbing energized bay by mistake	(1) Inform operators of nearby live equipment before start of work and provide supervision. (2) The operators shall take the designated channel and work in the barrier, and they are strictly prevented from moving and crossing the barrier without authorization. (3) Climbing operating equipment frames is strictly forbidden
3	Mechanical injury	Strictly abide by the regulations for use of general tools, carefully check the tools before use and do not use unqualified tools
4	Head, hands and feet injury due to narrow work space	(1) Wear safety helmet during work. (2) Provide unified command, ensure work and action in concert with one another

4. Precautions for safety in work

(1) Operators must know well the hazards, standard work instructions and work tickets before they can start work.

(2) Before start of work, the person in charge of work shall verify that the dual equipment name and number are consistent with the maintenance items listed in the work ticket before leading the team members into the site.

(3) If a team member leaves and returns to the site for a reason during work, equipment name and number shall be checked.

(4) Temporary power supply shall be connected from on-site power box, and leakage protector shall be connected. Connecting temporary power source from switch power box is strictly forbidden.

(5) Ladders shall be carefully inspected before use, and use of unqualified ladders is not allowed. During use, the ladder shall be held up by a specially assigned person, or other measures shall be taken to prevent the ladder from toppling.

3.5 Operation Procedures

1. Items to be checked before maintenance and standards for CJ6 electric operating mechanism

(1) Check motors, transmission gears, worm gears, worms, rotating shafts, auxiliary switches and motor control accessories for flexible and reliable operation.

(2) Check if travel switches, push buttons and AC contactors operate properly and auxiliary switches allow normal switchover; if motors have abnormal noise; if rotating parts are loose and if

electric and manual operations interlock.

2. CJ6 electric operating mechanism disassembly and maintenance process flow

(1) Preparations before CJ6 electric operating mechanism maintenance.

① Disconnect motor power supply and control power supply.

② Remove the bolts fixing the secondary element assembly terminal block and incoming cable conductor, remove the power cord connected to the motor, release cable clamps and pull incoming cable out of the mechanism box. Records shall be kept before removing incoming cables.

③ Remove the 4 bolts connecting the mechanism box to the foundation, remove the mechanism box and place it onto the maintenance platform.

④ Remove screws of the secondary wiring connecting the terminal block to the auxiliary switch and pull out the secondary wire.

⑤ Remove 2 fixing bolts from motor junction box and remove the cover. Unscrew the screws connecting it to secondary wiring between terminal boards and pull out the cable.

⑥ Unscrew the secondary connection screws on the opening/closing contactors in connection to travel switch, and remove secondary wiring.

(2) Disassembly of secondary elements of CJ6 electric operating mechanism.

① Remove the secondary wiring screws connecting the terminal board to the opening and closing push buttons, emergency stop buttons, opening and closing contactors, combination switches and knife switches (air switches), and remove secondary wiring.

② Remove terminal blocks, opening and closing buttons, emergency stop buttons, opening and closing contactors, combination switches and knife switches (air switches) from the L-shaped terminal blocks.

③ Remove secondary wiring fixing screws from auxiliary switches, remove secondary wiring and keep records before disassembly.

④ Remove the bolts fixing the auxiliary switch and the lower part of the gearbox base, remove auxiliary switch driving plate, take out auxiliary switch, unscrew the 2 screws for auxiliary switch rotating disc opening and closing switch block and take out opening and closing switch block.

(3) Requirements for maintenance process for secondary elements of CJ6 electric operating mechanism.

① Check travel switches, opening and closing buttons, emergency stop buttons and other parts for free movement, check contacts for burn, and use No. 00 emery cloth for treatment if there are burn marks. Replace the damaged one.

② Check secondary wiring terminals for tightness and check insulation.

③ Check terminal block number of terminal board and supply the lack. Replace damaged or cracked terminal blocks and rust-eaten crimping screws.

④ Check appearance of opening and closing contactors for damage, and replace the severely damaged contactors. Check actions of the contactors, adjust contact terminal gap and overtravel, and then check the contactors using multimeter for reliable closing and opening, check coils for burn and replace them as required.

⑤ Check contact terminals of contactor for burn marks and replace them as required.

⑥ Checks springs and leaves of travel switches, opening and closing buttons, emergency stop buttons, AC contactors, thermal relays, auxiliary switches; gently press the springs and leaves with hand to check reset status; replace them if there is permanent fatigue.

⑦ Check thermal relay, replace the damaged one and clean outer surface of the thermal relay with cleaning agent.

⑧ Make sure that thermal relay is correctly set.

⑨ Check that the heater is at good condition and automatic control devices operate accurately and reliably, check insulation resistance with 1000 V insulation resistance meter.

⑩ Check the L-shaped terminal block, remove rusts, correct deformation, provide rust prevention treatment, and replace the severely rust-eaten ones.

⑪ Clean all parts with cleaning agent, apply a small amount of neutral Vaseline oil on conductive contact surfaces of the elements.

(4) Reassembly after maintenance of secondary elements of CJ6 electric operating mechanism. Reassemble opening and closing contactors, travel switches, opening and closing buttons, combination switches, air circuit breakers and terminal boards in reverse order of disassembly. Pay attention to the following points during reassembly:

① Replace rust eaten fasteners and springs.

② Check travel switches, opening and closing, emergency stop buttons and ON/OFF status of contact terminals of contactor, and make sure that switchover is reliable and ON and OFF positions are correct.

③ Make sure that secondary wiring is correct after reassembly.

(5) Quality standard for maintenance of secondary elements of CJ6 electric operating mechanism.

① Closed position and open position of travel switches, opening and closing, emergency stop buttons, contactors and contacts shall be correct, flexible and free of jamming. The contacts have high contact performance, and the springs have high elasticity.

② Secondary circuit lead wires and cable heads removed shall be marked.

③ Terminal block numbers are clear and complete, and the terminal blocks are intact.

④ Insulation resistance of a secondary element measured with 1000 V insulation resistance meter shall be more than 2 MΩ.

(6) Reassembly after CJ6 electric operating mechanism maintenance. Reassemble in the reverse order of disassembly, and pay attention to the following points:

① Before reassembly, clean all parts and components with cleaning agent; after drying, apply molybdenum disulfide lithium on moving parts.

② During reassembling, pay attention to relative positions of auxiliary switch rotating disc and opening and closing switching blocks.

③ Check the connections and fixing bolts (screws) for tightness.

④ Check that stage-I and stage-II gears mesh at correct positions.

⑤ Rotate the mechanism with the handle to check if the transmission system operates flexibly and if there is axial movement of worm and intermediate shaft.

⑥ Recheck secondary wiring for correctness.

⑦ Operate the mechanism with the handle to check whether open and closed positions of the mechanism correspond to switching positions of the auxiliary switch and whether contact is reliable.

项目三 高压断路器检修

本项目的学习，注重学生在养成良好的安全生产意识和安全生产习惯的同时，培养团结协作的精神、一丝不苟、精益求精的工匠精神，树立勤业、乐业、爱岗敬业的职业道德。能按照标准化作业流程完成各项运检或测试工作及检修等，标准化作业流程执行力，能够准确地列出设备作用、工作原理、结构和主要运检项目。

模块一 高压断路器概述

一、高压断路器的定义

高压断路器是电力系统最重要的控制设备和保护设备，具有灭弧特性。通常将额定电压为 3 kV 及以上，能够接通和断开正常工作电流，并能快速切除过负荷电流和故障电流的开关电器称为高压断路器，它是开关电器中最为完善的一种设备。

高压断路器主要有控制和保护两大作用。

（1）控制作用：当电力系统正常运行时，根据其运行需要，将部分或全部电气设备，以及部分或全部线路投入或退出运行。

（2）保护作用：当电力系统发生故障时，它和继电保护装置相配合，将故障部分从系统中快速切除，减小停电范围，防止事故扩大，保护系统中各类设备不受损坏，保证系统无故障部分安全运行。

二、高压断路器的结构

高压断路器有很多种类型，但其基本结构类似，如图 3-1 所示，主要包括电路通断元件、绝缘支撑元件、操动机构、传动机构及基座等部分。

（1）通断元件：高压断路器的的关键部件，由接线端子、导电杆、动/静触头及灭弧室等组成，承担着接通或断开电路的任务，而灭弧能力的大小则决定了开关的开断能力。

（2）绝缘支撑元件：安装在基座上，起着支撑固定通断元件，并实现与各结构部分之间的绝缘作用。

（3）传动机构：把操动机构提供操作能量及发出的操作命令传递给通断元件。

（4）操动机构：起着控制通断元件的作用，当操动机构接到合闸或分闸命令时，向通断元件提供分、合闸操作的能量，经中间传动机构驱动动触头，实现断路器的合闸或分闸。

（5）基座：用于支撑、固定和安装开关电器的各结构部分，使之成为一个整体。

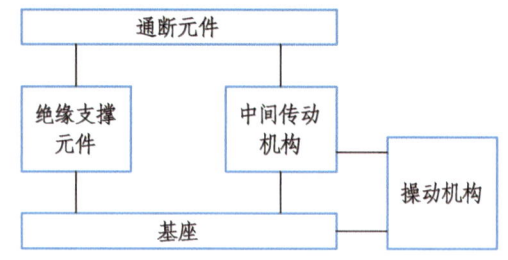

图 3-1　高压断路器原理结构示意图

三、高压断路器的类型

高压断路器按安装地点的不同，可以分为户内型和户外型两种；按灭弧介质和灭弧原理的不同，可以分为油断路器（又分为多油和少油）、真空断路器、六氟化硫（SF$_6$）断路器、空气断路器等。

（1）油断路器：采用绝缘油作为灭弧介质的断路器。

（2）空气断路器：采用压缩空气作为灭弧介质及操作机构能源的断路器。

（3）真空断路器：在真空中开断电流，利用真空的高绝缘强度来实现灭弧的断路器。

（4）六氟化硫（SF$_6$）断路器：采用具有优良灭弧性能的 SF$_6$ 气体作为灭弧介质的断路器。

国产高压断路器的型号表达和含义如下所示：

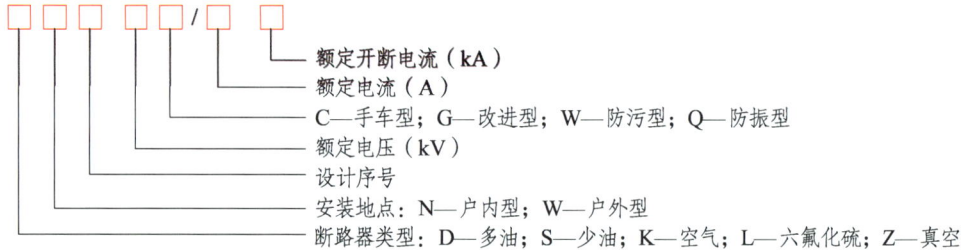

目前，在发电厂及变电站中最常用的是 SF$_6$ 断路器和真空断路器，有少量的少油断路器尚在运行中，而多油断路器和空气断路器在国内已趋于淘汰。

四、高压断路器的技术参数

（1）额定电压：表征断路器绝缘强度的参数。指断路器长期工作能承受的系统最高工作电压。

（2）额定电流：表征断路器通过长期电流能力的参数。指断路器允许连续长期通过的最大电流。

（3）额定开断电流：表征断路器开断电流能力的参数。指在额定电压下，断路器能保证开断的最大电流。

（4）额定关合电流：表征断路器关合电流能力的参数。指断路器能够可靠关合的电流最大峰值。

（5）动稳定电流：表征断路器通过短时电流能力的参数，反映断路器承受短路电流电动力效应的能力。指断路器在合闸状态或关合瞬间,允许通过的最大电流值。

(6)热稳定电流:表征断路器通过短时电流能力的参数,反映断路器承受短路电流热效应的能力。指断路器处于合闸状态下,在一定的持续时间内,所允许通过电流的最大周期分量有效值。

(7)合闸时间:表征断路器操作性能的参数。指断路器从接到合闸命令(合闸回路通电)起到断路器触头刚接触时所经过的时间间隔。

(8)分闸时间:表征断路器操作性能的参数,反映断路器开断过程的快慢。包括固有分闸时间和燃弧时间。

① 固有分闸时间 t_1:指断路器接到分闸命令起到灭弧触头刚分离时所经过的时间。

② 燃弧时间 t_2:指触头分离到各相电弧完全熄灭所经过的时间。

③ 全分闸时间 t_t:指断路器从接到分闸命令(分闸回路通电)起到断路器触头开断至各相电弧完全熄灭时所经过的时间间隔。它等于断路器固有分闸时间与灭弧时间之和。

分闸时间又称全分闸时间,三者的关系如图 3-2 所示,全分闸时间一般为 0.06~0.12 s,小于 0.06 s 的断路器,称为快速断路器。

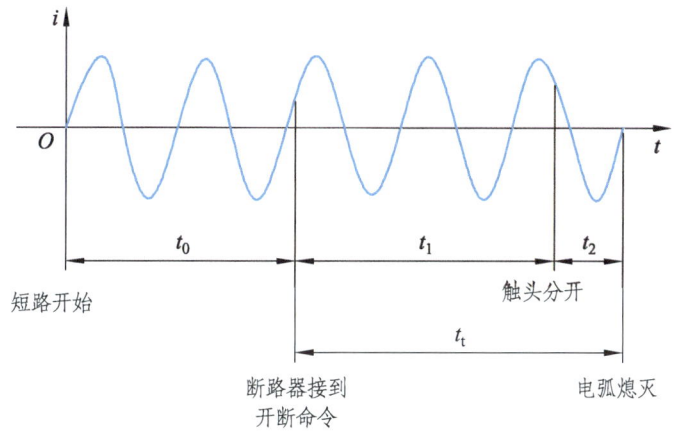

t_0—继电保护动作时间;t_2—燃弧时间;t_1—固有分闸时间;t_t—全分闸时间。

图 3-2 断路器开断电路时的各个时间

(9)额定操作循环:表征断路器操作性能的指标,反映断路器能承受一次或两次以上的关合、开断、或关合后立即开断的动作能力。指断路器根据实际运行需要,按一定时间间隔进行多次分、合的规定操作。额定操作循环分为两类:

① 自动重合闸操作循环:分—t'—合分—t—合分;

② 无自动重合闸操作循环:分—t—合分—t—合分。

其中,"分"表示分闸动作;"合分"表示合闸后立即分闸的动作。

"t'"表示无电流间隔时间,即断路器断开故障电路,从电弧熄灭到电路重新自动接通的时间,标准时间为 0.3 s 或 0.5 s,也即重合闸动作时间;"t"表示运行人员强送电时间,标准时间为 180 s。

Program 3 Maintenance of HV Circuit Breakers

This program aims to turn students into united collaborators, meticulous craftsmen who keep improving, and diligent, enthusiastic, and dedicated professionals while getting them to bear work safety in mind and develop a habit of paying heed to work safety. Through this program, students will be able to complete the operation and maintenance or tests according to the standardized operation flow, execute the standardized operation flow, and accurately list the functions, working principles, structures, and main operation and maintenance items of equipment.

Module 1 Overview of HV Circuit Breakers

3.1.1 Definition of HV Circuit Breaker

An HV circuit breaker is the most important control equipment and protection equipment in an electric power system, which can extinguish the arc. Usually, a switching device with a rated voltage of 3 kV and above that can make and break the normal working current and quickly cut off the overload current and fault current is called an HV circuit breaker. It is the most perfect among the switching devices.

An HV circuit breaker provides control and protection.

(1) Control: When the electric power system is operating properly, it can be used to put part or all of the electrical equipment and lines into or out of operation according to the operation needs.

(2) Protection: When the electric power system fails, it works in conjunction with the relay protection device to quickly remove the faulty part from the system. Thus, it narrows the scope of power interruption, prevents escalation, protects all kinds of equipment in the system from damage, and ensures that the trouble-free parts of the system can operate safely.

3.1.2 Structure of HV Circuit Breaker

There are many types of HV circuit breakers, but their basic structures are similar, as shown in Fig. 3-1, mainly including the circuit break-make element, insulating support element, operating mechanism, transmission mechanism, and base.

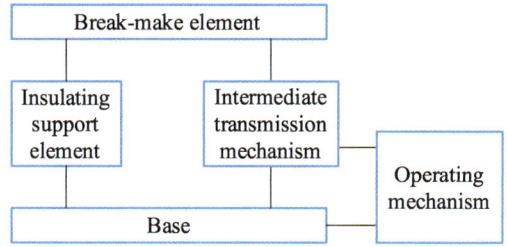

Fig. 3-1　Schematic Diagram of HV Circuit Breaker Structure

(1) Break-make element: It is a key component of the HV circuit breaker and consists of terminal block, conductive rod, moving/static contact terminal, and arc extinguishing chamber, etc. It can make or break a circuit, and the breaking capacity of a switch is determined by the arc extinguishing capacity.

(2) Insulating support element: It is installed on a base, and is able to support and secure the break-make element and isolate the structural parts.

(3) Transmission mechanism: It transmits the operating energy provided by the operating mechanism and the operation command issued by the operating mechanism to the break-make element.

(4) Operating mechanism: It controls the break-make element. When receiving a closing or opening command, the operating mechanism will provide the break-make element with the energy required for opening and closing, and after the intermediate transmission mechanism drives the moving contact terminals, the circuit breaker will be able to close or open.

(5) Base: It supports, fixes, and is fitted with the various structural parts of the switching device to make it a whole.

3.1.3　Types of HV Circuit Breakers

High-voltage circuit breaker can be divided into indoor and outdoor types as per different installation places; In view of different arc extinguishing mediums and principles, it can also be divided into oil circuit breaker (it can be further divided into high oil and less oil types), vacuum circuit breaker, sulfur hexafluoride (SF_6) circuit breaker, air circuit breaker and so on.

(1) Oil circuit breaker: The circuit breaker using insulating oil as arc extinguishing medium.

(2) Air circuit breaker: The circuit breaker using compressed air as arc extinguishing medium and energy for operating mechanism.

(3) Vacuum circuit breaker: The circuit breaker that provides vacuum breaking current, and uses high dielectric strength of vacuum to realize arc extinguishing.

(4) sulfur hexafluoride (SF_6) circuit breaker: The circuit breaker using SF_6 gas of high arc extinguishing performance as the arc extinguishing medium.

Model expressions and implications of high-voltage circuit breakers made in China are listed below:

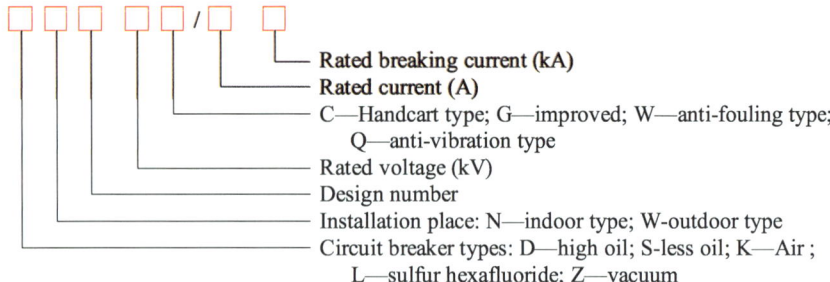

Presently, SF$_6$ circuit breaker and vacuum circuit breaker are most frequently used by power plants and substations. While a limited quantity of low oil circuit breakers are still being used, bulk oil circuit breakers and air circuit breakers are being eliminated in China.

3.1.4 Technical parameters for high-voltage circuit breaker

(1) Rated voltage: The parameter indicating dielectric strength of circuit breaker. It refers to the maximum system working voltage that can be withstood by circuit breaker during long-term operation.

(2) Rated current: The parameter indicating the capability of circuit breaker in passing through permanent current. It refers to the maximum continuous and permanent current permitted by the circuit breaker.

(3) Rated breaking current: The parameter indicating current breaking capacity of circuit breaker. It refers to the maximum current at which circuit breaker can ensure breaking under the rated voltage.

(4) Rated making current: The parameter indicating current making capacity of circuit breaker. It refers to the maximum peak value of current at which circuit breaker is available for reliable making.

(5) Dynamic stable current: The parameter indicating the capacity of circuit breaker in passing through short time current, and reflecting the capability of circuit breaker in withstanding electrodynamic effect from short-circuit current. It refers to the maximum current value that permitted by circuit breaker at closed status or at the moment of closing.

(6) Thermal stability current: The parameter indicating the capability of circuit breaker in passing through short time current, and reflecting the capability of circuit breaker in withstanding thermal effect from short-circuit current. It refers to the maximum valid value of periodic component of let-through current that permitted by the circuit breaker within a certain continuous time at closed status.

(7) Closing time: The parameter indicating operating performance of circuit breaker. It refers to the time interval commencing from reception of closing command (power-on of closing circuit) to contact with contact terminal of circuit breaker.

(8) Opening time: The parameter indicating operating performance of circuit breaker and reflecting breaking velocity of circuit breaker. This include inherent opening time and arcing time.

① Inherent opening time t_1: It refers to the time passed commencing from reception of opening command by the circuit breaker to the moment when arc extinguishing contact terminal is just separated;

② Arcing time t_2: It refers to the time passed commencing from the separation of contact terminal to complete extinguishing of electric arc at any phase.

③ Full opening time t_t: It refers to the time passed commencing from the reception of opening command (power-on to opening circuit) to breaking of contact terminal of circuit breaker and complete extinguishing of electric arc at any phase. It equal to the sum of inherent opening time of circuit breaker and arc extinguishing time.

The opening time is also known as full opening time. Relationship among the three is as shown in Fig. 3-2. Full opening time is normally 0.06–0.12 s. Any circuit breaker with opening time less than 0.06 s is called quick breaker.

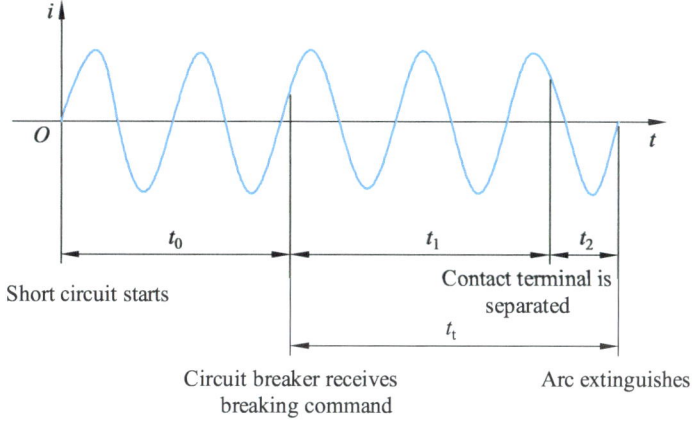

t_0—Relay protection actuation time; t_2—Arcing time; t_1—Inherent opening time; t_t—Full opening time.

Fig. 3-2　Circuit Breaking Time of Circuit Breaker

(9) Rated operation cycle: The indicator indicating operating performance of circuit breaker that reflects the capability of circuit breaker in withstanding closing and breaking for one time or over two times or actuating immediate breaking after closing. It refers to the specified operation of circuit breaker for repeated opening and closing at a certain time interval as per practical operation demands. Rated operation cycle falls into two categories:

① Auto reclosing operation cycle: opening–t'–closing/opening–t–closing/opening;

② Non-auto reclosing operation cycle: opening–t–closing/opening–t–closing/opening.

Wherein, "Opening" refers to opening action; "Closing/opening" refers to immediate opening action following closing. " t' " refers to no current interval time, namely the time from arc extinguishing to auto auto re-connection of circuit following disconnection of fault circuit by the circuit breaker. Standard time is 0.3 s or 0.5 s, or namely reclosing action time; "t" refers to the time of forced power transmission by operators. Standard time is 180 s.

模块二 真空断路器

真空断路器是利用真空的高介质强度来灭弧的断路器,因其灭弧介质和灭弧后触头间隙的绝缘介质都是高真空而得名。它具有触头开距短(一般为 10 mm)、熄弧快、体积小、重量轻、无爆炸危险、无污染、适用于频繁操作、灭弧不用检修等优点。真空断路器在中低压配电网中占据绝对优势,应用非常普及,特别在量大面广的 12 kV 等级中占到 99% 以上。

一、真空断路器的工作原理

由于真空间隙气体稀薄,气体分子的自由行程大,发生碰撞游离的机会少,击穿电压高,所以高真空度间隙的绝缘强度比灭弧介质的绝缘强度高得多。要满足真空灭弧室的绝缘强度要求,真空度一般要求在 $1.33 \times 10^{-3} \sim 1.33 \times 10^{-7}$ Pa 之间。

同时,在气体稀薄的真空间隙中,气体的游离不可能维持电弧的燃烧,所以真空间隙被击穿而产生电弧不是气体的碰撞游离的结果。实际上,真空间隙击穿产生的电弧,是在触头电极蒸发出来的金属蒸汽中形成的。在开断电流时,随着触头的分离,触头接触面积迅速减少,最后只有留下一个或几个微小的接触点,其电流密度非常大,温度急剧升高,使接触点的金属熔化并蒸发出大量的金属蒸汽。由于金属蒸汽温度很高,同时又存在很强的电场,导致强电场发射和金属蒸汽的电离,从而发展成真空电弧。因此,真空间隙击穿的主要因素除真空度外,还与触头的材料及其表面状况、剩余气体的种类、真空间隙长度以及电场的均匀程度有关。

真空断路器是利用在真空电弧中生成的带电粒子和金属蒸汽具有很高扩散速度的特性,在电弧电流过零,电弧暂时熄灭时,使触头间隙的介质强度能很快恢复而实现灭弧的。真空交流电弧的熄灭与其他交流电弧一样,主要决定于电流过零后弧隙介质强度的恢复。而弧隙绝缘强度的恢复,取决于带电粒子的扩散速度、开断电流的大小以及触头的面积、形状和材料等因素。在燃弧区域施加横向磁场和纵向磁场,驱动电弧高速扩散运动,可以提高介质强度的恢复速度,还能减轻触头的烧损程度,提高使用寿命。

真空断路器在开断大电流时,如果出现电弧的弧柱呈收缩状和阳极斑点(即呈现集聚性电弧,阳极表面有大而亮的斑点出现),阳极斑点会向弧柱喷射大量的金属蒸汽,造成弧柱压力加大,粒子大量扩散,此时真空断路器则完全丧失开断能力。如果要求电弧燃烧时在半个周波内不出现阳极斑点,只有采取提高真空中灭弧能力的措施,才能使真空断路器应用到大容量断路器的领域。

二、真空断路器的基本结构

真空断路器主要由真空灭弧室、支架和操动机构三部分组成。真空灭弧室是真空断路器的核心部件,具有开断、导电和绝缘的功能,主要由绝缘外壳、动静触头、屏蔽罩和波纹管等组成,其基本结构如图 3-3 所示。绝缘外壳是由绝缘筒、波纹管、动端盖板和静端盖板所组成的真空密封容器。灭弧室内的静触头固定在静导电杆,静导电杆穿过静端盖板并与之焊成一体;动触头固定在动导电杆的一端上,动导电杆的中部与波纹管的一个端口焊在一起,波纹管的另一端口与动端盖板的中孔焊接，动导电杆从中孔穿出外壳；在动、静触头和波纹管周围分别装有屏蔽罩和。由于波纹管在轴向上可以伸缩，因而这种结构既能实现从灭弧室外操动动触头作分合作用，灭弧室在无机械外力作用时，其动、静触头始终保持闭合位置，当外力使动导电杆向外运动时，触头才分离。真空灭弧室的性能主要取决于触头材料和结构，还与屏蔽罩的结构、灭弧室的材质以及制造工艺有关。

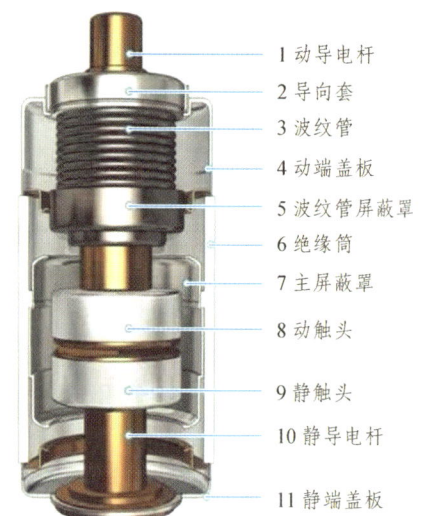

图 3-3　真空灭弧室结构示意图

1 动导电杆
2 导向套
3 波纹管
4 动端盖板
5 波纹管屏蔽罩
6 绝缘筒
7 主屏蔽罩
8 动触头
9 静触头
10 静导电杆
11 静端盖板

1. 绝缘外壳

真空灭弧室的绝缘外壳既是真空容器，又是动静触头间的绝缘体。其作用是容纳和支持灭弧室内的各种零件，与这些部件气密地焊接在一起，以确保灭弧室内的高真空度。为保证真空灭弧室工作的可靠性，一般要求在 20 年内，真空度不得低于规定值，所以需要严格密封。其次是要有一定的机械强度。绝缘筒用硬质玻璃、高氧化铝陶瓷或微晶玻璃等绝缘材料制成，以陶瓷为主。绝缘外壳常用硬质玻璃、氧化铝陶瓷或微晶玻璃制造。外壳的端盖常用不锈钢、无氧铜等金属制成。

波纹管的功能是用来保证灭弧室完全密封，同时使得断路器操动机构的运动得以传到动触头上。波纹管常用的材料有不锈钢、磷青铜等，以不锈钢性能最好，有液压成形和膜片焊接两种形式。波纹管允许伸缩量应能满足触头最大开距的要求。触头每次分、合，波纹管的波状薄壁就要产生一次大幅度的机械变形，使其很容易因疲劳而损坏。

2. 触头

真空断路器的触头，既是关合时的通流元件，又是开断时的灭弧元件。触头的材料和结构直接影响到灭弧室的开断容量、电气寿命、耐压强度、关合能力、截流过电压及长期导通电流能力等。由于纯金属触头缺乏真空触头良好的性能，故真空断路器触头材料使用了两种互不相同的合金组合材料，即铜铋材料（Cu/Bi）和铜铬材料（Cu/Cr）。

触头是真空灭弧室内最为重要的元件，其开断能力和电器寿命主要由触头结构来决定。目前真空断路器的触头系统一般都采用对接式，根据开断时灭弧的基本原理不同，其发展经历了三种典型结构型式，即非磁吹平板触头、横磁吹触头和纵磁吹触头，如图 3-4 所示。这些触头的共同特点是利用磁场力使真空电弧很快地运动，防止在触头上产生需要长时间冷却的受热区域。

（1）非磁吹平板触头。该触头的平板端面作为电接触和燃弧的表面，真空电弧在触头间燃烧时不受磁场的作用。开断小电流（瞬时值少于 10 kV）时，触头间的真空电弧为扩散型，燃弧后介质强度恢复快，灭弧性能好；开断电流较大时，真空电弧为集聚型，燃弧后介质强度恢复慢，因而开断可能失败。非磁吹平板触头容易加工，成本低，仅用于真空接触器和真空负荷开关等开断电流不超过 10 kA 的灭弧室中。

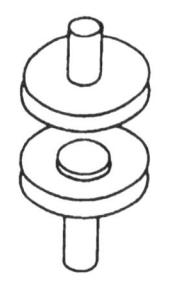  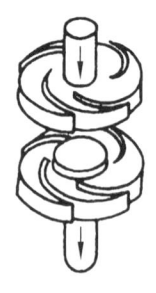  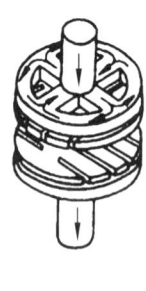

（a）平板非磁吹　　（b）杯状触头　　（c）螺旋触头　　（d）纵磁吹触头　　（e）纵磁吹触头
　　触头　　　　　　（横磁吹）　　　（横磁吹）

图 3-4　各种触头结构形状

（2）横磁吹触头。利用电流流过触头本身时所产生的横向磁场，驱使集聚型电弧不断在触头表面运动的触头称为横向磁场触头，主要可分为螺旋触头和杯状触头两种，如图 3-4（b）、（c）所示。螺旋触头用于开断小于 40 kA 的电流，近年来趋于淘汰；杯状触头可以开断 40 kA 以上的电流。

（3）纵磁吹触头。利用在磁场间隙中呈现的纵向磁场来提高开断能力的触头称为纵向磁场触头。纵向磁场的作用是削弱电弧自生磁场所产生的的磁收缩力，使真空电弧电流在电极间隙内及触头表面均匀分布，阻止阳极斑点的出现。起开断能力和抗电蚀性都强于横磁吹触头。

纵向磁场能约束带电质点，降低电弧电压，电弧不再聚集而呈现扩散形态。电弧在触头表面分成许多细弧，使电弧能量可均匀地输入触头的整个端面，触头表面均匀受热，不会造成触头表面局部的熔化，有利于提高开断电流和触头寿命，因而可用于开断更大的电流，其开断电流可以达到 70 kA，在实验室已高达 200 kA。

纵磁吹触头的结构基本上有两种，分为线圈型触头和杯状触头，触头结构如图 3-4（d）、（e）所示。线圈型触头靠在灭弧室外部装有线圈，当其被流过开关的电流所激励，将在触头间隙中形成一个相当均匀的纵磁场；杯状触头靠触头本身特殊结构产生纵磁场。后一种结构的优点是，产生磁场的元件是触头结构的一部分，不需要外加线圈。其缺点则是使得结构变得相当复杂，而且制作相当繁琐。研究表明，杯状纵向磁场触头开断能力大，触头磨损小，电气寿命长，结构简单，体积小，有利于真空断路器向大容量和小型化方向发展。

纵磁吹结构的真空断路器造价均高于横磁吹结构。

真空灭弧室技术已经有了很大的发展，重要标志是采用固封极柱真空灭弧室，即将真空灭弧室通过自动压力凝胶工艺包封在环氧树脂壳体内，形成固封极柱，避免了外力和外界环境对真空灭弧室及其他导电件的影响，增强了外绝缘强度，大大减少了装备工作量，并使得真空断路器小型化。

3. 屏蔽罩

真空灭弧室内常用的屏蔽罩有主屏蔽罩、波纹管屏蔽罩和均压屏蔽罩。主屏蔽罩的固定方式分为两大类，即固定电位式和悬浮电位式。悬浮电位式屏蔽罩有四种不同的固定方式：① 中间封接式；② 绝缘支柱式；③ 外屏蔽罩式；④ 绝缘端盖式。

主屏蔽罩采用导热性能好的材料制造，常用的材料为无氧铜、不锈钢和玻璃，其中铜是最常用的。在一定范围内，金属屏蔽罩厚度的增加可以提高灭弧室的开断能力，但通常其厚度不超过 2 mm。主屏蔽罩装设在触头的周围，一般固定在绝缘外壳内的中部，其主要作用是：

（1）防止燃弧过程中触头间产生的大量金属蒸汽和金属颗粒喷溅到绝缘外壳的内壁，导致外壳的绝缘强度降低或闪络。

（2）改善灭弧室内部电场分布的均匀性，降低局部电场强度，提高绝缘性能，促进真空灭弧室小型化。

（3）吸收部分电弧能量，冷却和凝结电弧生成物，有利于提高电弧熄灭后间隙介质强度的恢复速度，增大灭弧室的开断能力。

波纹管屏蔽罩包在波纹管的周围，用来保护波纹管免遭电弧生成物的烧损，防止电弧生成物凝结在波纹管表面上，影响波纹管的工作和降低其使用寿命。

均压屏蔽罩装设在触头附近，用于改善触头间的电场分布。

4. 波纹管

波纹管能保证动触头在一定行程范围内运动时，不破坏灭弧室的密封状态。波纹管通常采用不锈钢制成，有液压成形和膜片焊接两种。真空断路器触头每分合一次，波纹管便产生一次机械变形，长期频繁和剧烈的变形容易使波纹管因材料疲劳而损坏，导致灭弧室漏气而无法使用。波纹管是真空灭弧室中最易损坏的部件，其金属的疲劳寿命，决定了真空灭弧室的机械寿命。

三、真空断路器的整体结构与类型

1. 户内型真空断路器

真空断路器的整体结构可分为悬臂式和落地式两种基本形式。下面以 ZN28-12 系列真空断路器为例，介绍这两种典型结构。

ZN28-12 系列真空断路器采用中间封接式纵磁场真空灭弧室，根据灭弧室和操动机构布置方式的不同，大致可分为两类：一类是断路器本体与操动机构一起安装在箱形固定柜和手车柜中，称为整体式，即 ZN28-12 系列；另一类是断路器本体与操动机构分离安装在固定柜中，称为分体式，即 ZN28A-12 系列。分体式特别适合于旧柜无油化改造工程。

ZN28-12 真空断路器总体结构为落地式，如图 3-5 所示。每个真空灭弧室由一只落地绝缘子和一只悬挂绝缘子固定，真空灭弧室旁有一棒形绝缘子支撑。真空灭弧室上下铝合金支架既是输出接线的基座又兼起散热作用。在灭弧室上支架的上端面，安装有黄铜制作的导向板，使导电杆在分闸过程中对中良好。触头弹簧装设在绝缘拉杆的尾部。操动机构、传动主轴和绝缘转轴等部位均设置滚珠轴承，用于提高效率。断路器本体与操动机构一起安装在箱形固定柜和手车柜中，又称为整体式。

ZN28A-12 是固定开关柜专用真空断路器，总体结构为悬臂式，如图 3-6 所示。主导电回路、真空灭弧室与断路器机架前后布置。真空灭弧室用两只水平布置的悬臂绝缘子固定在机架的前面，主轴、分闸弹簧、缓冲器等部件安装在机架内。主轴通过绝缘拉杆、拐臂与真空灭弧室动导电杆连接，并从机架侧面伸出，与传动系统相连。断路器本体与操动机构分离安装在固定柜中，又称为分体式。

图 3-5　ZN28-12 真空断路器实物图　　　　图 3-6　ZN28A-12 真空断路器实物

2. 户外型真空断路器

户外真空断路器一般采用落地式结构,可分为箱式和支柱式。落地箱式是仿多油断路器结构,支柱式是仿少油断路器结构。ZW32-12 型户外支柱式真空断路器如图 3-7 所示,主要由真空灭弧室、上下绝缘罩、箱体、操动机构、隔离开关、电流互感器及驱动部件等组合而成。三相真空灭弧室分别封闭在三组绝缘罩内,绝缘罩固定在箱体上,箱体内安装弹簧操动机构,电流互感器安装在下出线端上,操作杠杆在箱体正面,箱体表面采用达克罗表面处理方式。

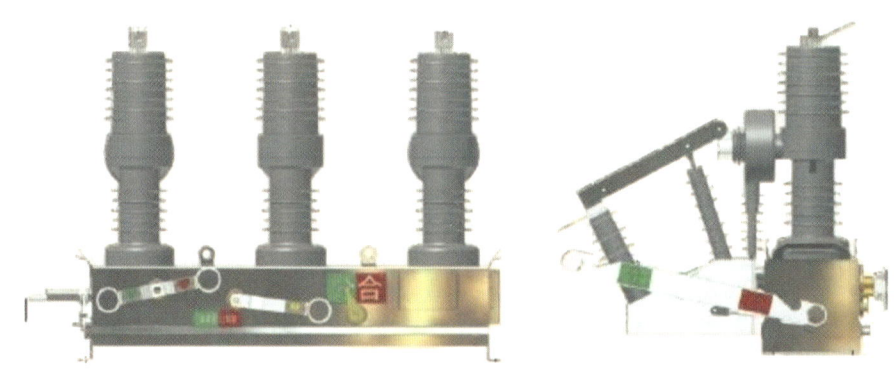

图 3-7　ZW32-12 型户外支柱式真空断路器

断路器同时具备电动和手动操作,并且可配置智能开关控制器,设有三段式过流保护、零序保护、重合闸、低电压、过电压保护等多种功能,它支持多种通信协议,允许选用多种通讯方式构成通信网,即可对开关进行本地手动或遥控操作,又可通过通信网实现远方控制。断路器为直立安装,三相分别对应线路三相,十分方便安装和线路接线。断路器支柱采用聚氨脂密封材料,内部采用新型的发泡灌封材料,增加了爬电距离,减少了体积。

3. 新型真空断路器

除了一般用途的标准型真空断路器(短路开断电流一般为 25~50 kA)被大量使用外,为满足特殊开断任务各种专用断路器不断被研制开发和使用。如用于发电机保护的特大容量真空断路器，短路开断电流高达 63~80 kA 及以上；用于开断感性负荷的低过电压真空断路器，不用加过电压吸收装置，而用新开发出的触头材料，将过电压限制至常规值的十分之一；用于投切电容的无重击穿真空断路器；用于频繁操作断路器，操作次数 5 万~6 万次，超频繁型真空断路器，操作次数 10 万~15 万次；用于一般场合的经济型真空断路器，开断电流 16~25 kA。

另外，具有多种功能和用途的新型真空断路器也在不断地被研制开发和使用。主要有以下三种：

（1）多功能真空断路器。

一般真空断路器为二工位，即合—分，完成关合与开断任务。现在的趋势是赋予真空断路器更多的功能，为此有两种做法，一是使真空断路器相柱在开断后移动或旋转，形成隔离和接地；另一是真空灭弧室内触头旋转完成隔离和接地。实现三工位（合—分—隔离）或四工位（合—分—隔离—接地）等功能。

（2）同步真空断路器。

同步真空断路器又叫选相真空断路器或受控真空断路器。其基本原理是使真空断路器在电压或电流最有利时刻关合或开断。与普通真空断路器相比，同步断路器具有的优势是，降低了电网瞬态过电压负荷；改善了电网供电质量；提高了断路器电寿命及性能；简化了电网设计，从而降低了整个系统费用。

（3）智能化真空断路器。

真空断路器智能化是建立在现代传感技术和数字化控制技术之上的。把计算机加入机械系统，使开关系统有了"大脑"，再加入"传感器"采集信息，用光纤传导信息，使开关系统有了"知觉"，大脑根据"知觉"做出判断与决定，使系统有了"智能"。

Module 2　Vacuum Circuit Breaker

Vacuum circuit breaker is a circuit breaker that uses high dielectric strength of vacuum for arc extinguishing. Its name comes from the fact that its arc extinguishing medium and the gap of contact terminal after arc extinguishing are high vacuum degree. It is provided with such advantages as narrow contact gap of contact terminal (normally 10 mm), quick arc extinguishing, small volume, light weight, prevention of explosion hazard and contamination, availability in frequent operation and maintenance free arc extinguishing. Vacuum circuit breaker enjoys an absolute advantage in medium and low voltage power distribution network, which has a highly popularized application. In particular, its proportion at 12 kV level of large quantity and extensive coverage is over 99%.

3.2.1　Working Principles of Vacuum Circuit Breaker

In view of rarefied gas at vacuum gap, long free path of gas molecules and limited opportunity for impact ionization and high breakdown voltage, insulating strength of high vacuum far exceeds that of arc extinguishing medium. To satisfy insulating strength as required by vacuum arc extinguishing chamber, vacuum degree is normally requested to be controlled at 1.33×10^{-3} to 1.33×10^{-7} Pa.

Meanwhile, in the vacuum gap where gas is rarefied, ionization of gas is unlikely to maintain arc combustion. Therefore, arc produced by breakdown of vacuum gap is not the consequence of gas impact ionization. Actually, arc produced by breakdown of vacuum gap is formed in the metallic vapor evaporated from electrode of contact terminal. Accompanied by separation of contact terminal at breaking current, contact area of contact terminal will witness a quick reduction, leaving one or several tiny contact points eventually. Moreover, extremely high current density and significant temperature rise will make the metal at contact point melted to evaporate a large quantity of metallic vapor. Extremely high temperature of metallic vapor and existence of extremely strong electric field may result in strong electric field emission and ionization of metallic vapor to the extent of developing into vacuum arc. Therefore, in addition to vacuum degree, major factors for breakdown of vacuum gap are also associated with contact terminal material and its surface conditions, varieties of residual gas, length vacuum gap and uniformity of electric field.

Vacuum circuit breaker makes use of the high diffusion rate of charged particles and metal vapor generated in vacuum arc to quickly recover dielectric strength of contact terminal gap for arc extinguishing in case of zero crossing of arc current and temporary arc extinguishing. Extinguishing of vacuum AC arc is the same as that of other AC arcs, which is mainly determined by recovery of arc gap dielectric strength following zero crossing of current. Meanwhile, insulating strength of arc gap depends on such factors as diffusion speed of charged particles, strength of breaking current as well as area, profile and material of contact terminal. By imposing transverse and longitudinal magnetic fields in arcing area to promote high-speed diffusion of arc, it is applicable to accelerate recovery of dielectric strength, alleviate burnout to contact terminal, and extend the service life.

Where arc column tends to contract, and is in the form of anode spot (in other words, it is in the form of agglomerating arc, and anode surface is provided with big and bright spots) in case of breaking of high current by vacuum circuit breaker, anode spot will spray a large quantity of metallic vapor to the arc column to the extent of resulting in increased pressure on arc column and massive diffusion of particles. In this case, the vacuum circuit breaker will completely lose its breaking capacity. Where it is requested to prevent occurrence of anode spot within a half cycle when the arc is combusting, the only way is to take measures to improve vacuum arc extinguishing capacity. Only in this way can vacuum circuit breaker be applied to the field of high-capacity circuit breakers.

3.2.2 Basic Structure of Vacuum Circuit Breaker

Vacuum circuit breaker mainly comprises three parts, namely vacuum arc extinguishing chamber, support and operating mechanism. Vacuum arc extinguishing chamber serves as the core part of vacuum circuit breaker, which is provided with such functions as breaking, conduction and insulation. It mainly comprises insulating housing, dynamic and static contact terminals, shield and bellows. Its basic structure is as shown in Fig. 3-3. Insulating housing is a vacuum-tight container that is composed of insulating cylinder, bellows, dynamic end cover plate and static end cover plate. Static contact terminal inside the arc extinguishing chamber is fixed to static conductive rod, and the static conductive rod penetrates through the static end cover plate, and is integrated with it through welding; Moving contact terminal is fixed to one end of moving conductive rod; the central part of moving conductive rod is welded to one port on the bellows; the other port on the bellows is welded to dynamic end cover plate; dynamic conductive rod penetrates through the housing via the central hole. Shield and is installed respectively at the periphery of moving and static contact terminals as well as bellows. As bellows is available for axial extension and contraction, such structure can realize manipulation of moving contact terminal outside of arc extinguishing chamber for closing and opening; where there is no mechanical force imposed on the arc extinguishing chamber, moving contact terminal and static contact terminal will remain closed, and will not separate until moving conductive rod moves outwards under external force. Performance of vacuum arc extinguishing chamber is mainly determined by material and structure of contact terminal, which is also related to the structure of shield, material of arc extinguishing chamber and manufacturing process.

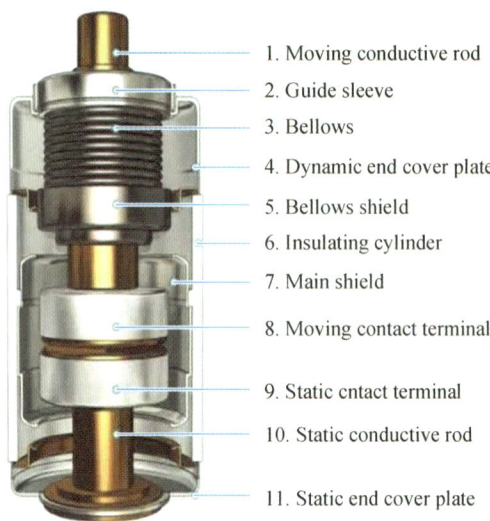

Fig. 3-3 Structural Diagram for Vacuum Arc Extinguishing Chamber

1. Moving conductive rod
2. Guide sleeve
3. Bellows
4. Dynamic end cover plate
5. Bellows shield
6. Insulating cylinder
7. Main shield
8. Moving contact terminal
9. Static cntact terminal
10. Static conductive rod
11. Static end cover plate

1. Insulating housing

Insulating housing of vacuum arc extinguishing chamber serves as both vacuum container and insulator between moving contact terminal and static contact terminal. Its function is to contain and support various parts inside arc extinguishing chamber. It is to be integrated with such parts through air tight welding to ensure high vacuum degree inside arc extinguishing chamber. To ensure operating reliability of vacuum arc extinguishing chamber, it is normally requested that vacuum degree shall not be lower than the specified value within 20 years. Therefore, strict sealing is required. Furthermore, it is essential to ensure a certain mechanical strength. Insulating cylinder is made of such insulating materials (mainly represented by ceramics) as hard glass, high alumina ceramics or microlite glass. Insulatinghousing is normally manufactured with hard glass, high alumina ceramics or microlite glass. End cover of the housing is normally manufactured with such metals as stainless steel and oxygen free copper.

The function of bellows is to ensure complete sealing of arc extinguishing chamber, and enable operating mechanism of circuit breaker to extend its motion to the moving contact terminal at the same time. Frequently used materials for bellows include stainless steel, phosphor bronze and so on. Stainless steel has the optimal performance among such materials, which is formed by hydroforming and diaphragm welding. Allowable extension and contraction of bellows shall be able to satisfy the maximum contact gap of contact terminal as required. Corrugated thin wall of bellows will subject to significant mechanical deformation every time when the contact terminal is opened and closed. As a result of it, it is extremely vulnerable to damage due to fatigue.

2. Contact terminal

Contact terminal on the vacuum circuit breaker serves as both a flow element for closing-opening and an arc extinguishing element for breaking. Material and structure of contact terminal has a direct impact on breaking capacity, electrical life, compressive strength, making capacity, cut-off over-voltage and long-term current conduction capacity of arc extinguishing chamber. As pure metallic contact terminal is in lack of the excellent performance of vacuum contact terminal, vacuum circuit breaker adopts two mutually different alloy composite materials, namely copper bismuth (Cu/Bi) and copper chromium (Cu/Cr).

Contact terminal serves as the most important element inside arc extinguishing chamber, of which breaking capacity and electrical life are mainly determined by structure of contact terminal. Presently, contact terminal system of vacuum circuit breaker normally adopts butt joint. In view of different basic principles on arc extinguishing for breaking, its development has witnessed three typical structural types, namely non-magnetic blow-out flat contact terminal, transverse magnetic blow-out contact terminal and longitudinal magnetic blow-out contact terminal. See Fig. 3-4. The common features of such contact terminals is the use of magnetic field force to ensure quick movement of vacuum arc, and prevent occurrence of heated area requiring long-time cooling on the contact terminal.

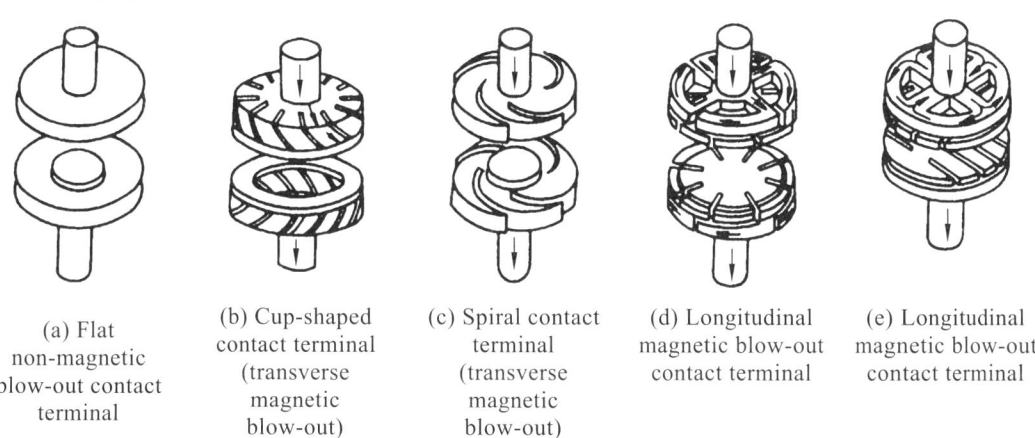

(a) Flat non-magnetic blow-out contact terminal   (b) Cup-shaped contact terminal (transverse magnetic blow-out)   (c) Spiral contact terminal (transverse magnetic blow-out)   (d) Longitudinal magnetic blow-out contact terminal   (e) Longitudinal magnetic blow-out contact terminal

Fig. 3-4 Structural Profile of Various Contact Terminals

(1) Non-magnetic blow-out flat contact terminal. Flat end surface of this contact terminal serves as the surface for electric contact and arcing. Vacuum arc combusting between each contact terminal is free of action of the magnetic field. Vacuum arc between each contact terminal is in diffusion pattern in case of breaking of low current (instantaneous value is less than 10 kV). It features in quick recovery of dielectric strength after arcing and excellent arc extinguishing performance; Vacuum arc is in agglomerating pattern in case of higher breaking current. However, slow recovery of dielectric strength after arcing may result in breaking failure. Non-magnetic blow-out flat contact terminal features in easy fabrication and low cost, which is only applied to arc extinguishing chamber where breaking current of vacuum contact, vacuum load switch and so on is no more than 10 kA.

(2) Transverse magnetic blow-out contact terminal. The contact terminal that makes use of transverse magnetic field produced by current passing through the contact terminal to enable continuous movement of agglomerating arc on the surface of contact terminal is called transverse

magnetic field contact terminal. It is mainly divided into spiral contact terminal and cup-shaped contact terminal as shown in Fig. 3-4 (b) and (c). Spiral contact terminal is used for breaking of current below 40 kA, which has been gradually eliminated in recent years; Cup-shaped contact terminal is available for breaking of current over 40 kA.

(3) Longitudinal magnetic blow-out contact terminal. The contact terminal that makes use of longitudinal magnetic field produced in the magnetic field gap is called longitudinal magnetic field contact terminal. The function of the longitudinal magnetic field is to weaken the magnetic contraction force generated by the self generated magnetic field of the arc to ensure uniform distribution of vacuum arc current inside electrode gap and on the surface of contact terminal, and prevent occurrence of anode spot. Its initial breaking capacity and electrical erosion resistance exceed that of transverse magnetic blow-out contact terminal.

Longitudinal magnetic field can restrict charged particles, and reduce arc voltage to ensure diffusion other than agglomerating of arc. The arc breaks down into numerous fine arcs on contact terminal surface to ensure uniform input of arc energy into the whole end surface of contact terminal and uniform heating of contact terminal surface to prevent local melting of contact terminal surface. As this is favorable for increase of breaking current and extension of service life of contact terminal, it can be used for breaking of higher current. Its breaking current can be up to 70 kA or 200 kA in the laboratory.

Longitudinal magnetic blow-out contact terminal is basically in two structures, which can be divided into coil contact terminal and cup-shaped contact terminal. Contact terminal structure is as shown in Fig. 3-4 (d) and (e). Coil contact terminal is installed with a coil adjacent to the external part of arc extinguishing chamber. Once it is excited by current passing through the switch, a relatively uniform longitudinal magnetic field is to be formed in the contact terminal gap; Cup-shaped contact terminal relies on its own unique structure to produce longitudinal magnetic field. Advantage of the later structure lies in the fact that the element producing magnetic field is a part of contact terminal structure, which requires no installation of additional coil. Its disadvantage lies in the fact that it may result in complicated structure and trivial details on fabrication. The research shows that cup-shaped longitudinal magnetic field contact terminal features in high breaking capacity, less wear to contact terminal, long electrical life, simple structure and small volume. It is favorable for development of vacuum circuit breaker in the orientation of high capacity and miniaturization.

Manufacturing cost of vacuum circuit breaker in longitudinal magnetic blow-out structure is higher than that of transverse blow-out structure.

Technologies on vacuum arc extinguishing chamber has witnessed an accelerated development. The significant sign is the use of embedded pole vacuum arc extinguishing chamber. In other words, the vacuum arc extinguishing chamber is enclosed in the epoxy resin housing based on automatic pressure gelation process to form an embedded pole. This can prevent impact from external force and external environment on vacuum arc extinguishing chamber and other conductive parts, enhance external insulation strength, significantly reduce work load of equipment, and realize miniaturization of vacuum circuit breaker.

3. Shield

Shields frequently applied to vacuum arc extinguishing chamber include main shield, bellows shield and pressure equalizing shield. Fixing of main shield falls into two categories, namely fixed potential type and suspended potential type. Shield of suspended potential type has four different fixing modes: (1) intermediate sealing-in, (2) insulating column, (3) external shield and (4) insulating end cover.

Main shield is made of materials with excellent thermal conductivity. Frequently used materials include oxygen free copper, stainless steel and glass. Among them, copper is used most frequently. Increased thickness of metallic shield within a certain scope can improve breaking capacity of arc extinguishing chamber. Nevertheless, its thickness is normally no more than 2 mm. Main shield is installed at the periphery of contact terminal, which is normally fixed to the internal center of insulating housing. Its major functions are stated as follows:

(1) Prevent massive metallic vapor and metallic particles produced between each contact terminal during arcing from spraying to inner wall of the insulating housing that may result in reduced insulating strength of housing or flashover.

(2) Improve uniformity of magnetic field distributed inside the arc extinguishing chamber, reduce local magnetic field strength, improve insulating performance, and promote miniaturization of vacuum arc extinguishing chamber.

(3) Absorb partial arc energy, and ensure cooling and condensation of arc products. It is favorable for acceleration of recovery of gap medium strength after arc extinguishing and improvement of breaking capacity of arc extinguishing chamber.

Bellows shield is enclosed at the periphery of the bellows to protect bellows from burnout by arc products, and prevent condensation of arc products on the bellows surface that may affect working performance of bellows, and reduce its service life.

Pressure equalizing shield is installed nearby the contact terminal to improve electric field distribution between each contact terminal.

4. Corrugated pipe

The bellows can make sure that the sealing state of the arc extinguishing chamber is not affected when the moving contact terminal moves within a certain range of travel. The bellows is normally made of stainless steel, which is formed through hydroforming and diaphragm welding. Every time the contact terminal of the vacuum circuit breaker is opened or closed, the bellows will subject to mechanical deformation. Long-term frequent and violent deformation is apt result in damage to the bellows due to fatigue to material to the extent of making arc extinguishing chamber unable for use due to leakage. The bellows is a part that is most vulnerable to damage insde the vacuum arc extinguishing chamber. Mechanical life of vacuum arc extinguishing chamber is determined by fatigue life of its metal.

3.2.3 Overall Structure and Types of Vacuum Circuit Breaker

1. Indoor vacuum circuit breaker

Overall structure of vacuum circuit breaker falls into cantilever and floor types. These two typical structures are introduced with the ZN28-12 series vacuum circuit breaker as an example in the following.

Vacuum arc extinguishing chambers of intermediate-sealed type with longitudinal magnetic field are adopted in the ZN28-12 series vacuum circuit breakers. According to the different arrangements of arc extinguishing chambers and operating mechanisms, circuit breakers can be roughly divided into two types. One type is that the circuit breaker body and the operating mechanism are installed together in the fixed cabinet of box shape and the handcart cabinet, which is called an integral type circuit breaker, namely the ZN28-12 series circuit breaker. Another type is that the circuit breaker body and the operating mechanism are installed separately in the fixed cabinet, which is called the separated type circuit breaker, namely the ZN28A-12 series circuit breaker. The separated type is particularly suitable for oil-free renovation projects of old cabinets.

The overall structure of the ZN28-12 vacuum circuit breaker is the floor-mounted type, as shown in Fig. 3-5. Each vacuum arc extinguishing chamber is fixed by a ground insulator and a suspension insulator. There is a rod insulator support next to the vacuum arc extinguishing chamber. The upper and lower aluminum alloy brackets of the vacuum arc extinguishing chamber are not only the base for output wiring but also serve as the heat dissipation. On the upper end face of the bracket on the arc extinguishing chamber, the guide plate made of brass is installed to ensure good alignment of the conductive rod during the opening process. The contact terminal spring is installed at the tail of the insulation tie rod. Ball bearings are installed in the operating mechanism, drive spindle, and insulated shaft to improve efficiency. The circuit breaker body and the operating mechanism are installed together in the fixed cabinet of box shape and the handcart cabinet, which is called an integral type.

ZN28A-12 is a dedicated vacuum circuit breaker for the fixed switch cabinet, with a cantilever type overall structure, as shown in Fig. 3-6. The main galvanic circle, vacuum arc extinguishing chamber, and circuit breaker frame are arranged in front and back. The vacuum arc extinguishing chamber is fixed to the front of the frame with two horizontally arranged cantilever insulators, and components such as the spindle, opening spring, and buffer are installed inside the frame. The spindle is connected to the moving conductive rod of the vacuum arc extinguishing chamber through the insulation tie rod and the crank arm, and extends from the side of the frame to be connected to the transmission system. The circuit breaker body and the operating mechanism are installed separately in the fixed cabinet, which is called the separated type.

Fig. 3-5　Physical Diagram of ZN28-12 Vacuum Circuit Breaker

Fig. 3-6　Physical Diagram of ZN28A-12 Vacuum Circuit Breaker

2. Outdoor vacuum circuit breaker

With the floor-mounted structure generally adopted, the outdoor vacuum circuit breakers can be divided into box type and strut type. The floor-mounted circuit breaker of the box type is a structure of simulated bulk oil circuit breaker, while the strut type is a structure of simulated low oil circuit breaker. The ZW32-12 outdoor vacuum circuit breaker of the strut type is as shown in Fig. 3-7, mainly composed of the vacuum arc extinguishing chamber, upper and lower insulating covers, box body, operating mechanism, disconnector, current transformer, and driving components. The three-phase vacuum arc extinguishing chambers are enclosed in three sets of insulating covers. The insulating cover is fixed on the box, and the spring operating mechanism is installed inside the box. The current transformer is installed on the lower outlet end, and the operating lever is on the front of the box. The box surface is treated with DACRO coating technology.

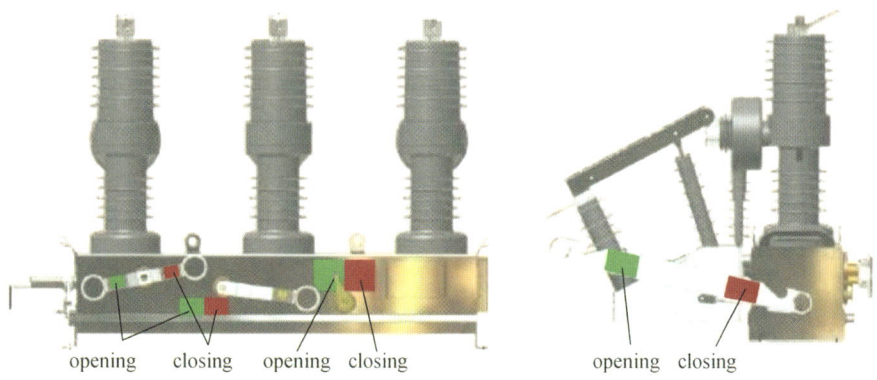

Fig. 3-7　ZW32-12 Outdoor Strut-type Vacuum Circuit Breaker

The circuit breaker has both electric and manual operation functions, and can be configured with the intelligent switch controller. It has multiple functions such as three-stage overcurrent protection, zero sequence protection, reclosing, low voltage protection and overvoltage protection. It supports multiple communication protocols and allows the use of multiple communication methods to form a communication network. It can perform local manual or remote-control operations on switches, as well as achieve the remote control through the communication network.

The circuit breaker is installed vertically, with three phases corresponding to the three phases of the circuit, which is very convenient for installation and wiring. The strut of the circuit breaker is made of polyurethane sealing material, and a new type of foam filling material is used internally, which increases the creepage distance and reduces the volume.

3. New type vacuum circuit breaker

In addition to the widespread use of standard vacuum circuit breakers for general purposes (short-circuit breaking current typically ranging from 25 to 50 kA), various specialized circuit breakers have been continuously developed and used to meet special breaking tasks. For example, with a short-circuit breaking current as high as 63-80 kA and above, the vacuum circuit breaker of extra-large capacity is used for the generator protection. Low overvoltage vacuum circuit breakers used to disconnect inductance loading do not require the addition of overvoltage absorption devices. With newly developed contact terminal materials, the overvoltage is limited to one tenth of the conventional value. The vacuum restrike-free circuit breaker is used for switching capacitors. Circuit breakers used for frequent operation can be operated up to 50,000 to 60,000 times. The vacuum circuit breaker of ultra-frequent type can be operated up to 100,000 to 150,000 times. The economical vacuum circuit breaker is used for general applications, with a breaking current of 16-25 kA.

In addition, new types of vacuum circuit breakers with a variety of functions and applications are constantly being developed and used. There are mainly the following three types:

(1) Multifunctional vacuum circuit breaker.

Generally, vacuum circuit breakers have two positions (namely closing-opening) to complete the closing and opening tasks. Now the trend is to give the vacuum circuit breaker more functions, and for this there are two practices. One is to complete isolation and grounding by making the phase column of the vacuum circuit breaker move or rotate after opening. The other is to complete isolation and grounding through the rotation of the contact terminal in the vacuum arc extinguishing chamber. Therefore, functions such as three positions (closing-opening-isolation) or four positions (closing-opening-isolation-grounding) are implemented.

(2) Synchronized vacuum circuit breaker.

Synchronized vacuum circuit breaker is also called vacuum circuit breaker of phase selection or controlled vacuum circuit breaker. Its basic principle is to make the vacuum circuit breaker close or open at the most favorable moment of voltage or current. Compared with the ordinary vacuum circuit breaker, the synchronized circuit breaker has the advantages of reducing transient overvoltage loads in the grid, improving the quality of power supply in the grid, improve the electrical life and performance of circuit breaker, and simplify the design of the power grid, thereby reducing the overall system costs.

(3) Intelligent vacuum circuit breaker.

The intelligent vacuum circuit breaker is produced based on the modern sensing technology and the digital control technology. Adding a computer to the mechanical system gives the switching system a "brain". Adding "the sensor" to collect information and using fiber optics to conduct information give the switching system "the perception". The brain makes judgments and decisions according to "the perception", which makes the system have "intelligence".

模块三　SF_6 断路器

SF_6 断路器是采用 SF_6 气体作为灭弧介质的断路器。SF_6 断路器的性能主要由断路器灭弧室结构决定。

SF_6 断路器最初是类似空气断路器的双压式断路器，用气泵将 SF_6 气体压入高压储气室，在开断电流时，SF_6 气体从高压储气室流入气压较低的灭弧室，从而产生气吹作用将电弧熄灭。这种高压断路器由于充气压力高，且 SF_6 气体在高压下容易液化，大大降低了灭弧性能；另外，还需要一套抽气系统，结构非常复杂，因而已被淘汰。

目前，SF_6 断路器多为单压式断路器，在开断短路电流时，由气缸与活塞之间的相对运动产生压气作用，使气缸内 SF_6 气体压强升高，气体从喷口排出，对电弧产生纵吹使其在电流过零时熄灭。单压式断路器结构简单，充气压强也较低，并且具有优越的开断性能，获得了广泛应用。

一、SF_6 断路器的灭弧室结构

（一）单压式 SF_6 断路器的灭弧室

单压式 SF_6 断路器的灭弧室是根据活塞压气原理工作的，故又称为压气式灭弧室。平时灭弧室只有一种压力为 0.3~0.7 MPa 的 SF_6 气体，起到绝缘作用。灭弧室的可动部分带有压气装置，在开断过程中，灭弧室所需的吹气压力由动触头系统带动压气缸对固定活塞相对运动产生。其 SF_6 气体是在封闭系统中循环使用，不能排向大气。这种灭弧装置结构简单、动作可靠，按灭弧室结构又分为变开距和定开距两种。

1. 定开距灭弧室结构和动作过程

图 3-11 为定开距灭弧室。断路器的触头由两个带嘴的空心静触头和动触头组成。图 3-8（a）中断路器处于合闸位置，这时动触头跨接于空心静触头之间，构成电流通路；动触头与压气缸在结构上连成一体，并与拉杆连接，操动机构可通过拉杆带动动触头和气压缸左右运动。固定活塞由绝缘材料制成，它与动触头、压气缸之间围成压气室。分闸时，操动机构通过拉杆带着动触头和压气缸向右运动，使压气室内的 SF_6 气体被压缩，压力约提高 1 倍左右，如图 3-8（b）所示。当动触头离开静触头时，将产生电弧，同时将原来被动触头所封闭的压气缸打开，高压 SF_6 气体迅速向两个静触头内腔喷射，对电弧进行强烈的对称双向纵吹，如图 3-8（c）所示。当电弧熄灭后，触头处在分闸位置，如图 3-8（d）所示。这种灭弧室中，断路器的弧隙由两个静触头保持固定的开距，故称为定开距灭弧室。由于 SF_6 气体的灭弧和绝缘能力强，所以开距一般不大，动作迅速。

定开距灭弧室的喷口采用耐电弧性能好的金属或石墨等导电材料制成。石墨能耐高温，在电弧作用下直接由固态变成气态，逸出功大，表面烧损轻。定开距灭弧室断口电场均匀，灭弧开距小，触头从分离位置到熄弧位置的行程很短，126 kV 的断路器只有 30 mm，电弧能量较小，熄弧能力强，燃弧时间短，可以开断很大的短路电流。但是压气室的体积较大。

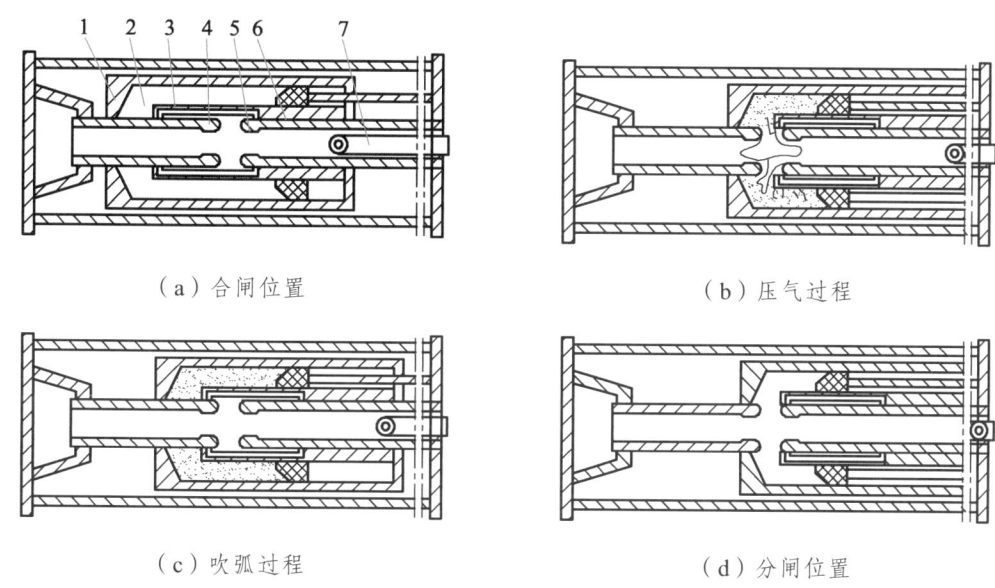

(a)合闸位置　　　　　　　　　(b)压气过程

(c)吹弧过程　　　　　　　　　(d)分闸位置

1—压气缸；2—压气室；3—动触头；4，5—静触头；6—固定活塞；7—拉杆。

图 3-8　定开距灭弧室动作过程

2. 变开距灭弧室结构和动作过程

由于在灭弧过程中，灭弧室内动、静触头间的开距，会随着压气室的运动而逐渐加大，即使电弧已被吹熄，动触头还将继续运动直至终止位置，即在吹弧过程中，触头开距不断加大，因此称为变开距灭弧室。变开距灭弧室按吹弧方式分为单向纵吹和双向纵吹，单吹式适用于中小容量断路器，高压大容量断路器采用双向纵吹居多。

断路器的导电体由工作触头和辅助触头两部分组成，合闸状态时，二者并联；分闸状态时，流经绝大部分电流的主触头先分离，电流转移到耐弧材料为铜钨合金做成的辅助触头上，随着操动机构运动，辅助触头打开而形成电弧，随即打开喷口间隙气吹熄电弧。工作触头放在外侧，可改善散热条件，提高断路器的热稳定性。

喷口用聚四氟乙烯或以聚四氟乙烯为主的填料制成的复合材料等绝缘材料制成，这类材料具有耐电弧、机械强度高、易加工、耐高温、直接受电弧短时作用不易炭化、烧损均匀、烧蚀量少、不受 SF_6 分解物侵蚀等特点。

变开距灭弧室的灭弧过程如图 3-9 所示。图 3-9（a）为断路器处于合闸位置。分闸时可动部分向右运动，压气室内的 SF_6 气体开始受压缩并提高压力。随着可动部分的运动，主静触头与主动触头首先分离，由于弧触头还未断开，所以此时不产生电弧，喷口也未形成，也无吹弧作用；直到可动部分向右移动到一定位置时，弧静触头与弧动触头开始分离，电弧产生，在喷嘴和弧动触头间形成喷口，SF_6 气体从两个方向吹向电弧，使电弧熄灭，如图 3-9（b）所示。电弧熄灭后，断路器处于分闸位置，如图 3-9（c）所示。

为了使分闸过程中压气室的气体集中向喷嘴吹弧，而在合闸过程中不致在压气室形成真空，故在固定活塞上设置了逆止阀。在分闸时，逆止阀堵住小孔，让 SF_6 气体集中向喷嘴吹弧。合闸时，逆止阀打开，使压气室与活塞的内腔相通，SF_6 气体从活塞小孔充入压气室，为下一次分闸做好准备。

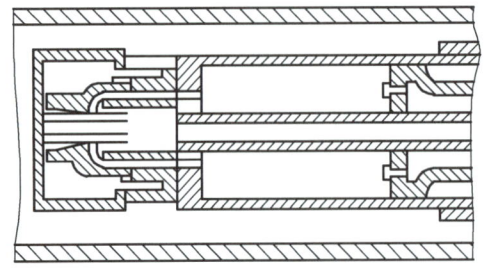 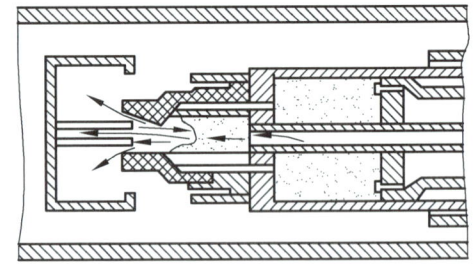

（a）合闸位置　　　　　　　　　　　（b）吹弧过程

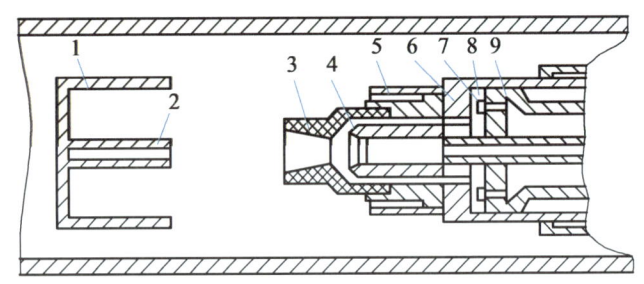

（c）分闸位置

1—主静触头；2—弧静触头；3—喷嘴；4—弧动触头；5—主动触头；
6—压气缸；7—逆止阀；8—压气室；9—固定活塞。

图 3-9　变开距灭弧室动作过程

3. 定开距与变开距灭弧室结构比较

变开距在吹弧过程中电极开距不断变大，破坏了气流场的死区；此外，虽然熄弧后有较大的的绝缘间隙，避免了熄弧后"电击穿"引发的电弧重燃，但由于燃弧时间增加，可能因介质强度恢复速度减慢导致发生"热击穿"，从而限制了变开距断路器产品的极限开断电流。

定开距结构的电场集中在固定的两喷口电极之间，开断过程中压气缸要耐受恢复电压，且在动触头出喷口瞬间，断口绝缘强度因短接一部分而降低，这使得定开距断路器产品的单元断口电压提高受限。

因此，由于 110 kV 断路器定开距灭弧室的单元断口电压不太高，开断电流较大，运用于开断较大电流时，宜采用定开距灭弧室的结构，而用于 220 kV 及以上的单断口 SF_6 断路器，以及用于发电厂出口反向开断时，采用变开距结构灭弧室较为有利。随着超高压断路器对提高单元断口电压的要求，变开距结构得到广泛运用。

（二）自能式 SF_6 断路器

压气式 SF_6 断路器要利用操作机构带动气缸与活塞相对运动来压气熄弧，因而操作机构负担很重，要求操作机构的操作功率大。

利用电弧自身的能量来熄灭电弧的自能式 SF_6 断路器，可以减轻操动机构的负担，减少对操动机构操作功率的要求，从而可以提高断路器的可靠性。自能式 SF_6 断路器代表了 SF_6 断路器发展的主流。自能式 SF_6 断路器按灭弧原理可分为旋转式、热膨胀式和混合式。

1. 旋弧式

旋弧式是利用设置在静触头附近的磁吹线圈在开断电流时自动地被电弧串接进回路，被开断电流流过线圈，在动、静触头之间产生磁场，电弧在磁场的驱动下高速旋转，电弧在旋转过程中不断地接触新鲜的 SF_6 气体，使电弧受到冷却而熄灭。按磁吹和电弧的运动方式不同可分为径向旋弧式和纵向旋弧式。

电磁驱动力随故障电流的减小而减小，所以旋弧式断路器灭弧能力受到较小的故障电流的限制。增加线圈匝数，就可以克服这一缺点，但线圈匝数的增加，受到机械强度的限制，因而在大的故障电流下，要承受大的电磁力。

旋弧式灭弧室结构简单，不需要大功率的操动机构，电弧高速旋转，触头烧损轻微，寿命长，在中压系统中使用比较普遍。

2. 热膨胀式

热膨胀式是利用电弧本身的能量，加热灭弧室内的 SF_6 气体，建立高压力，形成压差，并通过喷口释放，产生强力气流吹弧，从而达到冷却和吹灭电弧的目的。

灭弧室结构如图 3-10 所示，圆柱形的灭弧室被分成两个间隔，即密闭间隔和比密闭间隔大得多的排气间隔。在这两个间隔中都充有 SF_6 气体。当断路器处于合闸位置时，动触头通过触指连接到静触头，如中心线左部所示。分闸时，电流通过线圈，如中心线右部所示。当动触头运动一定距离后，在环状电极和动触头之间产生电弧。旋弧线圈产生与触头的同轴磁场，燃弧环中的电弧垂直于旋弧线圈的磁场，其间产生的电动力使电弧高速旋转，使电弧在 SF_6 气体中被拉长，旋转电弧不断接触新鲜的 SF_6 气体，释放热能，并将间隔中的气体加热，产生一个比排气间隔中较高的压力，当触头分开时，两个间隔经动触头中的喷嘴连通，此时，出现的气压差，被用来经过喷嘴形成纵向吹弧。在下一个电流过零点时，熄灭电弧。

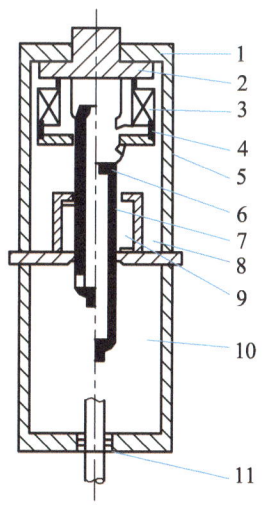

1—灭弧室圆筒；2—静触头；3—圆柱形线圈；4—触指；5—环状电极；6—喷嘴；7—动触头；8—密闭间隔；9—辅助吹气装置；10—排气间隔；11—对大气的密封。

图 3-10 热膨胀式灭弧室结构图

3. 混合吹弧式

无论是采用旋弧式灭弧，还是热膨胀式灭弧都能大大减轻操动机构的负担，提高断路器的性能价格比，但是任何一种灭弧室都有它的不足之处，为此往往将几种灭弧原理同时应用在断路器的灭弧室中。压气式加上自能吹弧的混合式灭弧有助于提高灭弧效能，不仅可以增大开断电流，而且可以明显减少操作功。混合吹弧式有多种方式，如旋弧+热膨胀，压气+热膨胀，压气+旋弧，旋弧+热膨胀+助吹。

二、SF_6断路器的附件

SF_6断路器的附件是指SF_6断路器及其操动机构配置的具有一定特殊功能的附属部件。如SF_6断路器上的压力表、压力继电器（也称压力开关）、安全阀、密度表、密度继电器、并联电容、并联电阻、净化装置、防爆装置等。它们虽然是附属部件，但是却起着非常重要的作用。

1. 压力表和压力继电器

SF_6气体压力是断路器绝缘、载流、开断与关合能力的宏观标志，运行中必须始终保持在规定的范围内。为监视SF_6气体压力的变化情况，应装设压力表和压力继电器。

（1）压力表。SF_6气体压力表起监视作用，按结构原理可分为弹簧管式、活塞式、数字式等。SF_6断路器一般采用弹簧管式压力表。

（2）压力继电器。压力继电器主要配置在断路器的操动机构上，带有多对电触点，用于控制操动机构电动机的起动、停止和输出闭锁断路器分闸、合闸、重合闸的指令以及发出相应的信号等。当气体压力升高或降低时，压力继电器使相应的行程开关电触点动作，以实现利用压力来控制有关指令和信号的输出。压力继电器起控制和保护作用。

（3）安全阀。安全阀是用于电动机油泵或空气压缩机系统的一种安全保护装置。它是压力继电器的一种特殊形式。与压力继电器不同之处是安全阀带不带电触点，且动作方式不同。当油压或气压超过规定的最高压力值时，安全阀内部机构装置动作，泄压至规定的压力值时自动关闭。

2. 密度表和密度继电器

气体密度表和密度继电器都是用来测量SF_6气体的专用表计，带指针及有刻度的称为密度表；不带指针及刻度的称为密度继电器。有的SF_6气体密度表也带有电触点，即兼作密度继电器使用。SF_6气体密度表起监视作用，密度继电器起控制和保护作用。

3. 并联电容和并联电阻

并联电容（也称均压电容）和并联电阻（也称合闸电阻）都是与断路器灭弧断口相并的、改善断路器分闸或合闸特性的重要附属元件。

为了降低断路器触头间弧隙的恢复电压速度，提高近区故障开断能力，在 63 kV 及以上电压等级的单断口SF_6断路器上也装设了并联电容。

为了限制合闸或分闸以及重合闸过程中的过电压，改善断路器的使用性能，采用在断口间并联电阻的方式。并联电阻片一般是由碳质烧结而成，外形与避雷器阀片很相似，但其热容量要大得多。

并联电阻的安装方式一般为两种：一种是并联电阻片与辅助断口均置于同一瓷套内，也可把并联电阻片布置在辅助断口的两侧，使电阻片在工作发热后更有利于热量扩散；另一种是合闸电阻片与辅助断口不在同一瓷套内，而是各自成独立元件，串联后并联在灭弧室两端。

选择并联电阻值的大小对限制合闸过电压影响很大。目前我国 500 kV 断路器上使用的并联电阻值一般为 400～450 Ω。

4. 净化装置

净化装置主要由过滤罐和吸附剂组成。吸附剂的作用是吸附 SF_6 气体中的水分和 SF_6 气体经电弧的高温作用后产生的某些分解物。

常用的吸附剂有以下几种。

（1）活性炭：是以果壳、煤、木材等为原料，经过炭化、高温活化等制成的吸附剂。

（2）分子筛：是一种人工合成的沸石，是具有四面骨架结构的铝硅酸盐。

（3）氧化铝：是一种由天然氧化铝或铝土矿经特殊处理而制成的多孔结构物质。

（4）硅胶：是一种坚硬多孔固体颗粒，以水玻璃为原料制成。

除了上述四种吸附剂外，还有漂白土、活性白土、吸附树脂、活性炭素纤维、炭分子筛、矾土、铝土、氧化镁、硫酸锶等数种吸附剂。目前，国内外 SF_6 开关设备上使用得最多的吸附剂主要是分子筛和氧化铝。

5. 压力释放装置

压力释放装置可分为两类：以开启压力和闭合压力表示其特征的，称为压力释放阀，一般装设在罐式 SF_6 断路器上；一旦开启后不能够再闭合的，称为防爆膜，一般装设在支柱式 SF_6 断路器上。

当外壳和气源采用固定连接时，所采用的压力调节装置不能可靠地防止过压力时，应装设适当尺寸的压力释放阀，以防止万一压力调节措施失效时外壳内部的压力过高。

当外壳和气源不是采用固定连接时，应在充气管道上装设压力释放阀，也可以装设在外壳本体上。

防爆膜的作用主要是当 SF_6 断路器在性能极度下降的情况下开断短路电流时，或其它意外原因引起的 SF_6 气体压力过高时，防爆膜破裂将 SF_6 气体排向大气，防止使断路器本体发生爆炸事故。防爆膜一般装设在灭弧室瓷套顶部的法兰处。

三、SF_6 断路器的整体结构与类型

SF_6 断路器的整体结构可分为瓷柱式和落地罐式两大类。

1. 瓷柱式 SF6 断路器

瓷柱式 SF_6 断路器的灭弧室安装在高强度瓷套中，用空心瓷柱支承和实现对地绝缘。灭弧室和绝缘瓷柱内腔相通，充有相同压力的 SF_6 气体，通过控制柜中的密度继电器和压力表进行控制和监视。穿过瓷柱的绝缘拉杆把灭弧室的动触头和操动机构的驱动杆连接起来，通过绝缘拉杆带动触头完成断路器的分合操作。

瓷柱式 SF_6 断路器系列性强，可以用不同个数的标准灭弧单元及支柱瓷套组成不同电压级的产品。这类断路器的结构简单，用气量少，运动部件少，价格相对便宜，是目前生产和使

用较多的一种。它具有单断口电压高、开断电流大、运行可靠性高和检修维护工作量小等优点。然而由于它重心高，抗震能力较差，且不能加装电流互感器，所以，使用场所受到一定限制。按照瓷柱式 SF$_6$ 断路器整体布置形式，可分为"I"形布置、"Y"形布置及"T"形布置三种形式。

（1）Y 形布置一般用于 220 kV 及以上的单柱双断口断路器，图 3-11 为 Y 形布置的 LW6-500 型 SF$_6$ 断路器一相结构。每相为单柱双断口，每个断口除了并联有电容外，还并联有合闸电阻；电容呈两侧对称排列，电阻呈水平布置；因电压较高，支柱有 3 个瓷套，其上端装有均压环；每相配有一台液压操动机构。

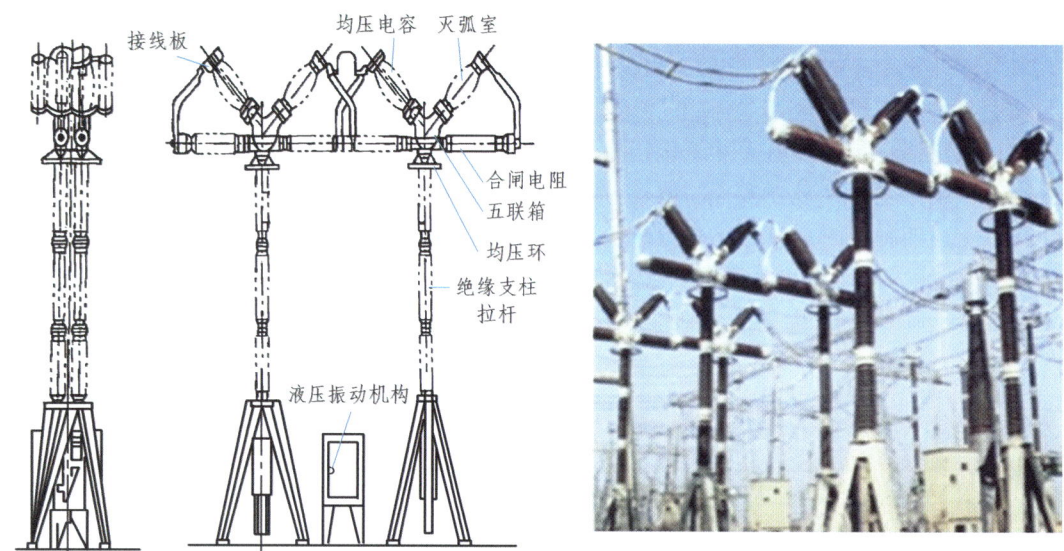

图 3-11　Y 形布置的 LW6-500 型 SF$_6$ 断路器

（2）T 形布置一般用于 220 kV 及以上特别是 500 kV 的单柱双断口断路器，图 3-12 为 T 型布置的 SFM-500 型 SF$_6$ 断路器一相结构。该断路器是我国与外企合作开发的产品，断路器每相有两个断口，灭弧室为变开距压气室结构，每相配一台气动操动机构，可进行单相操作及三相联动。图 3-13 为 ABB 公司开发的 HPL550 型 SF$_6$ 断路器，图 3-14 为阿尔斯通公司开发的 GL317 型 SF$_6$ 断路器。

（3）I 形布置一般用于 220 kV 及以下的单柱单断口断路器，三级安装在一个或三个支架上。图 3-15 为 I 形布置的 LW15-220 型 SF$_6$ 断路器一相结构。该型断路器为单断口结构，即每相只有一个断口。每相由灭弧室、支柱瓷套、机构箱组成。灭弧室采用变开距、双喷结构，支柱瓷套与灭弧室瓷套气室相通，支柱瓷套内的绝缘拉杆与灭弧室动触头相连，每相配一台气动操动机构，可进行单相操作及三相联动。

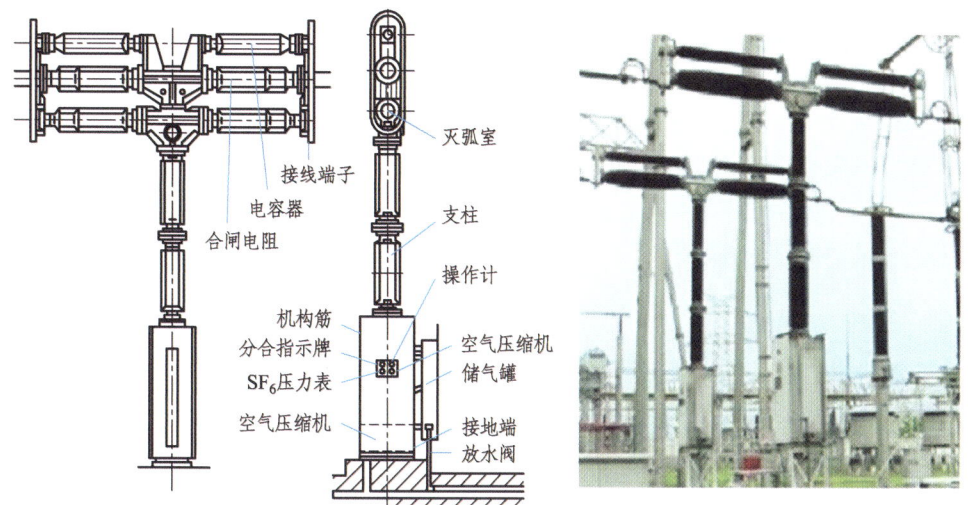

图 3-12 T 形布置的 SFM-500 型 SF$_6$ 断路器

图 3-13 T 形布置的 HPL550 型 SF$_6$ 断路器

图 3-14 T 形布置的 GL317 型 SF$_6$ 断路器

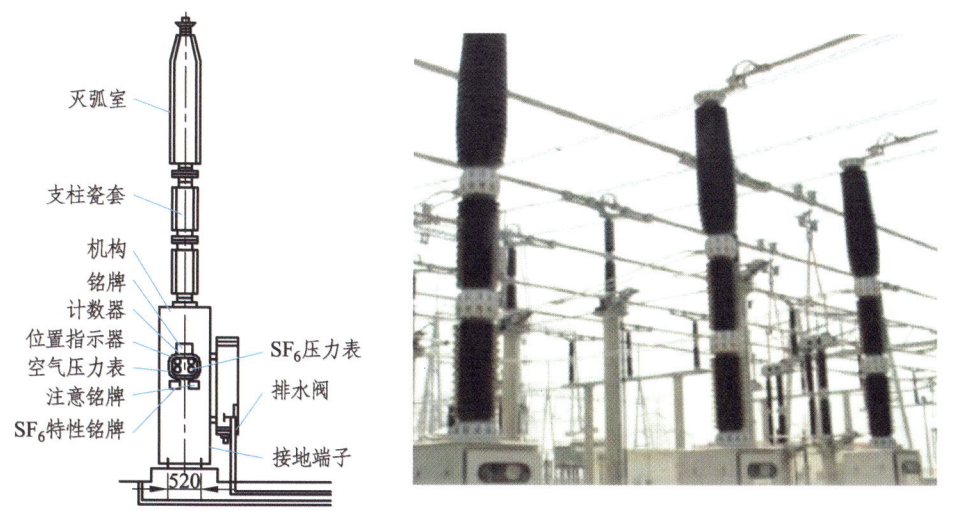

图 3-15 I 形布置的 LW15-220 型 SF$_6$ 断路器

项目三 高压断路器检修

2. 落地罐式 SF$_6$ 断路器

落地罐式 LM12-500 型 SF$_6$ 断路器如图 3-16 所示。其灭弧室安装在接地的金属罐中，高压带电部分用绝缘子支持，对箱体的绝缘主要靠 SF$_6$ 气体。绝缘操作杆穿过支承绝缘子，把动触头与机构驱动轴连接起来，在两个出线套管的下部都可安装电流互感器。

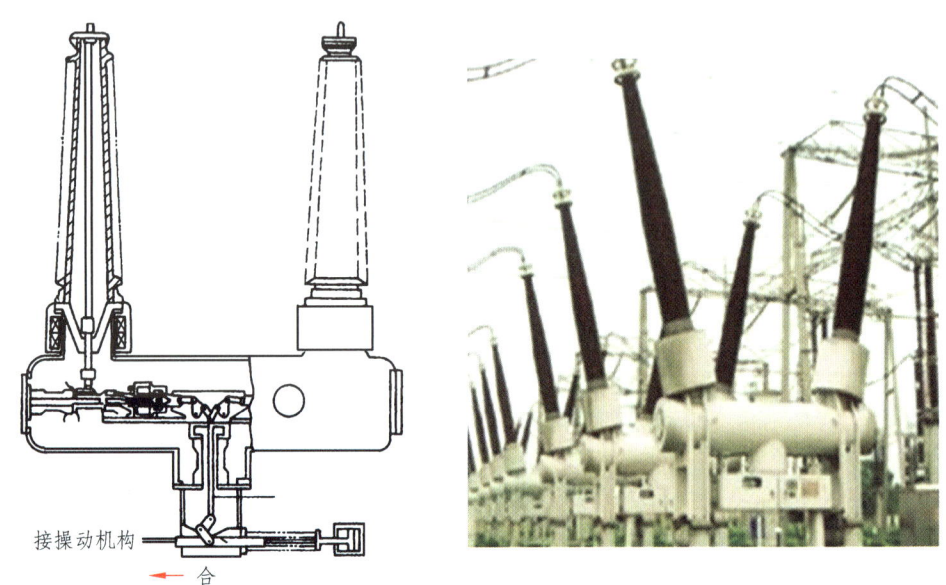

图 3-16 落地罐式 LM12-500 型 SF$_6$ 断路器

目前，110～500 kV 均有落地罐式 SF$_6$ 断路器，其外形基本相似，大多是引进日本三菱公司 SFMT 型或日立公司 OFPTB 技术的产品，如 OFPTB-500-50LA 型、国产 LW12 系列的 220 kV 和 500 kV 断路器。这种结构重心低，抗震性能好，灭弧断口间电场较好，断流容量大，可以加装电流互感器，还能与隔离开关、接地开关、避雷器等融为一体，组合成复合式开关设备。借助于套管引线，基本上不用改装就可以用于全封闭组合电器之中。但罐体耗用材料多，用气量大，系列性差，难度较大，造价比较昂贵。日本东芝、日立和三菱等公司已开发出 550 kV 63/50 kA 单断口罐式断路器。

Module 3 SF$_6$ Circuit Breaker

The SF$_6$ circuit breaker is a circuit breaker with the SF6 gas used as the arc extinguishing medium. The performance of SF$_6$ circuit breaker is mainly determined by the structure of the arc extinguishing chamber of the circuit breaker.

The SF$_6$ circuit breaker is initially a dual pressure circuit breaker similar to the air circuit breaker, with the air pump used to press the SF$_6$ gas into the high-pressure gas chamber. At the time of breaking the current, the SF$_6$ gas flows into the lower-pressure arc extinguishing chamber from the high-pressure gas chamber, thus generating a gas blowing effect to extinguish the arc. This high-voltage circuit breaker greatly reduces its arc extinguishing performance due to its high inflation pressure and the easy liquefaction of SF$_6$ gas under high pressure. In addition, a set of air extraction system is required, which is very complex in structure, so such circuit breaker has been phased out.

At present, SF$_6$ circuit breakers are mostly single pressure circuit breakers. At the time of breaking the short circuit current, the relative movement between the cylinder and the piston generates a compression effect, which increases the pressure of SF$_6$ gas in the cylinder. The gas is discharged from the nozzle, causing a longitudinal blasting to the arc and extinguishing it at the current zero-crossing. The single pressure circuit breaker is of simple structure, with low inflation pressure, and excellent breaking performance, which has been widely used.

3.3.1 Structure of Arc Extinguishing Chamber of SF$_6$ Circuit Breaker

1. Arc extinguishing chamber of single pressure SF$_6$ circuit breaker

The arc extinguishing chamber of a single pressure SF$_6$ circuit breaker operates based on the principle of piston compression. Hence it is also called a compressed arc extinguishing chamber. Normally, there is only one type of SF$_6$ gas with a pressure of 0.3–0.7 MPa in the arc extinguishing chamber, which serves as insulation. The movable part of the arc extinguishing chamber is equipped with the gas compression device. During the breaking process, the blowing pressure required for the arc extinguishing chamber is generated by the relative movement between the cylinder and the piston driven by the moving contact terminal system. Its SF$_6$ gas is recycled in the closed system and cannot be discharged to the atmosphere. This type of arc extinguishing device is simple in structure and reliable in action. According to the structure of the arc extinguishing chamber, arc extinguishing chambers can be divided into two types: variable clearance type and fixed clearance type.

(1) Structure and action process of arc extinguishing chamber with fixed clearance.

Fig. 3-8 shows the arc extinguishing chamber with fixed clearance. The circuit breaker is composed of two nozzle-equipped hollow static contact terminals and the moving contact terminal. In Fig. 3-8 (a), the circuit breaker is in the closed position. At this point, the moving contact terminal spans between the hollow static contact terminals, forming a current path; The moving contact terminal is structurally connected to the air cylinder and connected to the tie rod. The operating mechanism can drive the moving contact terminal and pneumatic cylinder to move left and right through the tie rod. The fixed piston is made of insulating material. Between it and the moving contact terminal and the air cylinder, a gas compression chamber is formed. At the time of opening, the operating mechanism moves the moving contact terminal and air cylinder to the right with the tie rod, compressing the SF_6 gas in the gas compression chamber, and increasing the pressure by about one time, as shown in Fig. 3-8 (b). When the moving contact terminal leaves the static contact terminal, an arc will be generated. At the same time, the air cylinder that was originally closed by the moving contact terminal is opened. High-pressure SF_6 gas rapidly sprays into the inner cavities of the two static contact terminals, exerting a strong symmetrical bidirectional longitudinal blasting on the arc, as shown in Fig. 3-8 (c). After the arc is extinguished, the contact terminal is in the open position, as shown in Fig. 3-8 (d). For this type of arc extinguishing chamber, the fixed clearance is kept by two static contact terminals for the arc gap of circuit breaker, hence such type of arc extinguishing chamber is called the arc extinguishing chamber with fixed clearance. Due to the strong arc extinguishing and insulation capabilities of SF_6 gas, the clearance is generally not large and the action is fast.

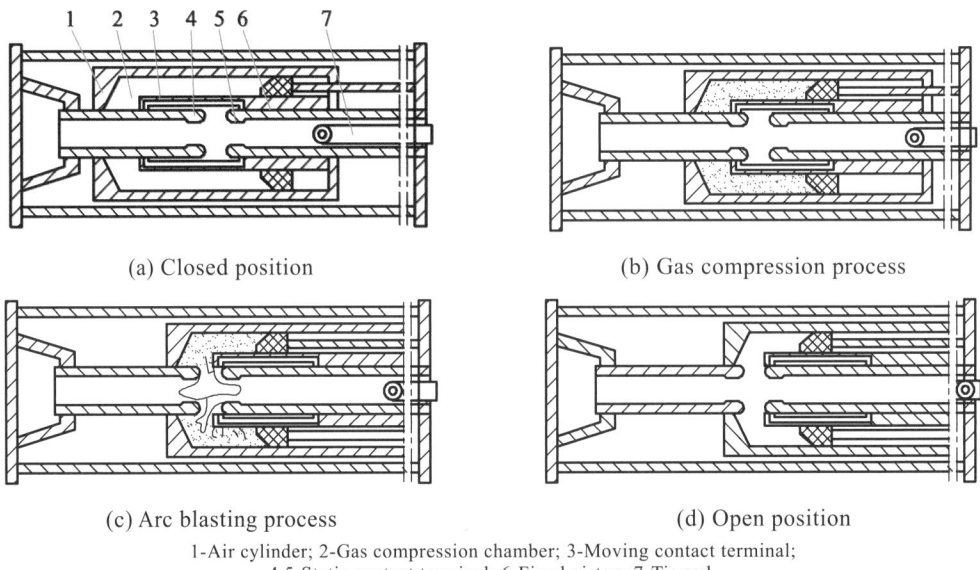

1-Air cylinder; 2-Gas compression chamber; 3-Moving contact terminal;
4,5-Static contact terminal; 6-Fixed piston; 7-Tie rod.

Fig. 3-8　Action Process of Arc Extinguishing Chamber with Fixed Clearance

The nozzle of the arc extinguishing chamber with fixed clearance is made of such conductive materials as metals or graphite with excellent arc resistance. Graphite can withstand high temperatures and directly changes from a solid state to a gaseous state under the action of an arc. It has a large work function and a light surface burning loss. With uniform electric field of open contact, small arc extinguishing clearance, very short stroke of the contact terminal from the separation position to the arc extinguishing position (only 30 mm for the circuit breakers of

126 kV), relatively small arc energy, strong capability of arc quenching and short time of arcing in the arc extinguishing chamber, it is possible for arc extinguishing chamber with fixed clearance to break very large short circuit current. However, the gas compression chamber has a relatively large volume.

(2) Structure and action process of arc extinguishing chamber with variable clearance.

During the arc extinguishing process, the clearance between the moving and static contact terminals in the arc extinguishing chamber will gradually increase with the movement of the gas compression chamber. Even if the arc has been blown out, the moving contact terminal will continue to move to the end position, that is, the contact terminal clearance continues to increase during the arc blasting process. So, such arc extinguishing chamber is called the arc extinguishing chamber with variable clearance. The arc extinguishing chamber with variable clearance is divided into unidirectional longitudinal blasting and bi-directional longitudinal blasting. The unidirectional blasting type is applicable to medium- and small-capacity circuit breakers, while the bi-directional longitudinal blasting is applicable to most high-pressure high-capacity circuit breakers.

The conductive body of the circuit breaker consists of two parts: the working contact terminal and the auxiliary contact terminal. When in the closed status, the two contact terminals are connected in parallel. When in the open status, the main contact terminal that flows through the majority of the current is first separated, and the current is transferred to the auxiliary contact terminal made of copper tungsten alloy, which is the arc resistant material. As the operating mechanism moves, the auxiliary contact terminal opens to form an arc, and then the nozzle gap is opened to blow out the arc. Putting the working contact terminal on the outside can improve heat dissipation conditions and enhance the thermal stability of the circuit breaker.

The nozzle is made of insulation materials such as polytetrafluoroethylene or composite materials made of fillers mainly composed of polytetrafluoroethylene. This type of material has characteristics such as arc resistance, high mechanical strength, easy processing, high temperature resistance, difficulty in carbonization under direct exposure to short-term arc action, uniform burning loss, low erosion, and resistance to SF_6 decomposition product corrosion.

The arc extinguishing process of the arc extinguishing chamber with variable distance is as shown in Fig. 3-9. Fig. 3-9 (a) shows the circuit breaker in the closed position. At the time of opening, the movable part moves to the right, and the SF_6 gas in the gas compression chamber begins to be compressed and the pressure is increased. With the movement of the movable part, the main static contact terminal 1 and the main moving contact terminal 5 first separate. Due to the fact that the arc contact terminal has not yet been disconnected, no arc is generated, the nozzle is not formed, and there is no arc blasting effect. Until the movable part moves to a certain position to the right, the arc static contact terminal 2 and the arc moving contact terminal 4 begin to separate. An arc is generated, and a spout between the nozzle and the arc moving contact terminal is formed. SF_6 gas is blown towards the arc from two directions, extinguishing the arc, as shown in Fig. 3-9 (b). After the arc is extinguished, the circuit breaker is in the open position, as shown in Fig. 3-9 (c).

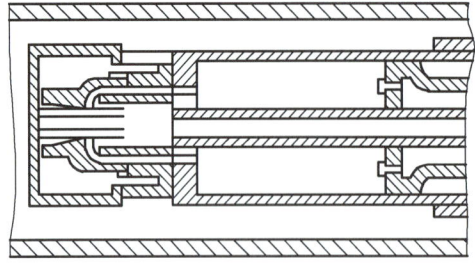
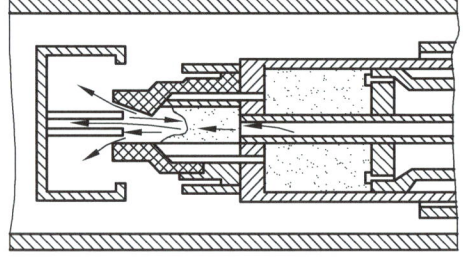

(a) Closed position　　　　　　　　　　(b) Arc blasting process

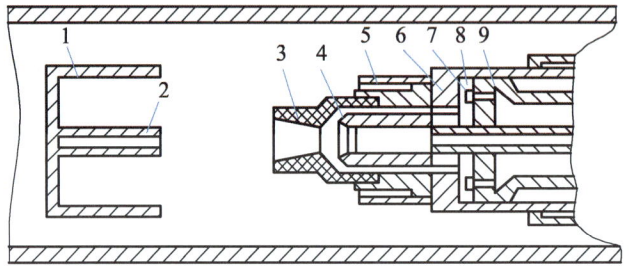

(c) open position

1-Main static contact terminal; 2-Arc static contact terminal; 3-Nozzle; 4-Arc moving contact terminal;
5-Main moving contact terminal; 6-Air cylinder; 7-Check valve;
8-Gas compression chamber; 9-Fixed piston.

Fig. 3-9　Action Process of Arc Extinguishing Chamber with Variable Clearance

A check valve is installed on the fixed piston to blast the arc centrally by the gas in the gas compression chamber towards the nozzle in the process of opening, while no vacuum is formed in the gas compression chamber in the process of closing. At the time of opening, the small hole is blocked by the check valve, allowing SF_6 gas to concentrate and blast the arc towards the nozzle. At the time of closing, the check valve opens to connect the gas compression chamber with the inner chamber of the piston, and SF_6 gas is charged into the gas compression chamber 8 from the small hole of the piston, preparing for the next opening.

(3) Comparison of structure between arc extinguishing chamber with fixed clearance and arc extinguishing chamber with variable clearance.

During the arc blasting process, the electrode clearance continuously increases, disrupting the dead zone of the airflow field. In addition, although there is a large insulation gap after arc extinguishing, the arc re-ignition caused by "electrical breakdown" after arc extinguishing is avoided. However, due to the increase in arcing time, the "thermal breakdown" may occur due to the slowdown of recovery rate of dielectric strength, thus limiting the limit breaking current of circuit breaker products with variable clearance.

The electric field of the structure with fixed clearance is concentrated between the two fixed nozzle electrodes. During the breaking process, the air cylinder needs to withstand the recovery voltage. At the moment when the moving contact terminal passes through the nozzle, the break insulation strength decreases due to a short circuit, which limits the increase in the unit break voltage of the circuit breaker product with fixed clearance.

Therefore, due to the low unit break voltage and large breaking current of the arc extinguishing chamber with fixed clearance of the 110 kV circuit breaker, the structure of the arc extinguishing chamber with fixed clearance should be used for breaking large current. However, for SF_6 circuit breaker with single break of 220 kV and above, as well as for reverse breaking at the outlet of power plants, it is more advantageous to use the arc extinguishing chamber with structure of variable clearance. With the demand for increasing unit break voltage in ultra-high voltage circuit breakers, structures with variable clearance have been widely used.

2. Self-energy SF_6 circuit breaker

For the compressed SF_6 circuit breaker, the operating mechanism is used to drive the relative movement between the cylinder and the piston. Therefore, the operating mechanism is heavily burdened and requires a high operating power.

The self-energy SF_6 circuit breaker which extinguishes the arc with the energy of the arc itself is able to reduce the burden on the operating mechanism and lower the requirements for the operation power of the operating mechanism, thus improving the reliability of circuit breaker. The self-energy SF_6 circuit breaker represents the mainstream of SF_6 circuit breaker development. The self-energy SF_6 circuit breakers can be divided into rotary, thermal expansion, and hybrid circuit breakers based on the arc extinguishing principle.

(1) Rotary arc type.

The circuit breaker with rotary arc is connected into the circuit in series by the arc automatically at the time of current breaking with the magnetic blow-out coil set around the static contact terminal, and the breaking current flows through the coil. The magnetic field is produced between moving and static contact terminals. The arc is driven by the magnetic field to rotate at a high speed, and it is constantly exposed to the fresh SF_6 gas during rotation, making the arc extinguished after being cooled. According to the different motion modes of magnetic blow-out and arc, it can be divided into radial and longitudinal rotary arc types.

The electromagnetic driving force decreases with the decrease of fault current, so the arc extinguishing ability of the circuit breaker with rotary arc is limited by the smaller fault current. Increasing the number of coil turns can overcome this drawback. However, the increase in the number of turns of the coil is limited by the mechanical strength, so large electromagnetic forces must be withstood under large fault currents.

The arc extinguishing chamber with rotary arc is simple in structure, with no need of a high-power operating mechanism. The arc rotates at high speed, with slight contact terminal damage and long service life, making such arc extinguishing chamber widely used in medium voltage systems.

(2) Thermal expansion type.

For the arc extinguishing chamber of thermal expansion type, the SF_6 gas in the arc extinguishing chamber is heated by the energy of the arc itself, and the heavy pressure is established, forming the differential pressure. Besides, release at the nozzle produces the arc blasting by strong airflow, thereby achieving the purposes of the cooling and the arc extinguishing by blasting.

The structure of the arc extinguishing chamber is as shown in Fig. 3-10. The cylindrical arc extinguishing chamber is divided into two compartments, namely the closed compartment and the exhaust compartment, which is much larger than the closed compartment. Both compartments are filled with SF_6 gas. When the circuit breaker is in the closed position, the moving contact terminal is connected to the static contact terminal through the contact finger, as shown on the left side of the centerline. At the time of opening, the current flows through coil, as shown on the right side of the centerline. When the moving contact terminal moves a certain distance, the arc is generated between the ring electrode and the moving contact terminal. The rotating arc coil generates a coaxial magnetic field with the contact terminal. The arc in the arcing ring is perpendicular to the magnetic field of the rotating arc coil, and the electric force generated during this process makes the arc rotate at high speed, causing the arc to be elongated in SF_6 gas. The rotating arc continuously contacts fresh SF_6 gas, releases heat energy, and heats the gas in compartment, generating a higher pressure than in the exhaust compartment. When the contact terminals are separated, the two compartments are connected through the nozzle in the moving contact terminal. At this point, the pressure difference that occurs is used to form a longitudinal arc blasting through the nozzle. The arc is extinguished at the next current zero-crossing.

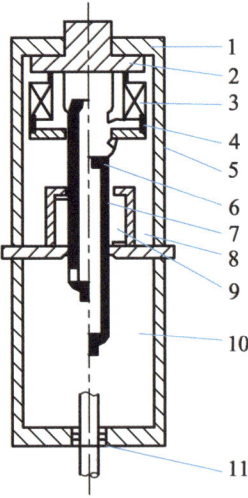

1-Barrel of arc extinguishing chamber; 2- Static contact terminal; 3-Cylindrical coil; 4-Contact finger; 5-Ring electrode; 6-Nozzle; 7- Moving contact terminal; 8-Closed compartment; 9-Auxiliary blasting device; 10-Exhaust compartment; 11-To atmosphere.

Fig. 3-10　Thermal Expansion Type Structural Diagram of Arc Extinguishing Chamber

(3) Hybrid arc extinguishing type.

Both arc extinguishing with rotary arc and arc extinguishing with thermal expansion can greatly reduce the burden on the operating mechanism and improve the performance price ratio of the circuit breaker. However, any type of arc extinguishing chamber has its shortcomings, so several arc extinguishing principles are often applied simultaneously in the arc extinguishing chamber of circuit breaker. The hybrid arc extinguishing method of compression and self-energy arc blasting can improve the arc extinguishing efficiency. Not only can it increase the breaking current, but it can also significantly reduce the operating power. There are various methods for hybrid arc extinguishing, such as rotating arc + thermal expansion, gas compression + thermal expansion, gas compression + rotating arc, and rotating arc + thermal expansion + auxiliary blasting.

3.3.2 Accessories for SF_6 Circuit Breaker

Accessories of SF_6 circuit breakers refer to the auxiliary components with certain special functions configured for SF_6 circuit breakers and their operating mechanisms, for example, pressure gauge, pressure relay (also known as pressure switch), safety valve, density gauge, density relay, parallel capacitor, parallel resistor, purification device, and explosion-proof device on SF_6 circuit breaker. Although they are accessory components, they play a very important role.

1. Pressure gauge and pressure relay

The SF_6 gas pressure is a macroscopic indicator of the insulation, current carrying capacity, breaking and closing capacity of the circuit breaker, and must always be maintained within the specified range during operation. To monitor the changes in SF_6 gas pressure, the pressure gauge and the pressure relay should be installed.

(1) Pressure gauge. The SF_6 gas pressure gauges play a monitoring role and can be divided into bourdon tube type, piston type, and digital type according to its structural principle. The bourdon tube pressure gauge is used generally for SF_6 circuit breaker.

(2) Pressure relay. The pressure relay is mainly configured on the operating mechanism of the circuit breaker, with multiple pairs of electrical contacts. It is used to control the motor startup and stop of the operating mechanism, to output the instructions for opening, closing and reclosing of locking circuit breaker, and to send corresponding signals. When the gas pressure increases or decreases, the pressure relay causes electrical contacts of the corresponding stroke switch to operate, in order to realize the use of pressure to control the output of relevant instructions and signals. The pressure relay plays a control and protection role.

(3) Safety valve. A safety valve is a safety protection device used in electric oil pump or air compressor system. It is a special form of pressure relay. The difference with a pressure relay is that the safety valve does not carry electrical contacts with different action method. When the oil pressure or air pressure exceeds the specified maximum value, the internal mechanism device of the safety valve acts. Such device is automatically shut down when the pressure is relieved to the specified value.

2. Density gauge and density relay

Gas density meters and density relays are specialized meters used to measure SF_6 gas. Meters with pointers and scales are called density gauges. Meters without pointers or scales are called density relays. Some SF_6 gas density meters also have electric contacts, and can be used as density relays. The SF_6 gas density meters play a monitoring role, and the density relay is for control and protection.

3. Parallel capacitors and parallel resistors

Parallel capacitors (also known as grading capacitors) and parallel resistors (also known as closing resistors) are important auxiliary components that are connected in parallel with the arc extinguishing break of the circuit breaker and improve the opening or closing characteristics of the circuit breaker.

In order to reduce the recovery voltage speed of the arc gap between contact terminals of circuit breaker and to improve the close-in fault breaking capacity, parallel capacitors are also installed in the single-break contact SF_6 circuit breakers with the voltage class of 63 kV and above.

In order to limit the overvoltage during closing, opening and reclosing and improve the performance of circuit breakers, a method of parallel connection of resistors between breaks is adopted. The parallel resistor is generally sintered from carbon. The shape of parallel resistor is very similar to the lightning arrester valve plate, but its heat capacity is much larger.

There are generally two installation methods for parallel resistors. One way is to place the parallel resistor disc and auxiliary break in the same porcelain bushing, or to arrange the parallel resistor disc on both sides of the auxiliary break, making it more conducive to heat diffusion after the resistor heats up during operation. The other way is that the closing resistor disc and the auxiliary break are not in the same porcelain bushing, but each form an independent component, which are connected in parallel at both ends of the arc extinguishing chamber after the connection in series.

The selection of parallel resistance values has a significant effect on limiting the closing overvoltage. At present, the parallel resistance value used on 500 kV circuit breaker in China is generally between 400 Ω and 450 Ω.

4. Purification device

The purification device is mainly composed of filtering tank and adsorbents. The function of the adsorbent is to adsorb moisture in SF_6 gas and certain decomposition products produced by SF_6 gas under high-temperature action of the arc.

The main adsorbents commonly used include the following.

（1）Activated carbon: It is an adsorbent made from raw materials such as fruit shells, coal, and wood through carbonization and high-temperature activation.

（2）Molecular sieve: It is an artificially synthesized zeolite and the aluminum silicate with a tetrahedral framework structure.

（3）Alumina: It is a porous material made of natural alumina or bauxite through special treatment.

（4）Silicone: It is a hard, porous solid particle made from sodium silicate.

In addition to the four adsorbents mentioned above, there are several other adsorbents such as bleaching earth, activated clay, adsorbent resin, activated carbonaceous fibers, charcoal molecular sieve, alumina, bauxite, magnesium oxide, and strontium sulfate. At present, the most widely used adsorbents on SF_6 switching devices at home and abroad are mainly molecular sieves and alumina.

5. Pressure relief device

Pressure relief devices can be divided into two types. Those characterized by opening pressure and closing pressure are called pressure relief valves, which are generally installed on SF_6 tank circuit breakers. Those that cannot be closed again once opened are called bursting disks and are generally installed on strut-type SF_6 circuit breakers.

When the casing and air source are connected in a fixed manner and the pressure regulating device used cannot reliably prevent the overpressure, the pressure relief valve of appropriate size should be installed to prevent excessive pressure inside the casing in case of failure of pressure regulating measures.

When the casing and air source are not connected in a fixed manner, the pressure relief valve should be installed on the inflation pipeline, or it can be installed on the casing body.

The role of the bursting disk is mainly to discharge SF_6 gas to the atmosphere with the rupture of the bursting disk, to prevent explosion accidents of the circuit breaker body in case that the SF_6 circuit breaker breaks the short-circuit current under extreme performance degradation, or when the SF_6 gas pressure is too high due to other unexpected reasons. The bursting disk is generally installed at the flange on the top of the arc extinguishing chamber porcelain bushing.

3.3.3 Overall Structure and Type of SF_6 Circuit Breaker

The overall structure of SF_6 circuit breakers can be divided into two types, namely live tank type and dead tank type.

1. Live tank SF_6 circuit breaker

The arc extinguishing chamber of live tank SF_6 circuit breaker is installed in the high-strength porcelain bushing and supported by the hollow knob insulators, and it is used for the insulation against ground. The arc extinguishing chamber is connected to the inner cavity of insulated knob insulator, filled with the SF_6 gas with the same pressure, and controlled and monitored through the density relay and pressure gauge in the control cabinet. The moving contact terminal of the arc extinguishing chamber is connected with the driving rod of the operating mechanism by the insulation tie rod passing through the knob insulator, and the circuit breaker is closed or opened by the contact terminal driven by the insulated tie rod.

The live tank SF_6 circuit breaker has a strong seriality, and can be composed of products of different voltage levels with different numbers of standard arc extinguishing units and supporting porcelain bushings. This type of circuit breaker is simple in structure, with low gas consumption, fewer moving parts, and relatively cheap price, making it widely produced and used at present. It has the advantages of high single break voltage, large breaking current, high operational reliability, and low maintenance workload. However, due to its high center of gravity, relatively poor shock resistance, and failure to install current transformers, its usage is limited to some extent. According to the overall layout of live tank SF_6 circuit breaker, they can be divided into "I" shape, "Y" shape, and "T" shape layouts.

(1) The "Y" shape layout is generally used for single-column circuit breaker of 220 kV and above with double breaks. Fig. 3-11 shows the one phase structure of LW6-500 SF_6 circuit breaker with "Y" shape layout. Each phase has a single column with double breaks, and each break is not only connected in parallel with the capacitor, but also with the closing resistor. Capacitors are arranged symmetrically on both sides, and resistors are arranged horizontally. Due to the high voltage, the strut has three porcelain bushings, and the upper end is equipped with the grading ring. Each phase is equipped with a hydraulic operating mechanism.

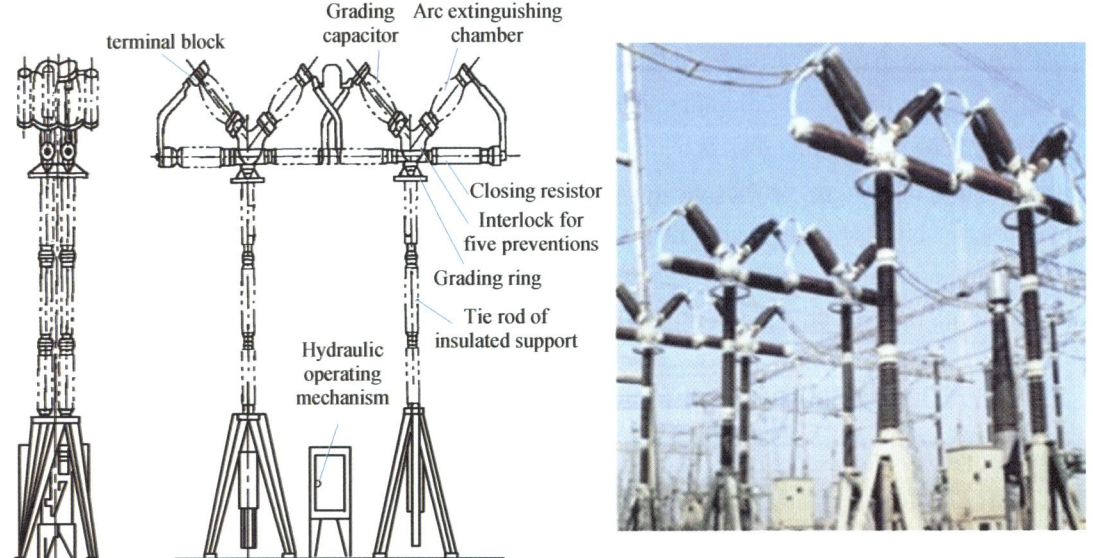

Fig. 3-11 "Y"-Shaped LW6-500 SF_6 Circuit Breaker

(2) The "T" shape layout is generally used for single-column circuit breaker of 220 kV and above (especially 500 kV) with double breaks. Fig. 3-12 shows the one phase structure of SFM-500 SF_6 circuit breaker with "T" shape layout. This circuit breaker is a product developed in cooperation between Chinese enterprises and foreign enterprises. The circuit breaker has two breaks per phase, and the arc extinguishing chamber is of the compression chamber structure with variable clearance. Each phase is equipped with the pneumatic operating mechanism, which can perform single-phase operation and three-phase linkage. Fig. 3-13 shows the HPL550 SF_6 circuit breaker developed by ABB, and Fig. 3-14 shows the GL317 SF_6 circuit breaker developed by Alstom.

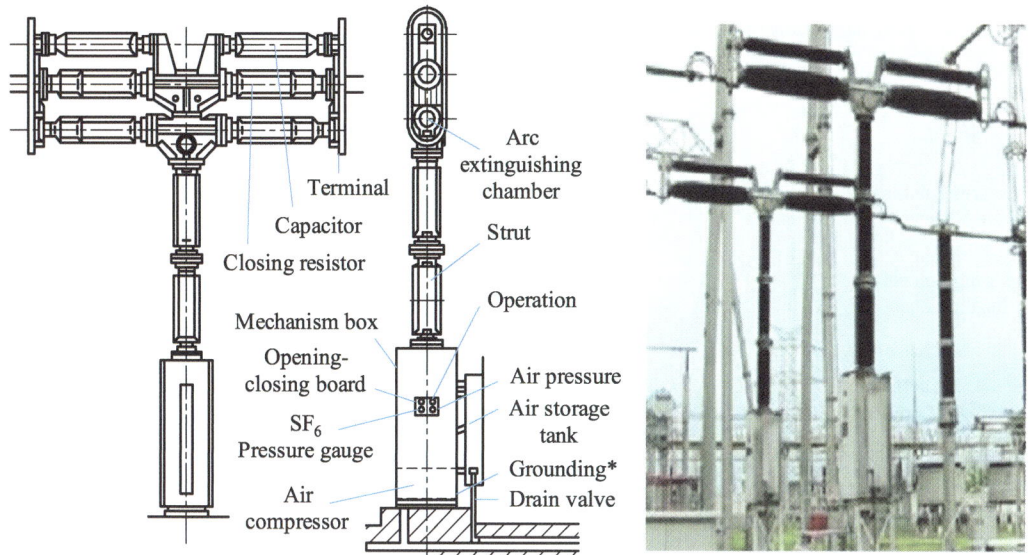

Fig. 3-12 "T"-Shaped SFM-500 SF_6 Circuit Breaker

Fig. 3-13 "T"-Shaped HPL550 SF_6 Circuit Breaker

Fig. 3-14 "T"-Shaped GL317 SF_6 Circuit Breaker

(3) The "I" shape layout is generally used for single-column circuit breaker of 220 kV and below with single break, with three stages of devices installed on one or three supports. Fig. 3-15 shows the one phase structure of LW15-220 SF_6 circuit breaker with "I" shape layout. This type of circuit breaker is of the single-break structure, which means there is only one break per phase. Each phase consists of the arc extinguishing chamber, the supporting porcelain bushing, and the mechanism box. For the arc extinguishing chamber, a variable clearance and dual-nozzle structure are adopted. The supporting porcelain bushing is connected to the gas chamber of the porcelain bushing of the arc extinguishing chamber, and the insulation tie rod inside the supporting porcelain bushing is connected to the moving contact terminal of the arc extinguishing chamber. Each phase

is equipped with the pneumatic operating mechanism, which can perform single-phase operation and three-phase linkage.

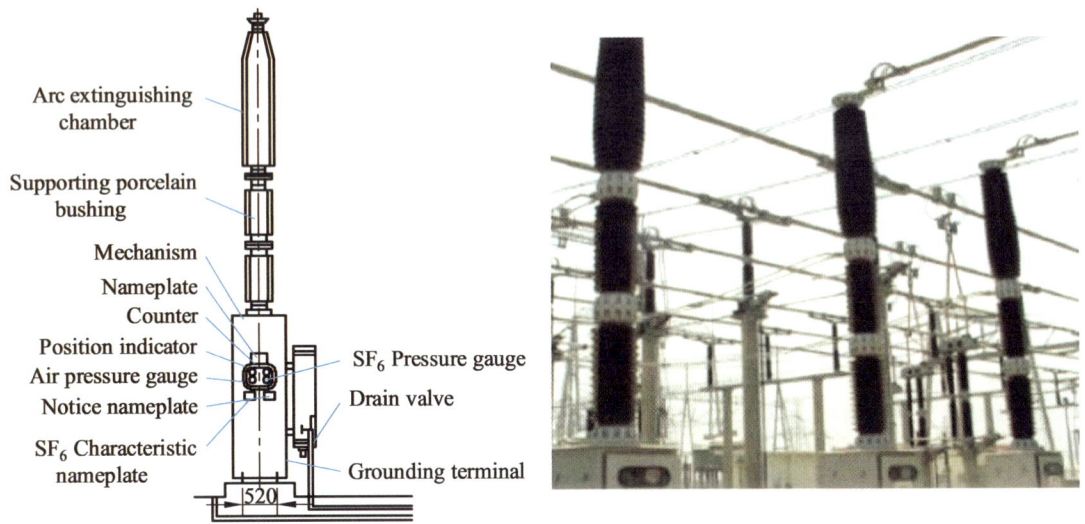

Fig. 3-15　"I"-Shaped LW15-220 SF$_6$ Circuit Breaker

2. Dead tank SF$_6$ circuit breaker

The LM12-500 dead tank SF$_6$ circuit breaker is as shown in Fig. 3-16. The arc extinguishing chamber is installed in the grounded metal tank, and HV live parts are supported by insulators. The box is insulated by the SF$_6$ gas mainly. The moving contact terminal is connected to the drive shaft of mechanism by the insulated operating lever through the supporting insulator, and the current transformers can be installed at the lower parts of two outgoing bushings.

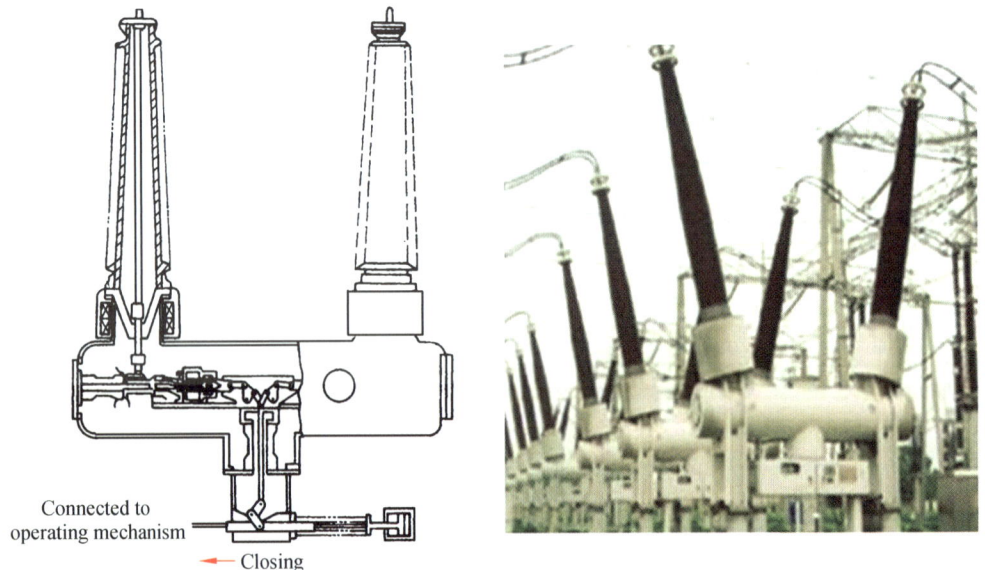

Fig. 3-16　LM12-500 Dead Tank SF$_6$ Circuit Breaker

At present, there are dead tank SF$_6$ circuit breakers with the range of 110 kV and 500 kV, with similar appearances. Most of them are imported products with Mitsubishi SFMT type or Hitachi OFPTB technology, such as the OFPTB-500-50LA type, and domestic 220 kV and 500 kV circuit breakers of LW12 series. With low center of gravity of structure, excellent shock resistance, good electric field between arc extinguishing breaks, and large breaking capacity, it can be retrofitted with current transformer, and integrated with disconnector, grounding switch and lightning arrester, forming the combined type switchgear. With the aid of bushing lead, it can be used in the fully enclosed GIS without being retrofitted basically. However, the tank body consumes a large amount of materials, with large gas consumption and poor seriation, and it is relatively difficult and expensive. Japanese companies such as Toshiba, Hitachi, and Mitsubishi have developed 550 kV 63/50 kA tank circuit breakers with single break.

模块四　高压断路器的操动机构

一、操动系统概述

高压断路器的操动系统包括操动机构、传动机构、提升机构、缓冲装置和二次控制回路等部分。主要部分的功能分述如下。

(一) 操动机构

操动机构是驱动断路器分合闸的重要配套设备，断路器的工作可靠性在很大程度上依赖于操动机构的动作可靠性。高压断路器的操动机构种类很多，结构差异很大，但基本上都是由操作能源系统、分闸与合闸控制系统、传动系统及辅助装置四个部分构成。

(二) 传动机构

1. 传动机构的作用和组成

传动系统是操动机构的做功元件与动触头之间相互联系的纽带，高压断路器的操动机构和本体在分、合闸过程中通过传动系统传递能量和运动，按照设计的性能要求完成分、合闸的操作。高压断路器的传动机构主要由操动机构中的传动元件、断路器中的提升机构和它们之间的传动机构三部分组成。

操动机构中的传动元件由连杆机构或液压、气动传动机构等构成，通过传动机构与断路器的提升杆相连。

传动机构是连接操动机构与提升机构的中间环节，起改变运动方向，增加行程并向断路器传递能量的作用。由于提升机构与操动机构总是相隔一定的距离，而且两者的运动方向也不一致，因此需要有传动机构，一般由连杆机构组成。传动机构主要有导轨型和平行运动型两类。导轨型的传动机构由于直连动静头的导杆受到导轨限制，受到与其运动方向相垂直的力，承受扭曲力矩易变形，因此动作可靠性受到影响。这种传动机构结构简单，价格便宜，常见于早期的断路器（如 SN1 型少油断路器）。平行运动型的传动机构可让动触头的导杆正确地沿直线方向移动，工作性能良好，在断路器中得到广泛应用。

提升机构是带动断路器动触头按一定轨迹运动的机构，通常它将操动机构传送的旋转运动或直线运动转变为动触头的直线运动，使断路器分、合闸，所以也叫变直机构。

以上三者之间相互关系见图 3-17。

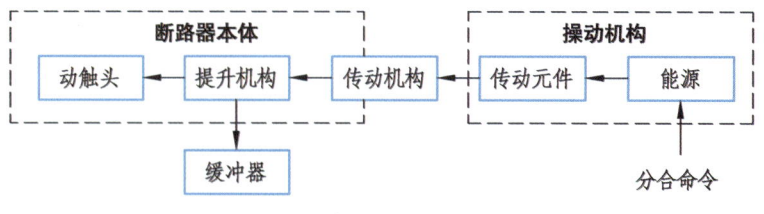

图 3-17　传动系统相互关系图

2. 传动机构的类型

传动系统形式很多，大致可分为以下几类：

（1）机械传动方式。常用的有杠杆、连杆机构、凸轮、齿轮等传动方式，其中以连杆机构使用最广泛。其优点是传动可靠，同步性好，加工简单，调整方便，维护容易，缺点是传递大功率时速度较低，冲击力大。

（2）压缩空气传动方式。一般使用在高压空气断路器及气动机构中，优点是反应较快、动作迅速。缺点是管道增长时动作时间随之增长，结构较复杂，加工及维护要求高。

（3）液压传动方式。多用于液压操动机构中，优点是动作平稳，传动力大，速度快，调整方便。缺点是结构复杂，加工难度大，传递速度受温度的影响。

（4）气压机械混合传动方式。多用于压缩空气断路器和少油断路器，这种传动方式是以杠杆代替部分管道和元件，优点是同步性好，传动快；缺点是结构复杂，维护要求较高。

（5）液压机械混合传动方式。多用于少油及 SF_6 断路器中，此种传动方式也是以杠杆代替部分管道和元件，优点是动作速度快，制造比液压机构简单；缺点是结构较复杂，冲击力大。

3. 连杆机构

连杆机构在高压断路器的传动系统中占有重要位置，各种传动机构大多都是由连杆机构组合而成。高压断路器及操动机构的连杆机构是比较复杂的，但是任何复杂的连杆都可以把它分解成几个四连杆机构，在有自由脱扣机构的操动机构中还会有一个五连杆机构。连杆机构的常用类型有以下几种。

（1）四连杆机构。

四连杆机构由三根活动连杆和一根固定连杆共组成，如图 3-18 所示，O_1 与 O_2 为固定轴销，A、B 为可动轴销，连杆 AB、AO_1、BO_2 为能往复摆动或转动可动的连杆，O_1O_2 可视为一根固定连杆。其中连杆 AO_1、BO_2 常称为拐臂，简称为臂，连杆 AB 简称为杆。若 AO_1 为主动臂，BO_2 则为从动臂，加在主动臂上的操作力产生的力矩 M 与主动臂的转动方向一致，而从动臂产生的力矩与从动臂转动的方向是相反的。用作图的与法可得到它们的运动轨迹和运动特性。改变四连杆机构各连杆的相对尺寸，可得到不同的机构型式。

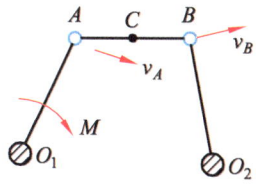

图 3-18 四连杆机构

（2）摇杆滑块机构。

摇杆滑块机构是四连杆机构的一种变形，常用作变直机构。如图 3-19 所示，O 为固定轴销，它没有从动臂，但有导向装置。当臂 OA 绕 O 摇动时，轴销 B 和滑块在导轨中作直线滑动。

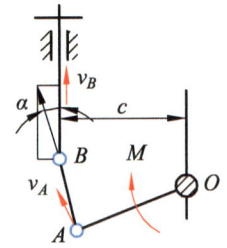

图 3-19　摇杆滑块机构

（3）准确椭圆机构。

如图 3-20 所示，O 为固定轴，且 $AB=AC=AO$。其中 OAC 相当于一个摇杆滑块机构，C 点在导轨内作直线运动，BC 是连杆，B 端限制在直线导轨里滑动。当滑块 C 在导轨中运动时，推动 A 点绕 O 旋转，这时 B 点作经过轴 O 的直线运动，而 BC 杆上除了 A、B、C 三点的其他任意点均作椭圆运动，故称准确椭圆直线机构。如果将断路器的导电杆或绝缘提升杆连接在 C 点，那么动触头（B 点）分、合闸都作直线运动。

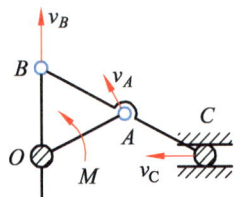

图 3-20　准确椭圆机构

（4）近似椭圆机构。

图 3-21 是由图 3-20 变化而来，即将图 3-20 导轨中的 C 点改在绕 O_2 摆动的摇杆端点上，这时若摇杆 O_2C 摆动不大，则 C 点轨迹为近似直线，B 点的轨迹也变为近似直线，而 BC 杆上除 A 点以外其他点的运动变为近似椭圆。

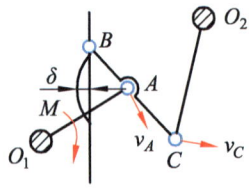

图 3-21　近似椭圆机构

（5）五连杆机构。

五连杆机构如图 3-22 所示，它有两个拐臂和两个连杆。其特点是主动臂与从动臂间没有确定的运动特点，即主动臂转过某一角度时，从动臂转过的角度可大可小，五连杆机构不能作传动机构，可在操作中用来实现自由脱扣。

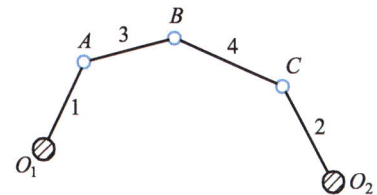

图 3-22　五连杆机构

（6）连杆式脱扣机构。

常用脱扣形式有折杆式、锁钩式、滚轮锁扣式三种。

① 折杆式脱扣机构。如图 3-23 所示，折杆式脱扣机构由连杆 4、5、6 组成。当操动机构接到分闸信号后，分闸电磁铁通电，电磁力 F_2 推动 C 点向上运动脱离死区后，断路器在分闸弹簧作用下自动分闸。

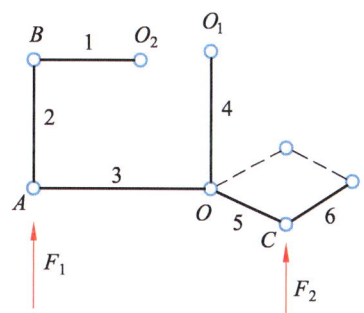

图 3-23　折杆式脱扣机构

② 锁钩式脱扣机构。如图 3-24 所示为锁钩式脱扣机构原理图，断路器在合闸位置时，分闸力产生的力矩 M 作用在连杆上，由于锁钩的阻挡，连杆不能被力矩 M 推动，因而可使断路器维持在合闸位置。断路器接到分闸信号后，分闸电磁铁通电，电磁力 F 推动锁钩反时针方向运动，当锁钩抬起一定距离后，连杆在力矩 M 的作用下，使断路器分闸。

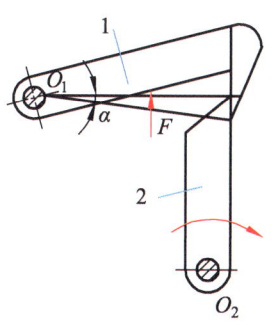

1—锁钩；2—连杆。

图 3-24　锁钩式脱扣机构

③ 滚轮锁扣式脱扣机构。如图 3-25 为滚轮锁扣式脱扣机构原理图，断路器在合闸位置时，滚轮被锁扣锁住，连杆上虽然有分闸力矩 M 作用，但无法使连杆转动，使断路器保持在合闸位置。当操动机构接到分闸信号后，分闸电磁铁通电，电磁力 F 推动锁扣向顺时针方向转动。

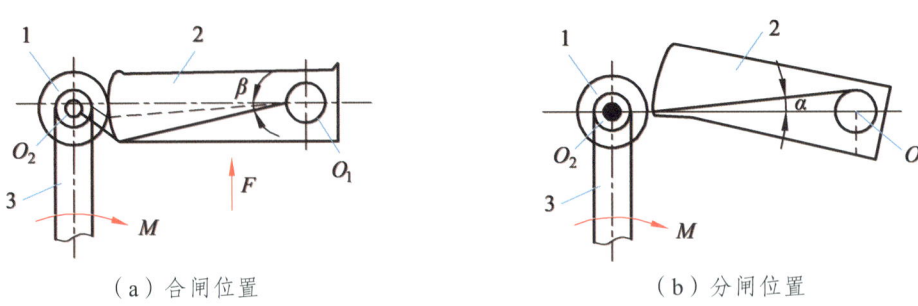

（a）合闸位置　　　　　　　　　　（b）分闸位置

1—滚轮；2—锁扣；3—连杆。

图 3-25　滚轮锁扣式脱扣机构

要使滚轮不被锁扣锁住，锁扣不仅要转过角 β，而且还要转过一个角度 α，使 A 点转出死区后，滚轮才能向顺时针方向转动，断路器分闸。

（三）缓冲装置

1. 缓冲装置的作用

断路器在操作过程中运动部件的速度很高，使得运动部件在运动即将结束时具有很大的动能，要使运动部件在较短的行程内停止下来，需要装置分、合闸缓冲器，使动作过程即将结束时的动能有控制地释放出来并转化为其他形式的能量，以保证在制动过程中吸收危及设备正常运行的冲击力，减少撞击，避免零部件变形损坏。缓冲器有时也可用于改变动作过程中的速度特性。缓冲器一般装设在提升机构旁。

2. 缓冲装置的类型

常用的缓冲装置有四类，即油缓冲器、弹簧缓冲器、气体缓冲器和橡皮缓冲器。

（1）油缓冲器。

油缓冲器一般用作分闸缓冲器，其原理结构如图 3-26 所示，它由油缸、活塞、撞杆、返回弹簧、端盖等组成，活塞与油缸壁之间有很小的间隙。当高速运动的部件撞击到缓冲器活塞上的撞杆后，活塞与运动部件一同向下运动，由于活塞下的油只能通过很小的缝隙向上流到活塞上方，使油流受阻，对活塞底部产生压力，阻碍活塞向下运动，形成对运动部件的缓冲。

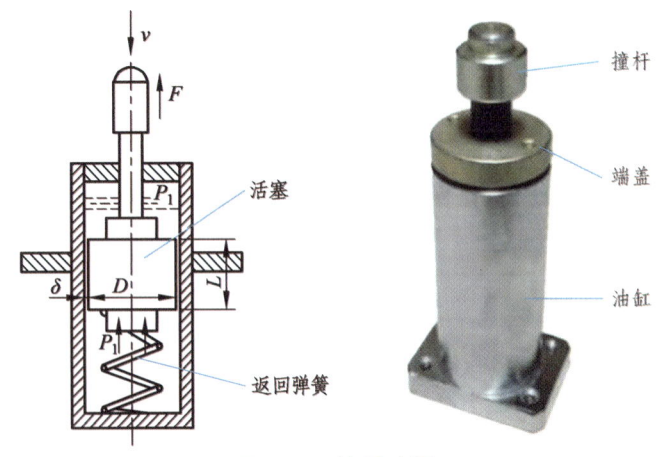

图 3-26　油缓冲器

一般油缓冲器多用于吸收分闸终了时的动能，因为分闸行程终了时传动系统的摩擦力很小，采用其他缓冲器易产生反弹跳跃或缓冲特性难以满足要求。

（2）弹簧缓冲器。

弹簧缓冲器利用运动部件撞击并压缩弹簧来吸收运动部件的动能而产生缓冲，其制动力与弹簧的压缩行程成正比。当需强力缓冲时必须增大弹力，在运动终了时，强力弹簧会使运动部件反弹，引起振动。弹簧缓冲器由弹簧、导杆、底座、撞杆组成，如图 3-27 所示。当运动部件与弹簧缓冲器相撞后，撞杆向上运动，弹簧被压缩，使运动部件的动能一部分变为弹簧的势能。

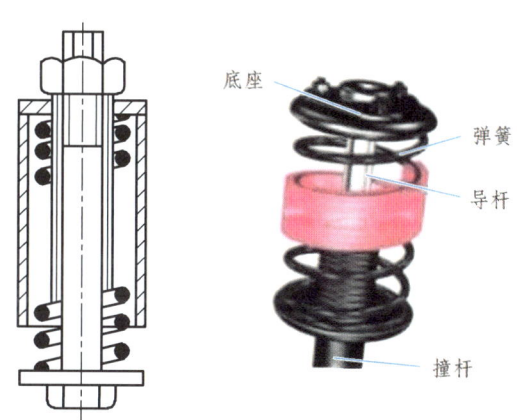

图 3-27　弹簧缓冲器

弹簧缓冲器结构简单，使用方便，特性不受温度影响，其缺点是有较大的反冲力。弹簧缓冲器多用作合闸缓冲器，因为合闸终了触头的摩擦阻力大，不易产生反弹振动，另外，被压缩的弹簧在分闸时释放能量还可以增加动触头的刚分速度。

（3）气体缓冲器。

气体缓冲器的原理与油缓冲器类似，只不过是缓冲器活塞运动压缩气体介质产生缓冲而已，但气体缓冲器在缓冲过程中产生的反弹力较大。气体缓冲器多用于压缩空气断路器及 SF_6 断路器中。

（4）橡皮缓冲器。

橡皮缓冲器在受到运动部件的撞击后，其动能消耗到压缩橡皮上，产生缓冲。橡皮缓冲器的优点是构造简单，反冲力不大，缺点是低温时橡皮弹性变坏，不耐油。一般用在缓冲能量不大的地方。

二、对操动系统的要求

断路器的全部使命，归根结底是体现在触头的分、合动作上，而分、合动作又是通过操动机构实现的。因此，操动机构工作性能的优劣，对高压断路器的工作性能和可靠性有着极为重要的影响。操动机构的动作性能必须满足断路器的工作可靠性的要求。

1. 具有足够的合闸功率

只有足够大的合闸功率的操动机构才能确保在各种规定条件下，按需求的合闸速度实现断路器可靠合闸，并维持在合闸位置，不产生误分闸。

在电网正常工作时，用操动机构使断路器关合的是工作电流，关合是比较容易的。但在电网事故情况下，如断路器关合到有预伏短路故障时，短路电流可达几万安以上，断路器导电回路受到的电动力达几千牛顿，从断路器导电回路的布置以及触头的结构来看，电动力的方向又常常是阻碍断路器关合的。因此，在关合有预伏短路故障的电路时，由于电动力过大，断路器有可能出现触头不能迅速、可靠地实现关合，从而引起触头严重烧伤。因此，操动机构必须具有足够的合闸功率，才能具有关合短裤故障的能力。

2. 能维持断路器处在合闸位置

由于合闸过程中，合闸命令的持续时间很短，而且操动机构的操作力也只能在段时间内提升，因此操动机构必须在完成断路器的合闸操作后，有保持合闸状态的机构，以保证在合闸命令和操作力消失后，断路器依旧能够维持在合闸位置。

3. 有可靠的分闸速度和足够的合闸速度

操动机构对断路器的分闸功率通常可以满足，这是由于断路器导电回路通过短路电流及触头结构所呈现的电动力方向，对于电路器的分闸起到加速作用。然而要求操动机构对断路器的分闸功能不仅能够电动对断路器分闸，在某些特殊情况下，还应可以进行手动分闸，而且要求断路器的分闸速度与操作人员的动作快慢和下达命令的时间长短无关。

4. 有自由脱扣装置与防跳跃措施

操动机构中自由脱扣装置的作用是，在断路器合闸过程中，若操动机构又接到分闸命令，则操动机构不应继续执行合闸命令，而应立即分闸，并保持在分闸位置。

当断路器关合在有预伏短路故障的电路时，断路器应自动分闸，此时若合闸命令还未解除，则断路器分闸后又将再次短路合闸，紧接着又会短路分闸。这样有可能会使断路器连续多次分、合短路电流，这一现象称为"跳跃"。出现跳跃时，断路器将连续多次合、分短路电流，造成触头严重烧损甚至引起爆炸事故。因此操动机构必须具有防止跳跃的能力，使得断路器关合短路电流而又自动分闸后，即使合闸命令尚未解除，也不会再次合闸。防跳跃可以采用机械的方法，不少操动机构中装设自动脱扣装置的目的就是为了防止跳跃。此外，尚有"电气防跳"措施，即在断路器的分、合闸操作的控制电路中，加装防跳跃继电器，防止跳跃的出现。

5. 复位与闭锁功能

断路器分闸后，操动机构中的各个部件应能自动地或通过简单的操作后，回复到准备合闸的位置，以保证操动机构的动作可靠。

同时，要求操动机构应具有以下的闭锁装置：

（1）分合闸位置连锁。保证断路器在合闸位置时，操动机构不能进行合闸操作；断路器在分闸位置时，操动机构不能进行分闸操作。

（2）低气（液）压与高气（液）压连锁。对于气动或液动操动机构，当气体或液体压力低于或高于额定值时，操动机构不能进行分、合闸操作。

（3）弹簧操动机构中的位置连锁。弹簧储能不到规定要求时，操动机构不能进行分、合闸操作。

三、操动机构的类型

根据所提供能源形式的不同,操动机构可分为手动操动机构(CS)、电磁操动机构(CD)、弹簧操动机构(CT)、气动操动机构(CQ)、液压操动机构(CY)以及新型操动机构等。其中,手动、电磁操动机构属于直动机构,弹簧、气动、液压操动机构属于储能机构。

在高压乃至特高压 SF_6 断路器中,配用的操动机构有三种:液压、气动以及弹簧操动机构。在中压真空断路器中,主要配用的操动机构为电磁操动机构和弹簧操动机构。此外,电压 10 kV、开断电流 6 kA 以下的轻型断路器常保留手动操动机构,用人力合闸,用已储能的分闸弹簧分闸。

(一)电磁操动机构

电磁操动机构是直接依靠电磁力合闸,可进行远距离控制和重合闸。其优点是结构简单、零件数少、工作可靠、制造成本低。其缺点是合闸线圈消耗的功率太大,合闸电流可达数百安,而且对二次操作电源可能形成一定冲击;要求配用 220 V/110 V 的大容量直流电源,因而辅助设施投资大,维护费用高;机构本身笨重,由于电磁时间常数影响,使合闸时往往有一定延迟,故在真空断路器中使用已逐渐减少。

电磁操动机构主要由作功元件(电磁系统)、连板系统、合闸维持、脱扣装置和缓冲系统等几部分组成。下面以 CD17 型电磁操动机构为例介绍电磁操动机构的结构及动作原理。

1. 结构和特点

CD17 型电磁操动机构是专为 ZN28-12 型真空断路器设计的操动机构,为平面五连杆机构,半轴脱扣,具备自由脱扣功能,脱扣功率小。机构左侧装有辅助开关,右侧装有分闸电磁铁,机构下部为合闸电磁铁,如图 3-28 所示。为了防止铁心吸合时黏附,合闸铁心加一黄铜垫和压缩弹簧,以保证铁心合闸终了时迅速落下。绕组和铁心间装有铜套,防止铁心运动时磨损绕组。

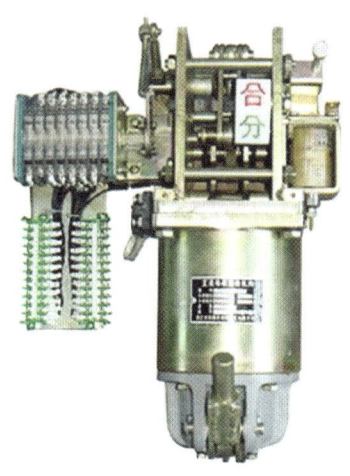

图 3-28 CD17 型电磁操动机构

合闸电磁铁下部由铸铁座和调整缓冲垫组成,座上装有合闸手柄供检修手动合闸用,橡胶调整缓冲垫不仅可起铁心缓冲作用,并可用于调整铁心顶杆与轮之间隙,以调整合闸速度。

2. 动作原理

（1）合闸操作。合闸前扣板处于复位状态，如图3-29（a）所示，合闸线圈通电后，铁心上扣板与半轴扣死，铁心推动顶轮上移，通过连杆带动输出轴转动约38°，通过机构外的传动杆，使断路器合闸，如图3-29（b）所示，此时断路器分闸弹簧储能，触头弹簧压缩。当铁心升到终点时，环与掣子出现（2±0.5）mm间隙，如图3-29（c）所示。这时，因主轴转动，带动辅助开关，使合闸回路常闭触点打开，切断合闸线圈电源，铁心落下，环被掣子撑住，完成合闸动作，如图3-29（d）所示。

（2）分闸操作。分闸线圈通电或用手力分闸时，半轴沿顺时针方向转动，扣板与半轴解扣，使环离开掣子，连杆系统变为不确定的五连杆机构，在分闸弹簧与触头弹簧的共同作用下，输出轴逆时针方向转动，完成分闸动作，同时带动辅助开关，使分闸回路常闭触点打开，切断分闸线圈的电源，如图3-29（e）所示。

（3）自由脱扣动作。合闸过程中，合闸铁心顶着轮向上运动，一旦接到分闸指令，使分闸铁心启动，使半轴与扣板解扣，在分闸弹簧与触头弹簧的共同作用下，轮从合闸铁心顶杆的端部滑下，实现自由脱扣，如图3-29（f）所示。

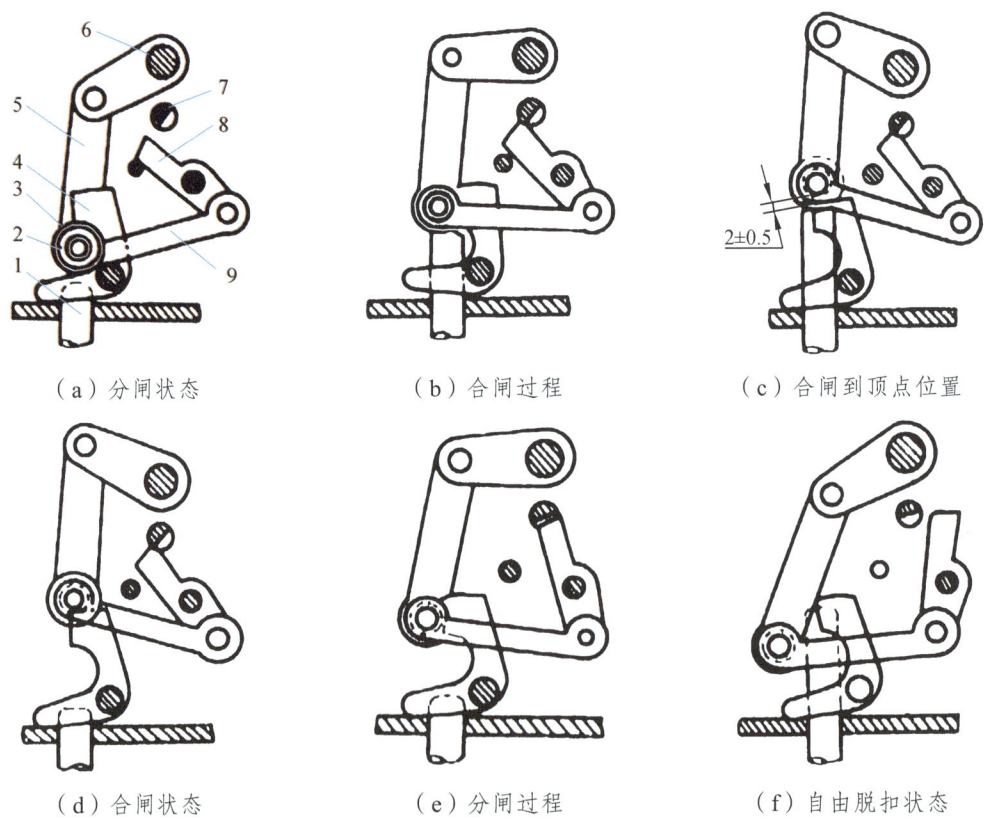

1—顶杆；2—滚轮；3—环；4—支架；5—连杆；6—输出轴；7—半轴；8—扣板；9—连板。

图3-29 CD17型操动机构动作状态图

（二）弹簧操动机构

弹簧操动机构是以弹簧作为储能元件的机械式操动机构。弹簧的储能借助电动机通过减速装置来完成,并经过锁扣系统保持在储能状态。其分合闸操作采用两个螺旋压缩弹簧实现。合闸时锁扣借助磁力脱扣,合闸弹簧释放的能量一部分用来合闸,另一部分用来给分闸弹簧储能。合闸弹簧一释放,储能电动机立刻给其储能,储能时间不超过 15 s,因而可实现断路器的快速自动重合闸。运行时分合闸弹簧均处于压缩状态。

1. 结构和特点

弹簧机构主要由合闸储能部分、传动部分和控制部分三部分组成。合闸储能部分包括电动机、减速装置、合闸弹簧、储能装置及保持释放装置。弹簧机构的合闸储能弹簧主要有压簧(也称螺旋弹簧)、拉簧(也称螺旋卷簧)和扭簧(也称蝶形弹簧)三种型式。按合闸弹簧储能所用的能源不同,弹簧机构可分为电动机储能弹簧机构和手力储能弹簧机构两种。合闸和分闸控制部分主要有脱扣器即脱扣机构。CT19 型弹簧操动机构结构如图 3-30 所示。

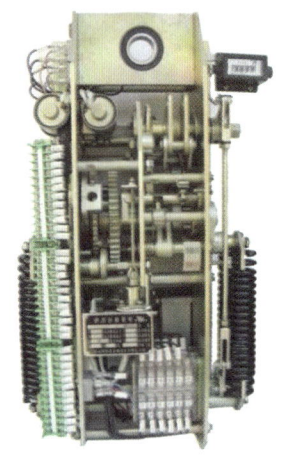

图 3-30　CT19 型弹簧操动机构

弹簧操动机构不需要专门的操作点源,储能电动机功率小,交直流两用,同时合闸弹簧的储能还可以通过人工手动完成,使用方便;但弹簧操动机构结构比较复杂,零件数量较多,成本较高,传动环节有出现故障的概率。

2. 动作原理

弹簧操动机构动作原理如图 3-31 所示。

在图 3-31(a)中,弹簧机构在合闸位置且分闸弹簧与合闸弹簧均已储能。拐臂和受分闸弹簧逆时针方向的力矩,此力矩被合闸保持掣子和分闸掣子阻挡。

断路器分闸时,分闸电磁铁的线圈接受分闸信号后带电,其铁心动作,冲击分闸掣子,分闸掣子顺时针方向旋转,释放合闸保持掣子,合闸保持掣子顺时针方向旋转,释放销子 A,拐臂和受分闸弹簧的推力,向逆时针方向旋转,拐臂通过连接的水平拉杆等传动元件和操作杆,使动、静触头快速分离,断路器分闸。

弹簧机构在分闸位置,合闸弹簧已储能时,棘轮轴承受连接在棘轮上的合闸弹簧逆时针方向的力矩,此力矩被储能保持掣子和合闸掣子销住。

断路器合闸时,合闸电磁铁的线圈接受合闸信号后带电,掣子动作,冲击合闸掣子,合闸掣子向顺时针方向旋转，释放储能保持掣子，储能保持掣子逆时针旋转释放 B 销，棘轮在合闸弹簧的作用下，逆时针方向旋转，同时带动棘轮轴旋转，使凸轮推动拐臂顺时针方向旋转，并带动拐臂轴上的拐臂顺时针方向旋转，同时压缩分闸弹簧储能。与拐臂相连接的水平拉杆和操作杆，使动触头快速合闸。

合闸操作完成后的机构状态如图 3-31（c）所示，A 销再次被合闸保持掣子锁住。

机构合闸操作完成后，合闸弹簧处于释放状态，棘爪轴通过齿轮与电机相连，断路器合闸到位后，对合闸弹簧进行储能。

合闸弹簧储能动作过程如下：

电动机启动，使棘爪轴旋转；偏心的棘爪轴上的两个棘爪，在棘爪轴的传动中与棘轮上齿交替进行啮合，使棘轮转动；棘轮逆时针方向旋转，带动拉杆使合闸弹簧储能，通过死点后，棘轮轴由合闸弹簧给以逆时针方向的转动力矩，此力矩通过 B 销被储能保持掣子锁住。

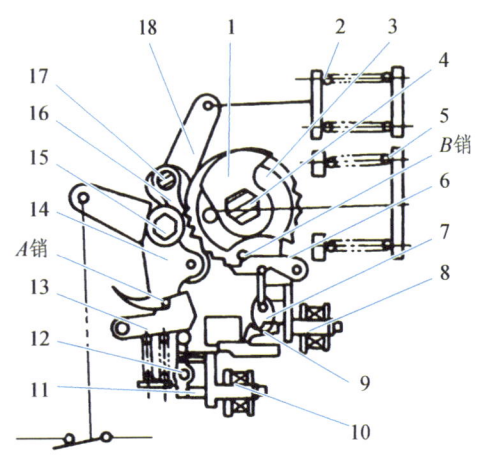

（a）合闸位置（合闸弹簧储能状态）

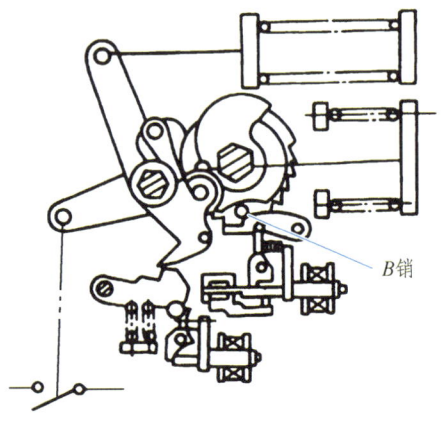

（b）分闸位置（合闸弹簧储能状态）

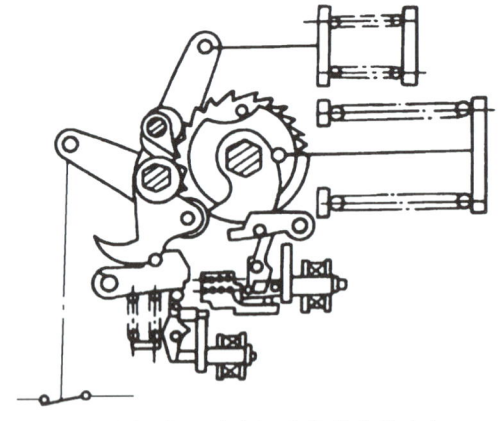

（c）合闸位置（合闸弹簧释放状态）

1—凸轮；2—分闸弹簧；3—棘轮；4—棘轮轴；5—合闸弹簧；6—储能保持掣子；7—合闸掣子；
8—合闸电磁铁；9—掣子；10—分闸电磁铁；11—铁芯；12—分闸掣子；13—合闸保持掣子；
14—拐臂；15—拐臂轴；16—棘爪；17—棘爪轴；18—拐臂。

图 3-31　弹簧操动机构工作原理图

(三)液压操动机构

液压操动机构将储存在储能器中的高压油作为驱动能传递媒体。储能器中的能量维持主要使用氮气,利用储能器中预存的能量,运用差动原理,间接推动操作活塞来实现断路器的分合闸操作。

1. 结构和特点

液压操动机构的主要构成元件有储能元件、控制元件、操动(执行)元件、辅助元件、电气元件五个部分,其结构如图 3-32 所示。

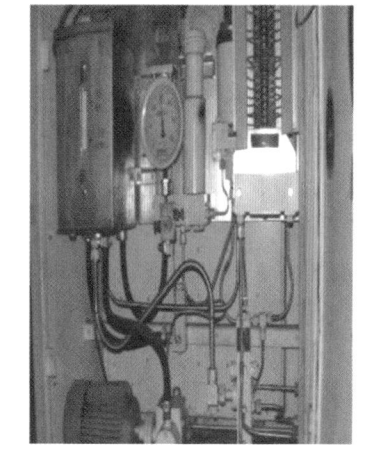

图 3-32 液压操动机构

(1)储能元件。包括储压器、滤油器和油泵等元件。当电动机驱动油泵时,油通过滤油器从油箱抽出打压送入储压器,压缩 N_2 储存能量。当操作时,气体膨胀对外做功,通过液压油传递给工作缸,转变成机械能,实现断路器分、合闸操作。

(2)控制元件。作为储能元件与操动元件的中间连接,发出分、合闸动作的液压脉冲信号,去控制操动元件。包括分、合闸电磁铁、分、合闸启动阀和二级阀等。

(3)操动元件。包括工作缸、压力开关、安全阀和放油阀等元件。工作缸借助连接件与断路器本体连接,受控制元件控制,驱动断路器实现分、合闸动作;压力开关用于控制电动油泵起动、停止和分、合闸闭锁;安全阀用于释放故障情况引起的过高压力,以免损坏液压元件;放油阀用于在调试和检修时释放油压,分为高压放油阀和低压放油阀。

(4)辅助元件。包括信号缸、油箱、排气阀、压力检测器和辅助储油器等元件。信号缸用于带动辅助开关切换电气控制回路,有的还带动分、合闸指示器及计数器;油箱是储油容器,平时与大气相通,操作时因工作缸排油,将会使它的内部压力瞬时升高;排气阀用以在液压系统压力建立之前排尽工作缸、管道内气体,以免影响动作时间和速度特性;压力检测器用于测量液压系统压力值;辅助储油器用于充分利用液压能量,减小工作缸分闸排油时的阻力,提高分闸速度。

(5)电子元件。包括分合闸线圈、加热器和微动开关等元件。分、合闸线圈分别用以操作分、合闸电磁阀(一级阀);加热器用于在外界低温时,保持机构箱内的温度,防止油液冻结和驱散箱内潮气,分为手动和自动两种;微动开关作为分、合闸闭锁触点和油泵起动、停止用触点,同时给主控室转换信号,以使起到监控作用。

液压操动机构的主要优点是操作平稳,无噪声,且需要控制的能量小,在不大的机构尺寸下就可以获得能量强大的操作力,以及液压元件质量轻且反应动作快,容易实现自动控制与各种保护,暂时失去电源时仍能操作多次,动作可靠性高,维修方便等,特别适用于 126 kV 及以上的高压、超高压特高压断路器。其主要缺点是加工工艺要求高,如果制造或装配不良,容易渗漏油,速度特性易受环境温度的影响。

2. 动作原理

液压操动机构采用差动原理,利用同一工作压力的高压油作用在活塞两侧的不同截面上测试作用力差,从而使活塞运动来驱动断路器进行分合闸操作。工作缸活塞和二级阀芯均按差压原理设计,一般在分闸侧常充有高压油,而在合闸侧则由阀系统进行控制,只有在合闸操作及合闸位置时才充入高压油,在分闸位置时与低压油箱连通。

液压操动机构系统的工作原理如图 3-33 所示。工作缸活塞右侧分闸腔与储压器直接连通,因此,不论是在合闸位置还是在分闸位置,都处在常高压状态。活塞左侧合闸腔则通过阀来控制。当合闸时,电磁铁线圈受电产生磁力,打开合闸电磁阀(一级阀),使高压油进入二级阀操纵活塞的合闸腔,操纵活塞推动二级阀的阀芯运动,于是关闭工作缸通往低压油缸的油路,打开高压阀口,使操纵活塞分闸腔的油从排油孔排出。储压器中的高压油进入工作缸活塞合闸侧,由于在活塞的合闸腔侧承压面积大于分闸腔侧承压面积,使活塞快速向右运动,

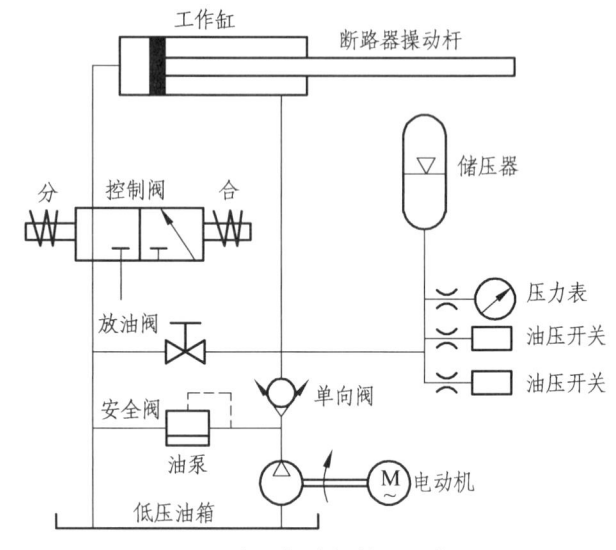

图 3-33　液压操动机构的工作原理

实现合闸。当分闸时,合闸腔中的高压油泄至低压油箱,同时在分闸腔内高压油的作用下,活塞向左运动,实现快速分闸。

液压操动机构是用油作为机械能传递的媒介,机械能是储存在储压筒内,目前,储压筒储存能量的方式主要有两种:① 利用氮气来储存能量,即在储压筒活塞的上部充入规定预充压力的氮气。氮气受压缩时就储存了能量。② 弹簧储能方式,即结构上使储压筒活塞与专用碟形弹簧相连。油泵打压时,被压的液压油推动储压筒活塞压缩碟形弹簧储能。

3. 闭锁防护功能

(1)油泵超时运转闭锁功能。一般油泵运转超过 3～5 min 时,时间继电器的常闭触点延时断开,切断电动机电源,停止打压。

(2)防慢分闭锁功能。

① 电气闭锁。当断路器和隔离开关处于合闸位置时,如果出现油压非常低,或降至零压时,将切断油泵电动机电源不会起动打压。

② 慢分阀。有三种方法:第一,将二级阀活塞锁住或加装防慢分装置;第二,在三级阀处设置手动阀,当油压降至零压时,将手动阀拧紧,使油压系统保持在合闸位置;当油压重新建立后松开此手动阀;第三,设置管状差动锥阀,该阀不论开关在分、合闸位置,只要系统一旦建立压力,不管压力有多大,该管状锥阀均产生一个维持在分、合闸位置的自保持力。

③ 机械法。利用机械手段将工作缸活塞杆维持在合闸位置上,待机械故障处理完毕后,即可拆除机械支撑。

(3)油压低闭锁功能。当油压降低至不足以保证断路器合闸或分闸时,利用微动开关使有关继电器励磁,断开合闸起动回路和分闸起动回路,从而实现分闸、合闸或重合闸闭锁功能。

4. CY 型液压操动机构

（1）结构简介。

常见的 CY 系列液压操动机构如图 3-34 所示，该机构主要由机构箱、储压筒、阀系统、工作缸、油泵、控制板等组成。CY3 型液压操动机构结构如图 3-35 所示，CYA8 型液压操动机构结构如图 3-36 所示。

① 机构箱。机构箱是整个机构的支承基架和保护外壳，箱内左右两侧和上方开有 3 个活门供检修用，上盖可以打开至 45°。下部装有加热器，当环境温度低于 0 ℃ 时投入运行。左侧门上开有监视压力表的孔。

② 储压筒。储压筒上部内腔中（活塞上方）预充有一定压力的氮气。储压筒活塞杆处装有微动开关，分别起到油泵启动与停止、合闸闭锁、分闸闭锁、重合闸闭锁、压力降低发告警信号等作用。

（a）CYA5 系列

（b）CYA6 系列

（c）CYA8 系列

图 3-34　CY 型液压操动机构

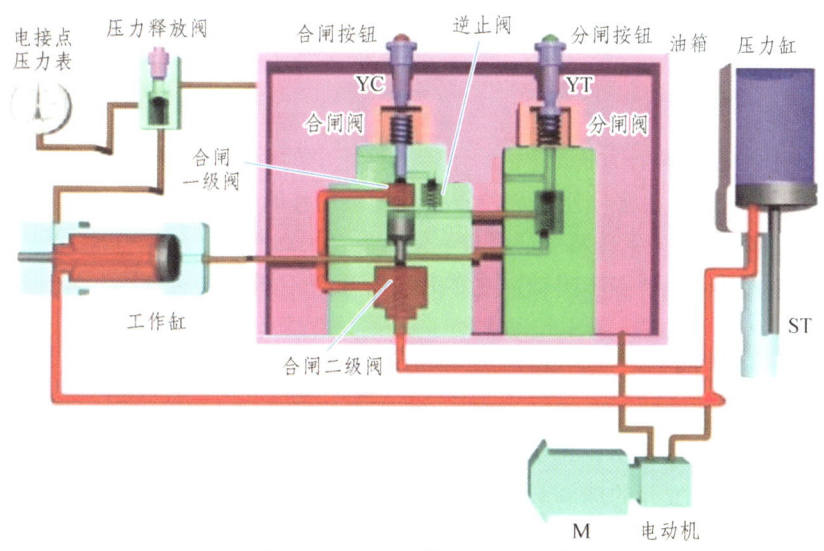

图 3-35　CY3 型液压操动机构

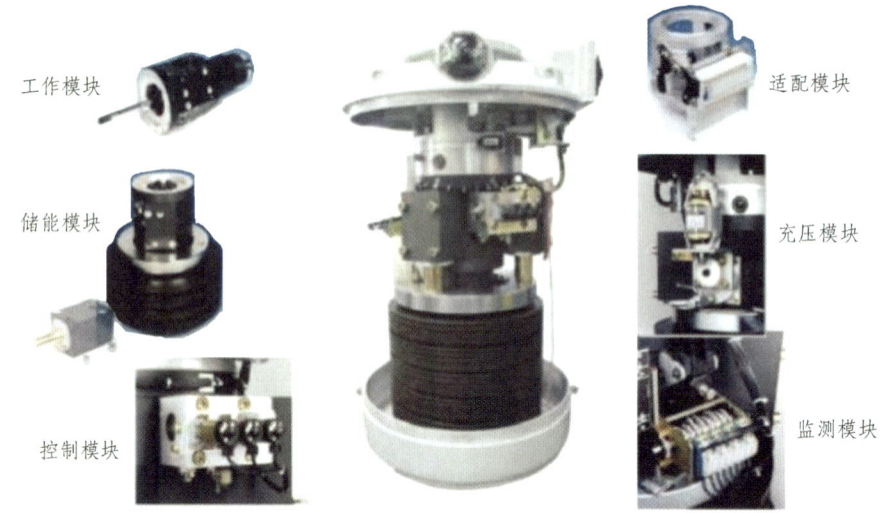

图 3-36　CY8A 型液压操动机构

③ 阀系统。阀系统由滤油器、放油阀、操动阀组成。操动阀放在油箱中，箱内有 10 号航空液压油。

④ 工作缸。工作缸和活塞为高强度耐磨元件，活塞左端接高压油，运行时，活塞左端始终保持常高压，活塞右端接操作管，高压油经合闸回路进入活塞右端。根据差动原理，活塞向左运动，通过水平拉杆带动断路器合闸。当右侧合闸回路中压力释放后，活塞在常高压作用下，向右运动，带动断路器分闸。

⑤ 油泵。油泵是双柱塞式的，柱塞是打压元件，柱塞与阀座为滑动配合，电动机带动油泵工作时，低压油经单向阀进入阀座内腔，曲轴转动时，滚珠轴承推动柱塞向阀座运动，于是腔内的油被挤压，并将另一侧单相阀（高压侧）打开，油经阀口进入高压油管，单相阀打开，液压油经管道进入储压筒中储存起来。曲轴转动一周，两柱塞各工作一次。

⑥ 控制板。控制板上装有辅助开关、接触器、中间继电器、电接点压力表，接线端子排等。

（2）动作原理。

① 机构储能。

油泵电机的电路接通后，油泵开始始运转，油箱中的低压油经滤油器进入油泵，高压油进入储压筒内。当储压筒活塞杆使停泵位置微动开关动作时，油泵停止运转，储能过程结束。运行中，当储压筒活塞杆向下位移至起泵位置使微动开关动作时，油泵自动运转补压。

由于在电路中没有油泵零压闭锁，在零表压时油泵不能自启动打压，为此须人为地将闭锁回路临时短接一下（或者人为地按动）。

② 合闸及合闸自保持。

当合闸电磁铁接到合闸命令或者手按铁心按钮时，合闸电磁铁的可动铁心向下运动，推动合闸一级启动阀的阀杆运动，先堵住阀座下排油孔，然后打开一级球阀，于是从合闸控制油管来的高压油通过被推开的球阀阀口，并经内部通道推交流接触器的动铁心。当油压高于低压力异常闭锁压力时，油泵的起停才能自动进行。开合闸保持逆止阀。高压油进入二级阀的锥阀心的上部，推动该锥阀心高速向下运动。它首先预封住阀座上排油孔的通路，然后推开下球阀心。上锥阀心利用其锥面密封住二级阀的上阀口，并堵住排油回路。从合闸进油管来的高压油经二级阀的下阀口和管路进入工作缸的合闸腔（无活塞杆的一侧）。这时，工作缸活塞两端均受到相同压强的高压油作用，由于合闸侧受压面积大得多，使该活塞向合闸方向运动，直到合闸终了为止。此时，辅助开关也完成了切换，合闸电磁铁失电，合闸一级阀在其复位弹簧力作用下返回，合闸一级阀腔中的高压油通过排油孔泄放。同时，由于高压油自保持回路的作用，操动机构得以保持在合闸状态。

合闸二级阀的锥阀芯所处的位置，决定了断路器是处于合闸还是分闸状态。为了保持合闸状态，必须使二级阀锥阀芯处于合闸位置，即在合闸操作后，该锥阀芯上部必须始终保持有高压油作用，以保证使该二级阀的下阀口打开，而上阀口关闭。

为了防止由于慢性渗漏使锥阀芯上部的油压降到零，机构设置了自保持的高压油补充回路。它是由油管、带 0.5 mm 节流孔的接头、保持阀和油管路等部件组成，借此来实现合闸自保持。

③ 分闸。

当分闸电磁铁接到分闸命令或者手按分闸按钮时，分闸电磁铁的动铁心推动分闸一级阀的阀杆向下运动，从而使分闸阀钢球阀口打开。合闸保持回路的高压油经管路和分闸一级阀阀口，通过阀座上排油孔排放到低压油箱中。二级阀芯在其下部高压油作用下立即向上返回，先将上阀口打开，工作缸合闸侧腔内的高压油经管路和上阀口以及阀座上的排油孔排泄到低压油箱中。工作缸活塞在分闸侧高压油作用下向分闸方向运动，最终完成分闸操作。同时，辅助开关也完成切换，将分闸电磁铁的电路断开。

在二级阀上阀口打开的同时，二级阀钢球在其复位弹簧作用下，迅速将下阀口关闭，使高压油不会过多地被泄放掉。

④ 合闸闭锁。

当储能装置的油压已不够合闸时，储压筒活塞下移，当降到闭锁合闸的微动开关时，微动开关触点闭合，合闸闭锁继电器动作，切除了合闸控制回路。另外当储压筒氮气漏气，氮气压力异常降低或由于某种原因油压不正常升高时，压力表接点闭合，启动有关继电器，从而切除合闸回路及油泵电动机电源回路。

⑤ 分闸闭锁。

当储能装置的油压降低已不满足分闸要求时，储压筒活塞杆下移，当降到闭锁分闸的微动开关时，微动开关接点闭合，分闸闭锁继电器随之动作，从而切除分闸控制回路。如上所述，如油压异常升高或降低以及氮气漏失，则电接点压力表动作，同样切除分闸回路及油泵电动机电源。

5. 集成化液压操动机构

集成化液压操动机构是为满足断路器需要操动机构提供较小操作功率而设计的，它不仅体积缩小，而且实现了功能元件的模块化。机构外部无配油管，元件小型化，各主要部件集成化，减少了泄漏点，实现了产品高质量，机构无外漏，操作噪声低，更方便检修。

集成化液压操动机构的动作原理如图 3-37 所示。

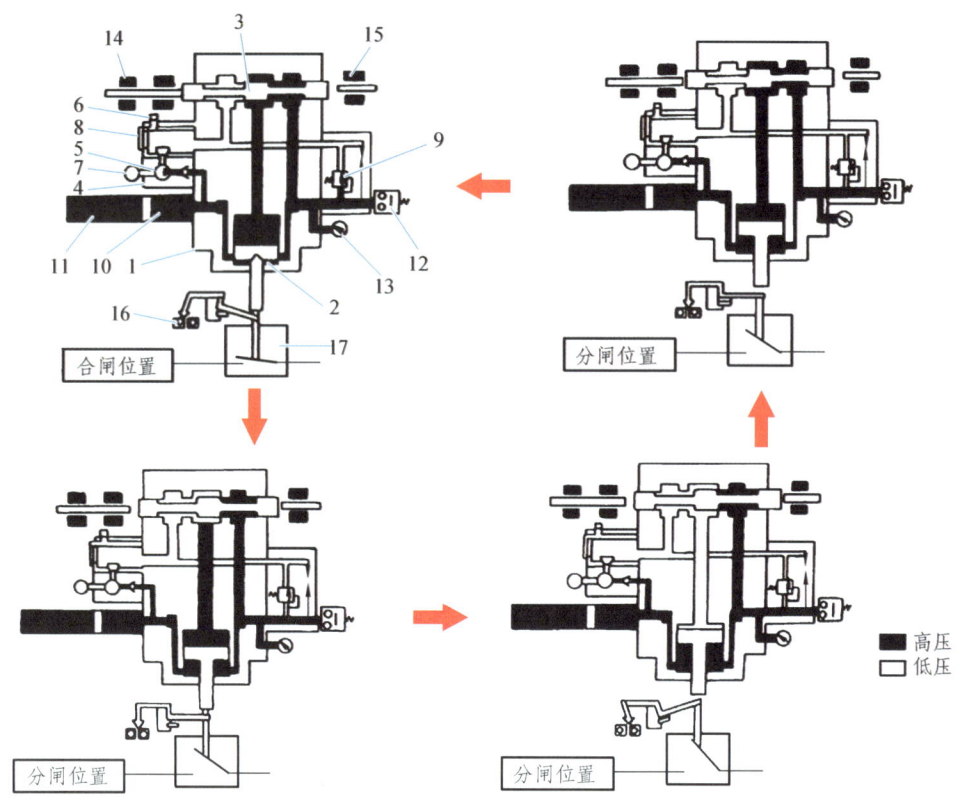

1—工作缸；2—活塞；3—换向阀；4—液泵组件；5—液压泵；6—油箱；7—电动机；8—油位计；
9—安全阀；10—蓄能器；11—高压氮气；12—油压开关；13—油压表；14—分闸线圈；
15—合闸线圈；16—分合指示器；17—断路器。

图 3-37 集成化液压操动机构的动作原理

（1）分闸动作。

接到分闸命令，分闸线圈受电励磁，换向阀向右移动。由于换向阀的移动，活塞上侧的高压油排出。蓄能器向活塞下侧充入高压油，借助油压作用下，活塞向分闸方向高速移动，进行分闸动作。活塞两侧的压力差使活塞可靠地保持在分闸的位置。

（2）合闸动作。

接到合闸命令，合闸线圈受电励磁，换向阀向左移动。换向阀的移动，使蓄能器的高压油流入活塞的上侧，进行合闸动作。尽管活塞上下两侧都充满了高压油，由于活塞杆上下两侧横截面积差导致的压力差，活塞向下快速移动，完成合闸动作，并可靠地保持在合闸位置（具有固有的失压防慢分功能）。

（四）气动操动机构

气动操动机构是利用压缩空气作为能源产生推力的操动机构。由于以压缩空气作为能源，因此气动操动机构不需要大功率的直流电源，独立的储气罐能供气动机构多次操作。断路器的分合闸全部依靠压缩空气，并依靠压缩空气的推力将断路器维持在分闸或合闸位置。气动操动机构在国内主要用于 220 kV 和 500 kV 的高压断路器。

1. 结构和特点

单一的压缩空气作动力的气动操动机构已淘汰，断路器当前采用的是以压缩空气作动力进行分闸操作，辅以合闸弹簧作为合闸储能元件的气动操动机构。压缩空气靠操动机构自备的压缩机进行储能，分闸过程中通过气缸活塞给合闸弹簧进行储能，同时经过机械传递单元使触头完成分闸操作，并经过锁扣系统使合闸弹簧保持在储能状态。合闸时，锁扣借助磁力脱扣，弹簧释放能量，经过机械传递单元使触头完成合闸操作。

气动操动机构的缺点是体积较大，零部件的加工准确度比电磁操动机构还高，同时需要配备压缩空压装置及压缩空气罐，对空气的气密性要求很高，因此活塞和气缸维护的要求高。加之气动操动机构中能量的传递是压缩空气，操作过程中会发生动作延迟，因而在特高压断路器上较少使用。

2. 动作原理

气动合闸、弹簧分闸的气动机构原理，如图 3-38 所示。合闸时，电磁铁通电，铁心推动启动阀的顶杆把阀口打开，使压缩空气通过保持阀及复归阀而将中间阀打开。压缩空气通过调节螺钉上端的孔口进入工作阀活塞的下面，推动活塞向上运动而使断路器合闸。在合闸过程中，若合闸信号提前撤除，则保持阀的球向上封住阀口而保证合闸过程继续进行到底。合闸完毕时，工作阀的活塞打开通向复归阀的孔口，使复归阀的活塞向右运动，中间阀活塞上部的压缩空气通过复归阀排向大气，中间阀复位，工作阀活塞下的压缩空气通过阀下部的排气口排出，活塞下落复位。

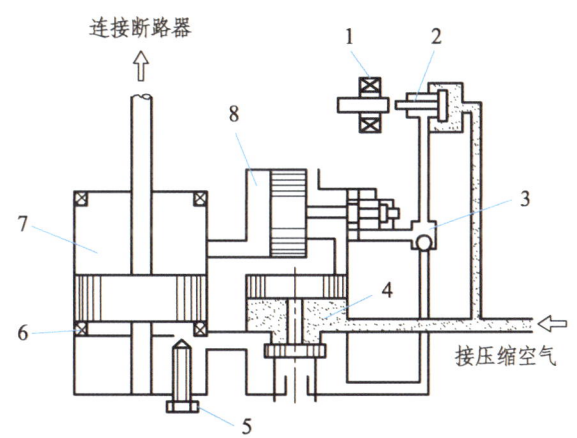

1—电磁铁；2—启动阀；3—保持阀；4—中间阀；5—调节螺钉；
6—缓冲垫；7—工作阀；8—复回阀。

图 3-38 气动机构原理图

3. CQ6 型气动操动机构

CQ6 型气动操动机构是由活塞和气缸组成的驱动机构，还包括控制压缩空气补给的控制阀、由电信号操纵的合闸和分闸电磁铁，以及合闸弹簧、缓冲器、分闸保持掣子、脱扣器等其他零件组成，如图 3-39 所示。

CQ6-I 型气动操动机构配用于 252 kV SF_6 断路器或 GIS，每极断路器均装有一个机构箱，机构箱装有气动操动机构和压缩空气罐，各极断路器的压缩空气罐之间用 22 mm 的铜管连通，以维持压力一致，压缩空气由中间极机构箱内空气压缩机组提供。每台断路器既可进行单相操动，又可进行三相电气联动。

图 3-39　气动操动机构

CQ6-II型气动操动机构配用于 126 kV SF$_6$ 断路器或 GIS，每台断路器在中间相装设一个机构箱。机构箱装有气动操动机构、空气压缩机组和压缩空气罐。断路器三相动作靠机构箱内的传动部件实行三相机械联动。

（1）分闸操作。

断路器在合闸位置时，圆柱阀被凸轮拐臂压在最低端且由掣子 1 锁住，从而将压缩空气封闭在储气罐中。由于装在圆柱阀上的弹簧的作用，掣子 1 受到一个逆时针的力矩，但其同时又被掣子 2 锁住，如图 3-40（a）所示。

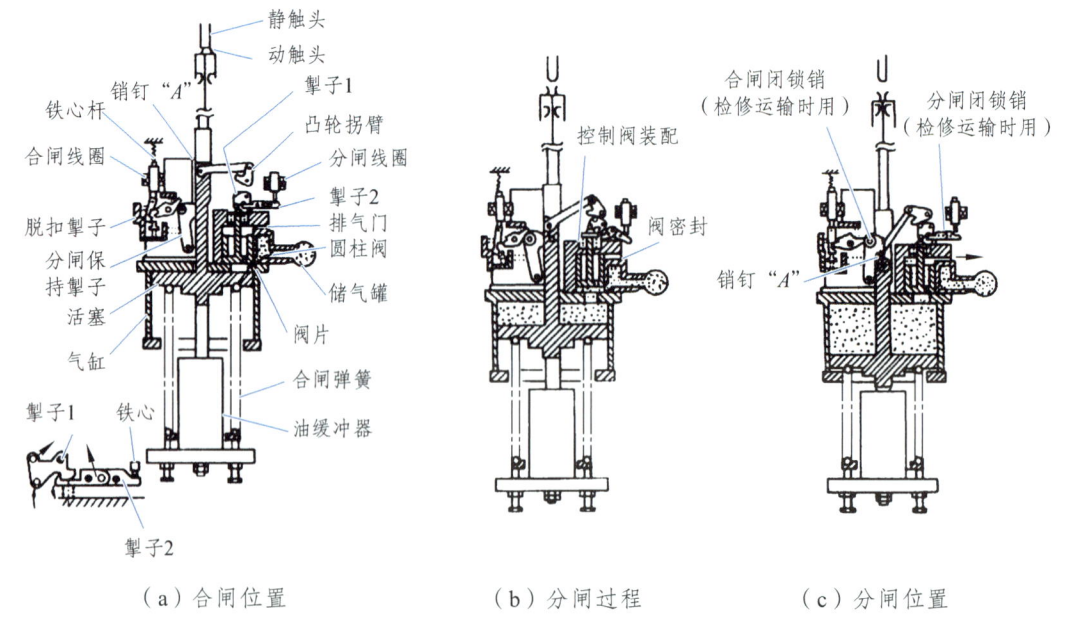

图 3-40　CQ6（AM）操动机构动作原理

分闸操作的动力是由储存在储气罐中的压缩空气供给的。打开圆柱阀，压缩空气进入气缸，气缸内的活塞向下运动，使触头分开。

分闸动作过程如下：

分闸线圈通电，分闸线圈的铁心杆向下运动，撞击掣子2，掣子2是由二个连杆和三个销钉组成，标白点的一个销钉连接两个连杆，标黑点的二个销钉分别将两个连杆固定在机架上。掣子2的右侧的连杆在分闸线圈铁心杆的撞击下将顺时针转动，左侧的连杆则逆时针转动，则掣子1与掣子2间的啮合被释放，如图3-40（b）所示。当掣子1被解脱时，圆柱阀便不受约束从而靠弹簧力打开。储气罐中的压缩空气流入气缸。压缩空气使活塞向下运动，带动触头分闸，如图3-40（c）所示。

在分闸过程的末期，圆柱阀又被与活塞连在一起的凸轮拐臂压回最低端并被掣子1锁住。这样，圆柱阀便返回到合闸位置，气缸中的压缩空气通过排气口排出。最后，销钉A被分闸保持掣子锁住，断路器被保持在分闸状态。

在分闸位置时，断路器是通过连接在机架上的分闸保持掣子在机械上锁住。分闸保持掣子受到由合闸弹簧力产生的逆时针方向的力矩作用，同时，又与脱扣掣子和自身轴销构成"死点"结构产生顺时针方向的力矩，保持断路器的分闸位置。

在分闸过程中，合闸弹簧被压缩储能，为下一次的合闸做好了准备。

（2）合闸操作。

合闸操作的动力是由合闸弹簧供给的。分闸保持掣子解脱后，活塞在合闸弹簧的作用下向上运动，带动触头合闸。

合闸动作过程：合闸线圈通电，合闸线圈的铁心杆向下运动，撞击脱扣掣子。脱扣掣子与分闸保持掣子的啮合被解脱。分闸保持掣子逆时针转动，销钉A从分闸保持掣子的约束中释放。活塞和触头在合闸弹簧力的作用下向上运动完成合闸。

（3）防跳装置。

在CQ6型操动机构上装有一个机械防跳装置。防跳装置的原理如图3-41所示，其动作构成如下：

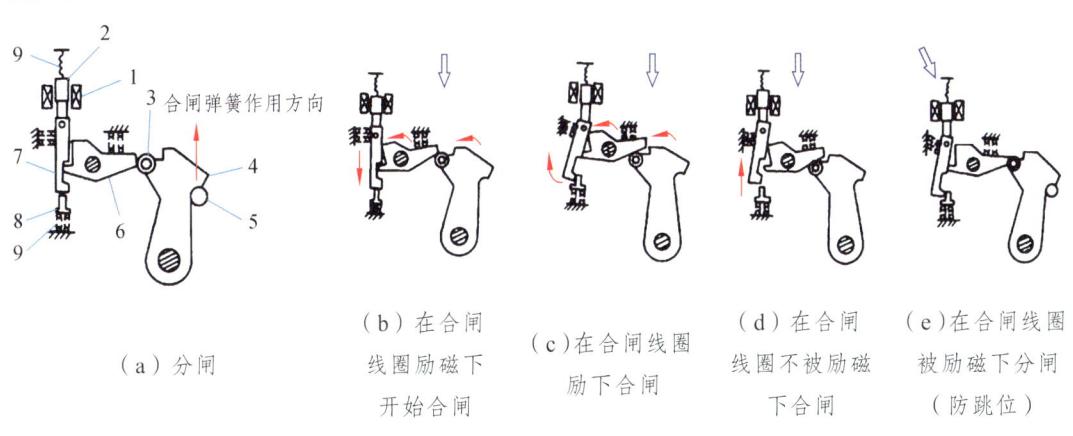

图 3-41 防跳装置

分闸保持掣子锁住销钉A，使断路器保持在分闸位置。销钉A与操作杆连在一起，合闸弹簧的反力作用在其上，这样，销钉A便给分闸保持掣子一个逆时针的转矩，但同时该掣子还被脱扣掣子通过滚轮锁住。

当合闸线圈被合闸信号励磁时，铁心杆带动脱扣杆撞击脱扣掣子，使它逆时针方向转动，解脱了对分闸保持掣子的约束，分闸保持掣子便在合闸弹簧的反力作用下逆时针转动，销钉 A 被解脱，断路器合闸。同时，铁心杆通过脱扣杆压下防跳销钉。

滚轮推动脱扣掣子的回转面，使其进一步逆时针转动。从而，脱扣掣子使脱扣杆顺时针转动，如图 3-41（b）所示，从防跳销钉上滑脱，而防跳销钉使脱扣杆保持倾斜状态，如图 3-41（c）所示。

如果断路器此时得到了意外的分闸信号而开始分闸，销钉 A 便会向下运动，分闸保持掣子在复位弹簧作用下顺时针转动锁住销钉 A。然后，分闸保护掣子本身又被脱扣掣子锁住。

在这一过程中，只要合闸信号一直保持，脱扣杆由于防跳销钉的作用始终是倾斜的，从而铁心杆便不能撞击脱扣掣子，因此，断路器不能重复合闸操作，实现防跳功能。如图 3-41（e）所示。

当合闸信号解除时，合闸电磁铁失磁，铁心杆通过电磁铁内弹簧返回，则铁心杆和脱扣杆均处于图 3-41（a）的状态，为下次合闸作好了准备。

AM 型与 CQ6 型气动操动机构的结构和动作原理完全相同。

（五）新型操动机构

1. 永磁操动机构

永磁操动机构不同于电磁机构和弹簧机构，它是一种崭新的用于真空断路器的操动机构。它利用电磁铁操动，永久磁铁锁扣，电容器储能，电子器件控制，其结构如图 3-42 所示。

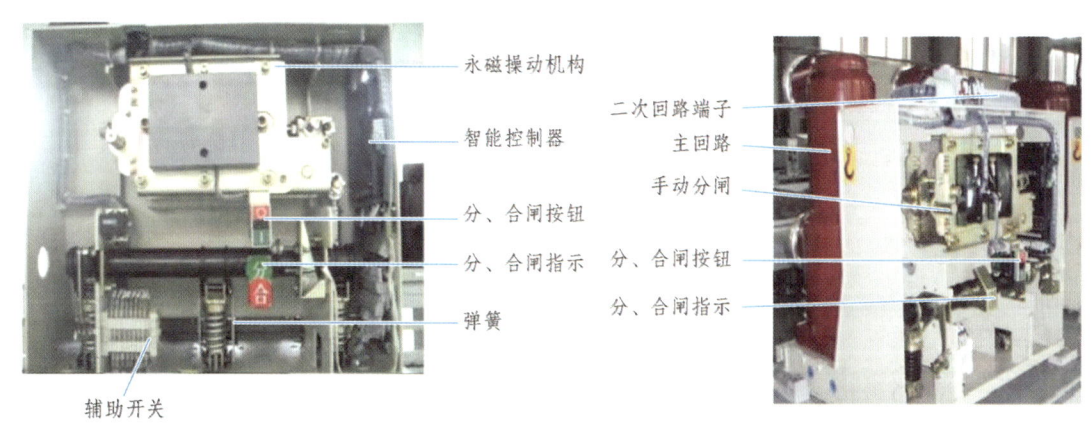

图 3-42　永磁操动机构

在进行分、合闸操作时，储能电容器向分闸或合闸线圈放电，使其受电励磁，从而产生电磁吸力驱动动铁心运动。利用永久磁铁产生保持力，而不需任何机械能，就可将真空断路器保持在分、合闸位置上。电容器提供的巨大脉冲能量，够一次重合闸之用。

永磁机构的控制采用现代的电力电子技术构成电子控制单元。一般采用接近开关来检测开关的分、合闸状态。

永磁机构大致可分三种：双线圈式永磁机构、单线圈式永磁机构及分磁路式永磁机构。

（1）双线圈式永磁机构。双线圈式永磁机构的特点是采用永久磁铁使真空断路器分别保持在分闸和合闸的极限位置上,使用励磁线圈将机构的铁心从分闸位置推动到合闸位置,使用另一励磁线圈将机构的铁心从合闸位置推动到分闸位置。

其特点:① 由于机构在进行合闸时,不需给分闸提供能量,合闸能量较小,合闸线圈线径较细,需要的电源电流小;② 机构在合闸位置时,永久磁铁只需提供克服触头弹簧的力,而不包括分闸弹簧的力。

(2)单线圈式永磁机构。单线圈式永磁机构也是采用永久磁铁使真空断路器分别保持在分闸和合闸的极限位置上,但分闸、合闸共用一个励磁线圈。合闸的能量主要来自励磁线圈,分闸的能量主要来自分闸弹簧。

其特点:① 分闸是靠分闸弹簧和触头弹簧释放的能量,可以通过调整分闸弹簧来调整分闸特性。分闸的弹簧输出特性可与断路器所要求的速度特性一致。② 分、合闸共用一个操作线圈,结构较简单,体积较小,更适合户外封闭式箱体内安装。

(3)分离磁路式永磁机构。分离磁路式永磁机构就是把分闸、合闸和保持磁路分开,使用这种方法能优化磁路,永久磁铁只用于保持合闸位置。

其特点:该类型机构在永久磁铁所产生的合闸保持力方面表现出了优越的性能。由于双工作气隙较短的导磁回路,使该机构能用较少的永磁材料提供较大的合闸保持力,但其结构较复杂,要求加工精度较高,加工和装配难度较大。

以上三种不同型式的永磁机构各有其利弊,应根据不同型式的断路器要求来选取。单线圈式机构的结构较简单,分闸速度可调,但开关合闸时需要给分闸弹簧储能,合闸消耗能量较大,适用于需要合闸功较小的开关,如柱上真空断路器;双线圈式机构则适合于合闸功较大的开关;若开关的合闸保持力较大,可考虑采用分离磁路式永磁机构。

2. 电动机操动机构

电动机操动机构完全不同于液压、气动、弹簧及液压弹簧操动机构,它利用先进的数字技术与简单、可靠、成熟的电动机相结合,不仅满足断路器操动机构的所有核心要求,而且在性能和功能方面具有许多新优势。

电动机操动机构的基本原理是一台用电子器件控制的电动机,去直接操动断路器的操作杆。该电动机操动机构由一些单元组成,主要包括能量缓存单元、充电单元、变换器单元、控制单元、电动机与解算器单元及输入/输出单元。电动机由能量缓存单元经变换器供电,能量缓存单元由充电单元(电源单元)来充电。基于微处理器的控制单元控制速度和监视。电动机操动机构的操动通过输入/输出(I/O)来实现。各单元之间的连接如图 3-43 所示。

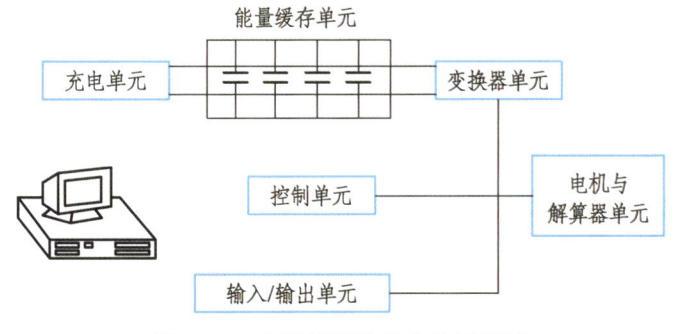

图 3-43　电动机操动机构的示意图

Module 4　Operating Mechanism of High-voltage Circuit Breaker

3.4.1　Overview of Operating System

The operating system of high-voltage circuit breaker includes several parts, including the operating mechanism, transmission mechanism, lifting mechanism, buffer device, and secondary control circuit. The functions of the main parts are as follows.

3.4.1.1　Operating mechanism

The operating mechanism is an important supporting equipment for driving the opening and closing of the circuit breaker, and the working reliability of the circuit breaker largely depends on the action reliability of the operating mechanism. There are many types of operating mechanisms for high-voltage circuit breakers, with significant structural differences. However, the operating mechanism is basically composed of four parts: operating energy system, opening and closing control system, transmission system, and auxiliary devices.

3.4.1.2　Transmission mechanism

1. Functions and composition of transmission mechanism

The transmission system is the link between the working components of the operating mechanism and the moving contact terminals. The operating mechanism and body of the high-voltage circuit breaker transmit energy and complete the motion through the transmission system during the opening and closing processes, and complete the opening and closing operations according to the designed performance requirements. The transmission mechanism of the high-voltage circuit breaker mainly consists the transmission components in the operating mechanism, the lifting mechanism in the circuit breaker, and the transmission mechanism between them.

The transmission component in the operating mechanism is composed of the link gear or hydraulic or pneumatic transmission mechanism, and is connected to the lifting rod of the circuit breaker through the transmission mechanism.

The transmission mechanism is the intermediate link connecting the operating mechanism and the lifting mechanism, which plays a role in changing the movement direction, increasing the stroke, and transmitting energy to the circuit breaker. Due to the fact that the lifting mechanism and the operating mechanism are always separated by a certain distance, and their movement directions are also inconsistent, a transmission mechanism is required, which is generally composed of the link gear. There are two main types of transmission mechanisms: guide rail type and parallel motion type. Due to the limitation of the guide rod directly connected with moving and static terminals by the guide rail and the force perpendicular to the movement direction, it is prone to deformation when subjected to twisting torque, thus affecting the reliability of the action of the transmission mechanism of guide rail type. This transmission mechanism is simple in structure, with cheap price. It is common in early circuit breakers (such as SN1 type low oil circuit breaker). The transmission mechanism of parallel motion type allows the guide rod of the moving contact terminal to move correctly in a straight direction, with good working performance. It is widely used in the circuit breaker.

The lifting mechanism is a mechanism that drives the moving contact terminals of a circuit breaker to move along a certain trajectory. Usually, it converts the rotary or linear motion transmitted by the operating mechanism into the linear motion of the moving contact terminal, causing the circuit breaker to open and close. So it is also called the straightening mechanism.

The interrelationship between the above three is as shown in Fig. 3-17.

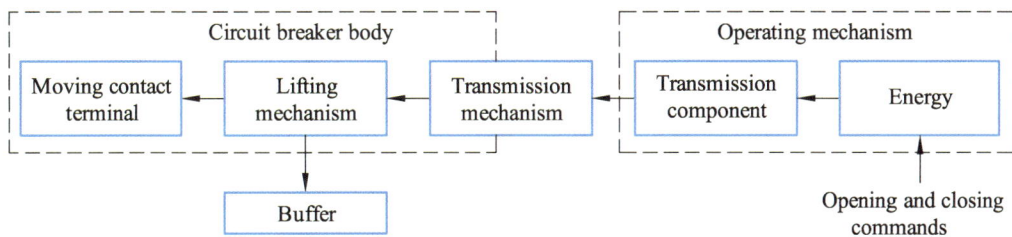

Fig. 3-17 Interrelationship Diagram of Transmission System

2. Types of transmission mechanism

There are many forms of transmission systems, which can be roughly divided into the following types:

(1) Mechanical transmission. Commonly used transmission methods include lever, link gear, cam, and gear. Among them, the link gear is the most widely used. Advantages are reliable transmission, good synchronization, simple processing, convenient adjustment, and easy maintenance. While disadvantages are low speed and large impact force when transmitting large power.

(2) Compressed air transmission. It is generally used in the high-voltage air circuit breaker and pneumatic mechanisms, with the advantages of fast response and quick action. While disadvantages are the action time increasement as the pipeline growth, more complex structure, and high processing and maintenance requirements.

(3) Hydraulic transmission It is mostly used in the hydraulic operating mechanism, with advantages of smooth operation, large power transmission, fast speed, and convenient adjustment. While disadvantages are complex structure, great processing difficulty, and the influence of temperature on transmission speed.

(4) Hybrid pneumatic and mechanical transmission. It is mostly used for the compressed-air circuit breaker and the low oil circuit breaker. For this transmission method, a lever is used to replace some pipes and components. Advantages are good synchronization and fast transmission, while disadvantages are complex structure and high maintenance requirements.

(5) Hybrid hydraulic and mechanical transmission. It is mostly used for the low oil circuit breaker and the SF_6 circuit breaker. For this transmission method, a lever is also used to replace

some pipes and components. Advantages are fast action speed and simpler manufacturing compared to hydraulic mechanisms, while disadvantages are complex structure and big impact force.

3. Link gear

The link gear plays an important role in the transmission system of the high-voltage circuit breaker, and most of the various transmission mechanisms are composed through the combination of link gears. Link gears of high-voltage circuit breakers and operating mechanisms are relatively complex. However, any complex connecting rod can be decomposed into several four-bar linkages, and there is also a five-bar linkage in the operating mechanism with a trip-free mechanism. There are the following common types of link gears.

(1) Four-bar linkage.

The four-bar linkage consists of three movable connecting rods and one fixed connecting rod, as shown in Fig. 3-18. O_1 and O_2 are fixed pins, A and B are movable pins, connecting rods AB, AO_1, and BO_2 are movable connecting rods that can swing or rotate back and forth, and O_1O_2 can be regarded as a fixed connecting rod. The connecting rods AO_1 and BO_2 are often referred to as crank arms (abbreviated as arms), and the connecting rod AB is abbreviated as the rod. If AO_1 is the master arm, BO_2 is the slave arm. The direction of torque M generated by the operating force applied to the master arm is consistent with the direction of rotation of the master arm, while the direction of the torque generated by the slave arm is opposite to the direction of rotation of the slave arm. With the drawing method, their motion trajectories and characteristics can be obtained. Changing the relative dimensions of each connecting rod in a four-bar linkage can result in different types of mechanisms.

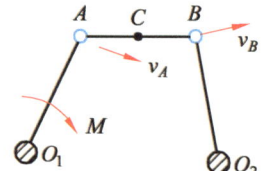

Fig. 3-18　Four-bar Linkage

(2) Slider rocker mechanism.

The slider rocker mechanism is a deformation of the four-bar linkage, commonly used as a straightening mechanism. As shown in Fig. 3-19, O is a fixed pin with no slave arm. But this pin has a guide device. When the arm OA swings around O, the pin B and the slider slide in a straight line in the guide rail.

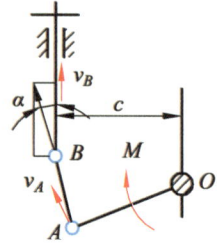

Fig. 3-19　Slider Rocker Mechanism

(3) Accurate elliptical mechanism.

As shown in Fig. 3-20, O is the fixed axis and $AB=AC=AO$. OAC is equivalent to a slider rocker mechanism, and point C moves in a straight line within the guide rail. BC is the connecting rod, and end B is limited to sliding within the linear guide rail. When slider C moves in the guide rail, point A is pushed to rotate around O, and then point B moves in a straight line through axis O. Any point on the BC rod, except for points A, B, and C, undergoes the elliptical motion, hence it is called the accurate elliptical linear mechanism. If the conductive rod or insulation lifting rod of the circuit breaker is connected to point C, the moving contact terminal (point B) will move in a straight line when opening and closing.

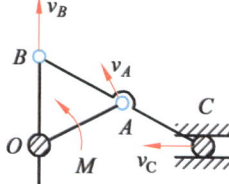

Fig. 3-20　Accurate Elliptical Mechanism

(4) Approximate elliptical mechanism.

As shown in Fig. 3-21, it is a change from Fig. 3-20, that is, the point C in the guide rail in Fig. 3-20 is changed to the endpoint of the rocker swinging around O_2. At this point, if the swing of the rocker O_2C is not significant, the trajectory of point C is approximately a straight line, the trajectory of point B also becomes approximately a straight line, and motion trajectories of other points on the BC bar except for point A becomes an approximately ellipse.

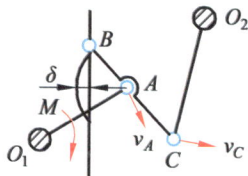

Fig. 3-21　Approximate Elliptical Mechanism

(5) Five-bar linkage.

The five-bar linkage is as shown in Fig. 3-22, which has two crank arms and two connecting rods. Its characteristic is that there is no definite motion characteristic between the master arm and the slave arm. When the master arm rotates at a certain angle, the angle that the slave arm rotates can be large or small. The five-bar linkage cannot be used as the transmission mechanism and can be used to achieve the trip-free during operation.

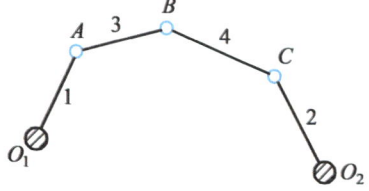

Fig. 3-22　Five-bar Linkage

(6) Tripping mechanism of connecting rod type.

There are three commonly used tripping types: folding rod type, lock hook type, and roller lock latch type.

① Tripping mechanism of folding rod type As shown in Fig. 3-23, the tripping mechanism of folding rod type consists of connecting rods 4, 5, and 6. When the operating mechanism receives the opening signal, the opening electromagnetic iron is energized. After the point C is pushed upwards by electromagnetic force F_2 to break away from the dead zone, the circuit breaker automatically opens under the action of the opening spring.

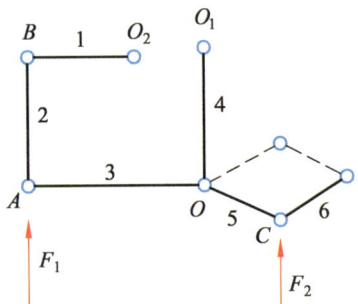

Fig. 3-23　Tripping Mechanism of Folding Rod Type

② Tripping mechanism of lock hook type The schematic diagram of the tripping mechanism of lock hook type is as shown in Fig. 3-24. When the circuit breaker is in the closed position, the torque M generated by the opening force acts on the connecting rod. Due to the obstruction of the lock hook, the connecting rod cannot be pushed by the torque M, thus maintaining the circuit breaker in the closed position. After the circuit breaker receives the opening signal, the opening electromagnetic iron is energized. The lock hook is pushed counterclockwise by electromagnetic force F. When the locking hook is lifted a certain distance, the connecting rod causes the circuit breaker to open under the action of torque M.

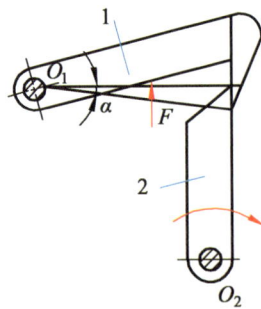

1-Hood; 2-Rod.

Fig. 3-24　Tripping Mechanism of Lock Hook Type

③ Roller lock latch type tripping mechanism The schematic diagram of the tripping mechanism of roller lock latch type is as shown in Fig. 3-25. When the circuit breaker is in the closed position, roller is locked by the latch. Although there is an opening torque M on the connecting rod, such torque cannot cause the connecting rod to rotate, thus keeping the circuit breaker in the closed position. When the operating mechanism receives the opening signal, the opening electromagnetic iron is energized, and the lock latch is pushed by electromagnetic force F to rotate clockwise.

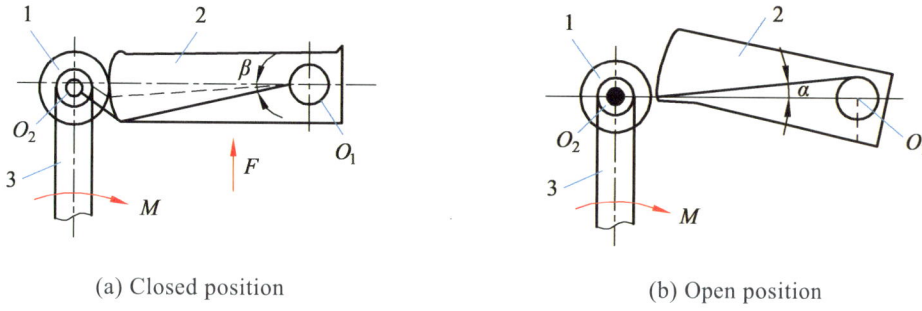

(a) Closed position   (b) Open position

1-Roller; 2-Lock latch; 3-Connecting rod.

Fig. 3-25   Roller Lock Latch Type Tripping Mechanism

To keep the roller from being locked by the latch, the latch must be turned not only through angle β, but also through an angle α. After point A is turned out of the dead zone, the roller can be turned in a clockwise direction, and the circuit breaker opens.

3.4.1.3 Buffer device

1. Functions of buffer device

The very high speed of the moving parts of the circuit breaker during operation makes the moving parts have a lot of kinetic energy when the movement is about to end. To make the moving parts stop within a short stroke, it is necessary to install opening and closing buffers. So the kinetic energy at the end of the action process can be controlled to be released and converted into other forms of energy to ensure that the impact force that endangers the normal operation of the equipment is absorbed during the braking process, reduce impact, and avoid component deformation and damage. The buffer can sometimes be used to change the speed characteristics during the action process. The buffer is generally installed next to the lifting mechanism.

2. Types of buffer device

There are four types of commonly used buffer devices, namely oil buffer, spring buffer, gas buffer, and rubber buffer.

(1) Oil buffer.

The oil buffer is generally used as an opening buffer, with its principle structure is as shown in Fig. 3-26. It is composed of oil cylinder, piston, impact rod, return spring, and end cover. There is a small gap between the piston and the cylinder wall. When the high-speed moving component hits the impact rod on the buffer piston, the piston and the moving component move downwards together. Due to the fact that the oil under the piston can only flow up to the top of the piston through a very small gap, the oil flow is blocked, creating the pressure on the bottom of the piston, hindering the piston downward movement, and forming a buffer for the moving component.

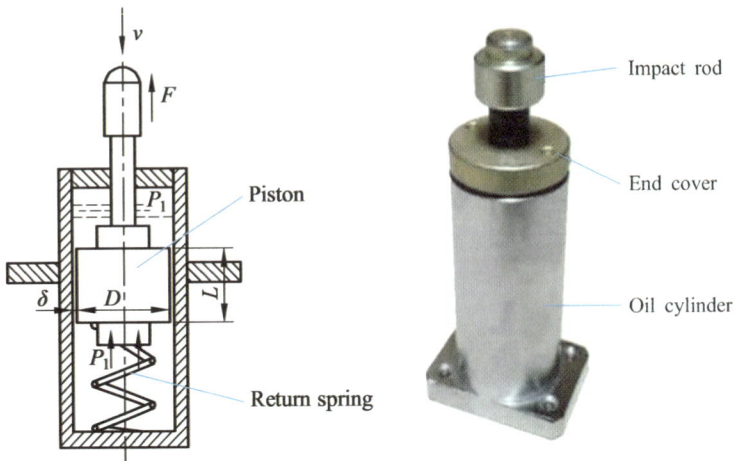

Fig. 3-26   Oil Buffer

Because the friction force of the transmission system at the end of the opening stroke is very small, using of other buffers can easily cause rebound and jumping, or the buffering characteristics are difficult to meet the requirements, generally oil buffers are mostly used to absorb the kinetic energy at the end of the opening stroke.

(2) Spring buffer.

With moving parts to impact and compress the spring to absorb the kinetic energy of the moving parts, the spring buffer generates buffering. The braking force is proportional to the compression stroke of the spring. When strong buffering is required, the elastic force must be increased. At the end of the movement, the strong spring will cause the rebound of moving parts and cause vibration. The spring buffer consists of a spring, guide rod, base, and impact rod, as shown in Fig. 3-27. When the moving component collides with the spring buffer, the impact rod moves upwards, and the spring is compressed, converting a portion of the kinetic energy of the moving component into the potential energy of the spring.

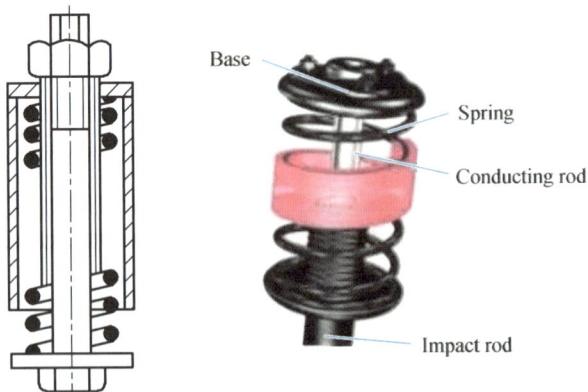

Fig. 3-27   Spring Buffer

The spring buffer is simple in structure, and easy to use. Its characteristics are not affected by temperature. However, its disadvantage is that it has a large recoil force. The spring buffer is often used as a closing buffer because the frictional resistance of the contact terminal at the end of closing is high, making it difficult to generate the rebound vibration. In addition, the compressed spring can release energy during opening and increase the rigid opening speed of the moving contact terminal.

(3) Gas buffer.

The principle of the gas buffer is similar to that of the oil buffer, except that the movement of the buffer piston compresses the gas medium to create the buffer. However, the rebound force generated by the gas buffer during the buffering process is relatively large. Gas buffers are commonly used in the compressed air circuit breaker and the SF_6 circuit breaker.

(4) Rubber buffer.

After being impacted by the moving component, the kinetic energy of the rubber buffer is consumed on the compressed rubber, creating buffering. The advantage of rubber buffer is its simple structure and low recoil force. The disadvantage is that the rubber elasticity deteriorates at low temperatures and the rubber is not resistant to oil. The rubber buffer is generally used in areas with low buffering energy.

3.4.2 Requirements for Operating System

The entire mission of a circuit breaker is ultimately reflected in the opening and closing actions of contact terminals, which are achieved through the operating mechanism. Therefore, the performance of the operating mechanism has a crucial impact on the working performance and reliability of high-voltage circuit breakers. The operating performance of the operating mechanism must meet the requirements for the reliability of the circuit breaker.

1. Sufficient closing power

Only an operating mechanism with sufficient closing power can ensure reliable closing of the circuit breaker according to the required closing speed under various specified conditions, and maintain it in the closed position without causing false opening.

When the power grid in the normal working state, the operating mechanism is used to close the circuit breaker with the working current, which is relatively easy to close. However, in case of a power grid accident, if the circuit breaker closes a circuit with a pre voltage short circuit fault, the short circuit current can reach tens of thousands of amperes or more, and the conductive circuit of the circuit breaker receives an electrodynamic force of several thousand newtons. From the arrangement of the conductive circuit of the circuit breaker and the structure of the contact terminals, the direction of the electrodynamic force often hinders the closing of the circuit breaker. Therefore, when closing a circuit with a pre voltage short circuit fault, the circuit breaker may not be able to quickly and reliably close the contact terminals due to excessive electrodynamic force, resulting in severe burns to the contact terminals. Therefore, the operating mechanism must have sufficient closing power to obtain the ability to close shorts faults.

2. Keeping the circuit breaker in the closed position

During the closing process, the duration of the closing command is short, and the operating force of the operating mechanism can only be improved in a short period of time. Therefore, the operating mechanism must have a mechanism to maintain the closed status after completing the closing operation of the circuit breaker, to ensure that the circuit breaker can still be kept in the closed position after the closing command and operating force disappear.

3. Reliable opening speed and sufficient closing speed

Due to the short circuit current and the electrodynamic force direction presented by the contact terminal structure in the conductive circuit of the circuit breaker, usually the opening power of the operating mechanism on the circuit breaker can be met, which plays an accelerating role in the opening of the circuit breaker. However, for the opening function of the operating mechanism on the circuit breaker, it is required to not only electrically open the circuit breaker, but also to manually open it in certain special circumstances. Moreover, the opening speed of the circuit breaker is required to be independent of the speed of the operator's actions and the length of time required to give commands.

4. Trip-free device and anti-jump measures

The function of the trip-free device in the operating mechanism is that during the closing process of the circuit breaker, if the operating mechanism receives the opening command again, the operating mechanism should not continue to execute the closing command, but should immediately open and maintain the open position.

When the circuit breaker is closed in a circuit with a pre voltage short circuit fault, the circuit breaker should automatically open. If the closing command is not yet removed, the short-circuit closing occurs again after the opening of the circuit breaker. Immediately after that, there will be a short-circuit opening again. This may cause the circuit breaker to continuously open and close the short-circuit current for several times, which is called "jumping". When there is a jump, the circuit breaker will continuously close and open the short-circuit current for many times, which will cause serious burns on the contact terminals and even cause an explosion. Therefore, the operating mechanism must have the ability to prevent jumping, so that the circuit breaker will not close again after the circuit breaker closes the short-circuit current and automatically opens, even if the closing command has not been removed. Mechanical methods can be used to prevent jumping, and many operating mechanisms are equipped with automatic tripping devices to prevent jumping. In addition, there are "electrical anti-jump" measures, that is, the installation of anti-tripping relays in the control circuits for the opening and closing operations of the circuit breakers to prevent the occurrence of jumping.

5. Reset and locking functions

After the circuit breaker is opened, each component in the operating mechanism should be able to return to the position where it is ready to close automatically or through simple operation, to ensure the reliable operation of the operating mechanism.

At the same time, it is required that the operating mechanism should have the following locking devices:

(1) Interlocking of open and closed positions. It should be ensured that the operating mechanism cannot perform the closing operation when the circuit breaker is in the closed position. The operating mechanism cannot perform the opening operation when the circuit breaker is in the open position.

(2) Interlocking of low gas (liquid) pressure with high gas (liquid) pressure. For pneumatic or hydraulic operating mechanisms, when the gas or liquid pressure is below or above the rated value, the operating mechanism cannot perform opening or closing operations.

(3) Position interlocking in the spring operating mechanism. The operating mechanism cannot perform opening and closing operations when the spring energy storage does not meet the specified requirements,

3.4.3 Type of Operating Mechanism

According to the different forms of energy provided, operating mechanisms can be divided into manual operating mechanisms (CS), electromagnetic operating mechanisms (CD), spring operating mechanisms (CT), pneumatic operating mechanisms (CQ), hydraulic operating mechanisms (CY), and new type operating mechanisms. Among them, manual and electromagnetic operating mechanisms are straight line mechanisms, while spring, pneumatic, and hydraulic operating mechanisms are energy storage mechanisms.

There are three types of operating mechanisms used in high-voltage and even ultra-high voltage SF_6 circuit breakers: hydraulic, pneumatic, and spring operating mechanisms. For vacuum circuit breakers of medium voltage, the operating mechanisms mainly used are electromagnetic operating mechanisms and spring operating mechanisms. In addition, for light circuit breakers with a voltage of 10 kV and a breaking current of 6kA or less, manual operating mechanisms are often retained, with manual closing and energy stored opening springs used for opening.

3.4.3.1 Electromagnetic operating mechanism

The electromagnetic operating mechanism directly relies on electromagnetic force to close, and can perform remote control and reclosing. Its advantages are simple structure, fewer parts, reliable operation, and low manufacturing cost. However, its disadvantages are too large power consumed by the closing coil, hundreds of amperes of the closing current, and a certain impact on the secondary operating power supply. It is required to use a 220/110 V high-capacity DC power supply, resulting in high investment in auxiliary facilities and high maintenance costs. The mechanism itself is heavy, and there is often a certain delay when closing due to the influence of electromagnetic time constant, so its use in vacuum circuit breakers has gradually decreased.

The electromagnetic operating mechanism mainly consists of several parts, including the working element (electromagnetic system), connecting plate system, closing maintenance, tripping device, and buffer system. In the following section, the structure and operating principle of the electromagnetic operating mechanism will be introduced with the CD17 electromagnetic operating mechanism as an example.

1. Structure and characteristics

The CD17 electromagnetic operating mechanism is the operating mechanism designed specifically for the ZN28-12 vacuum circuit breaker. It is a planar five-bar linkage with half shaft tripping, with trip-free function and low tripping power. The left side of the mechanism is equipped with the auxiliary switch, the right side is equipped with the opening electromagnet, and the lower part of the mechanism is a closing electromagnet, as shown in Fig. 3-28. In order to prevent the adhesion when the core is closed, a brass pad and compression spring are added to the closing core to ensure that the core quickly falls when the closing is completed. The copper bush is installed between the winding and the core to prevent the winding from being worn out during the movement of the core.

Fig. 3-28　CD17 Electromagnetic Operating Mechanism

The lower part of the closing electromagnet is composed of the cast iron seat and the adjustable buffer pad. The seat is equipped with the closing handle for manual closing during the maintenance. The adjustable buffer pad not only serves as a buffer for the core, but also can be used to adjust the gap between the top rod of the core and the wheel to adjust the closing speed.

2. Action principle

(1) Closing operation. Before closing, the buckle plate is in a reset state, as shown in Fig. 3-29 (a). After the closing coil is energized, the upper buckle plate of the core and the half shaft are locked. The top wheel is pushed upwards by the core, and the output shaft is driven to rotate by about 38 ° through the connecting rod. The circuit breaker is closed through the transmission rod outside the mechanism, as shown in Fig. 3-29 (b). At this point, the opening spring of the circuit breaker stores energy and the spring of the contact terminal compresses. When the core reaches the end point, there is a gap of 2±0.5 mm between ring and latch, as shown in Fig. 3-29 (c). At this point, due to the rotation of the spindle, the auxiliary switch is driven to open the normally closed contact of the closing circuit. As the power supply of the closing coil is cut off, the core falls and the ring is supported by the latch, the closing action is completed, as shown in Fig. 3-29 (d).

(2) Opening operation. When the opening coil is powered on or manually opened, the half shaft rotates in a clockwise direction, and the buckle plate releases from the half shaft, causing the

ring to leave the latch. The linkage system becomes an uncertain five-bar linkage. Under the joint action of the opening spring and contact terminal spring, the output shaft rotates counterclockwise to complete the opening. At the same time, the auxiliary switch is driven to open the normally closed contact of the opening circuit, cutting off the power supply of the opening coil, as shown in Fig. 3-29 (e).

(3) Trip-free action. During the closing process, the closing core moves upwards against wheel. Once the opening command is received, the opening core is activated to release the half shaft from the buckle plate. Under the joint action of the opening spring and contact terminal spring, wheel slides down from the end of the top rod of closing core to achieve the trip-free, as shown in Fig. 3-29 (f).

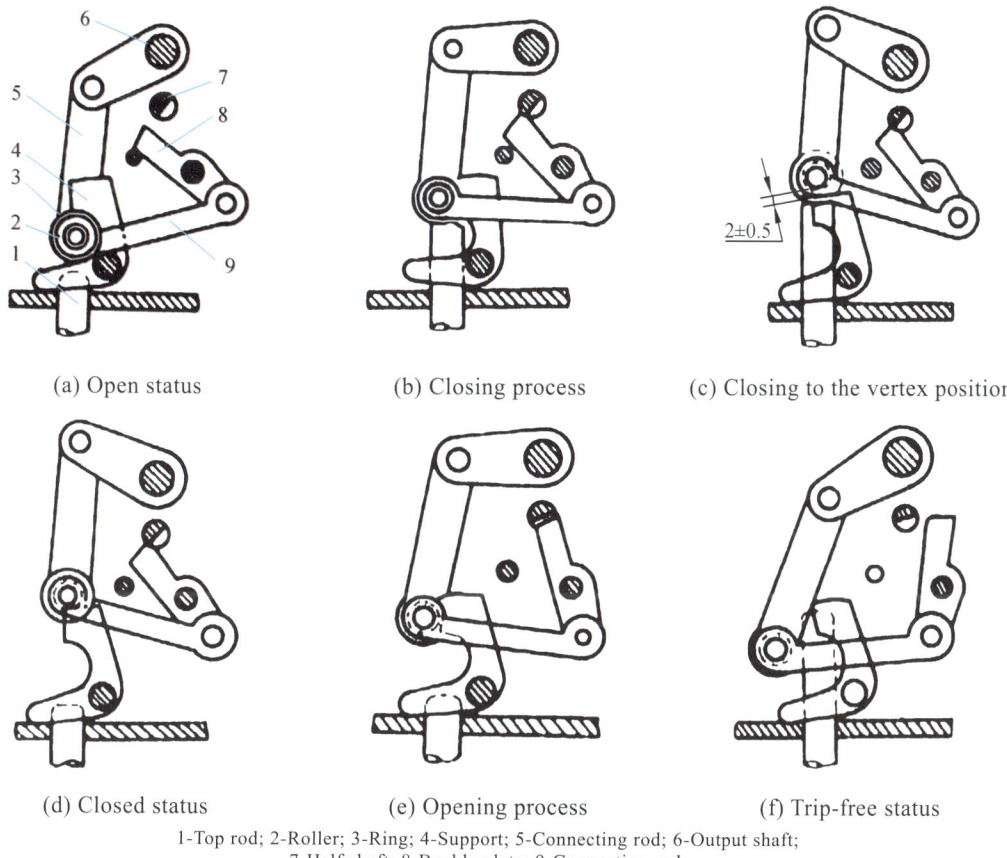

(a) Open status (b) Closing process (c) Closing to the vertex position

(d) Closed status (e) Opening process (f) Trip-free status

1-Top rod; 2-Roller; 3-Ring; 4-Support; 5-Connecting rod; 6-Output shaft;
7-Half shaft; 8-Buckle plate; 9-Connecting rod.

Fig. 3-29 Action Status Diagram of CD17 Operating Mechanism

3.4.3.2 Spring operating mechanism

The spring operating mechanism is a mechanical operating mechanism with springs used as energy storage components. The energy storage of the spring is achieved with the electric motor through the deceleration device, and is maintained in the energy storage state through the latch system. The opening and closing operations are achieved with two spiral compression springs. At the time of closing, the latch tripping is realized with the help of magnetic force. Part of the energy released by the closing spring is used for closing, while the other part is used to store energy for the opening spring. As soon as the closing spring is released, the energy storage motor immediately stores energy for it, with a storage time of no more than 15 seconds, thus achieving fast automatic reclosing of the circuit breaker. During the operation, the opening and closing springs are in a compressed state.

1. Structure and characteristics

The spring mechanism mainly consists of the closing energy storage part, the transmission part, and the control part. The closing energy storage part includes the electric motor, the reduction device, the closing spring, the energy storage device, and the holding release device. There are three main types of closing energy storage springs for spring mechanisms, namely compression springs (also known as spiral springs), tension springs (also known as spiral coil springs), and torsional springs (also known as butterfly springs). According to the different energy sources used for energy storage by the closing spring, spring mechanisms can be divided into spring mechanisms for electric motor energy storage and spring mechanisms for manual energy storage. The closing and opening control part mainly consists of the release, which is a tripping mechanism. The structure of the CT19 spring operating mechanism is as shown in Fig. 3-30.

Fig. 3-30 CT19 spring operating mechanism

A specialized operating point source is not required for the spring operating mechanism, and the energy storage motor has low power and can be used for both AC and DC currents. At the same time, the energy storage of the closing spring can also be manually completed, making the operating mechanism convenient to use. However, the structure of the spring operating mechanism is relatively complex, with a large number of parts and high costs. There is a probability of failure in the transmission link.

2. Action principle

The action principle of the spring operating mechanism is as shown in Fig. 3-31.

In the Fig. 3-31 (a), the spring mechanism is in the closed position and energy is stored for both the opening spring and closing spring. The crank arms are subjected to the counterclockwise torque from the opening spring, which is blocked by the closing holding latch and opening latch.

In case of the circuit breaker opening, the coil of the opening electromagnet is energized after receiving the opening signal. The core acts and impacts the opening latch. The opening latch rotates clockwise to release the closing holding latch. The closing holding latch rotates clockwise to release the pin A. The crank arms rotate counterclockwise under the thrust of the opening spring. The crank arm quickly separates the moving and static contact terminals through transmission components such as connected horizontal tie rods and operating rods, and the circuit breaker opens.

When the spring mechanism is in the open position and the closing spring has stored energy, the ratchet shaft bears the counterclockwise torque of the closing spring connected to the ratchet. This torque is pinned by the energy storage holding latch and the closing latch.

In case of the circuit breaker closing, the coil of the closing electromagnet is energized after receiving the closing signal. The latch acts and impacts the closing latch. The closing latch rotates clockwise to release the energy storage holding latch. The energy storage holding latch rotates clockwise to release the pin B. Under the action of the closing spring, the ratchet rotates counterclockwise and drives the ratchet shaft to rotate, causing cam to push the crank arm to rotate clockwise, and driving the crank arm on the crank arm shaft to rotate clockwise, while compressing the store energy of the opening spring. The horizontal tie rod and operating rod connected to the crank arm quickly close the moving contact terminal.

After the closing operation is completed, the mechanism status is as shown in Fig. 3-31 (c), and the pin A is once again locked by the closing holding latch.

After the closing operation of the mechanism is completed, the closing spring is in a released state, and the pawl shaft is connected to the motor through gears. After the circuit breaker is closed in place, store energy is conducted for the closing spring.

The energy storage process of the closing process is as follows:

The electric motor starts, causing the pawl shaft to rotate. The two pawls on the eccentric pawl shaft engage alternately with the upper teeth of the pawl wheel during the transmission of the pawl shaft, causing the pawl wheel to rotate. The ratchet rotates counterclockwise, driving the tie rod to store energy in the closing spring. After passing through the dead point, the ratchet shaft is given a counterclockwise rotational torque by the closing spring. This torque is locked by the energy storage holding latch through the pin B.

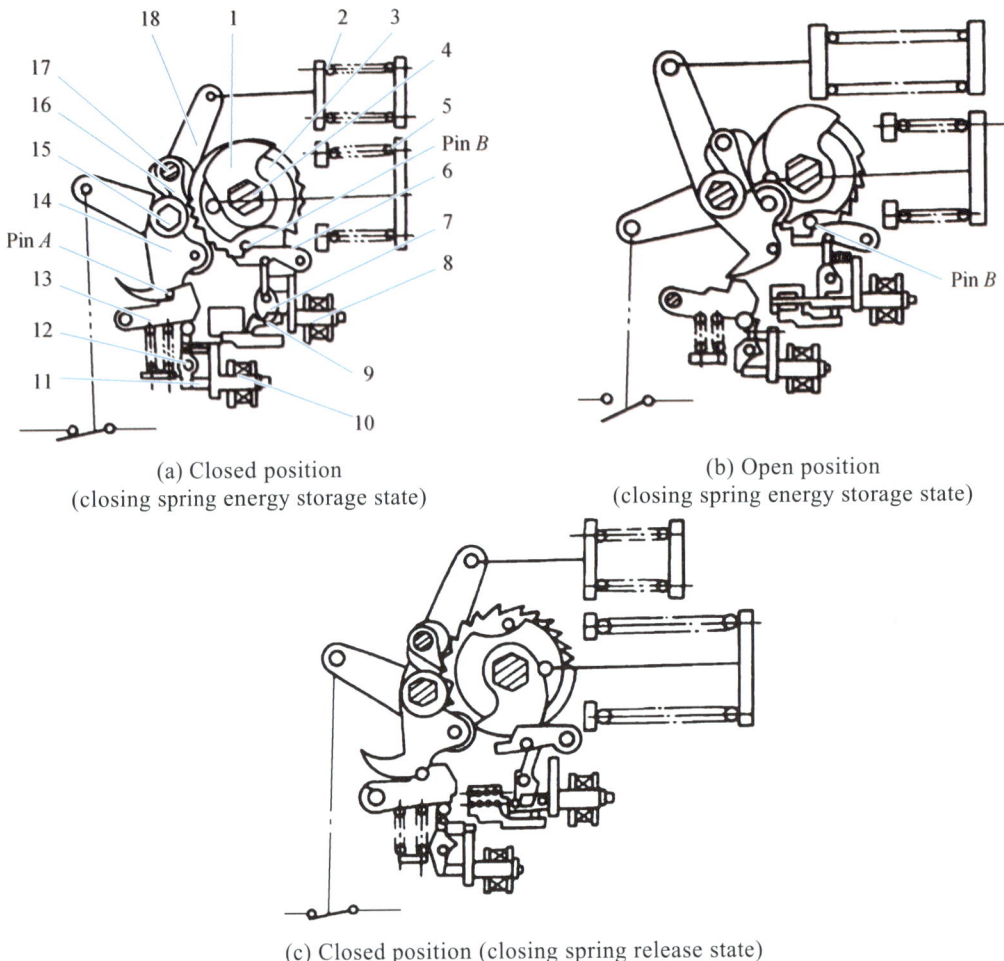

(c) Closed position (closing spring release state)

1-Cam; 2-Opening spring; 3-Ratchet; 4-Ratchet shaft; 5-Closing spring; 6-Energy storage holding latch; 7-Closing latch; 8-Closing electromagnet; 9-Latch; 10-Opening electromagnet; 11-Core; 12-Opening latch; 13-Closing holding latch; 14-Crank arm; 15-Crank arm shaft; 16-Pawl; 17-Pawl shaft; 18-Crank arm.

Fig. 3-31　Working Principle Diagram of Spring Operating Mechanism

3.4.3.3　Hydraulic operating mechanism

The hydraulic operating mechanism uses high-pressure oil stored in the energy storage device as the driving energy transmission medium. The nitrogen gas is mainly used for the energy maintenance in the energy storage device. By utilizing the pre stored energy in the energy storage device according to the differential principle, the operating piston is indirectly pushed to achieve the opening and closing operations of the circuit breaker.

1. Structure and characteristics

The main components of the hydraulic operating mechanism include energy storage component, control component, operating (executing) component, auxiliary component, and electrical component. Its structure is as shown in Fig. 3-32.

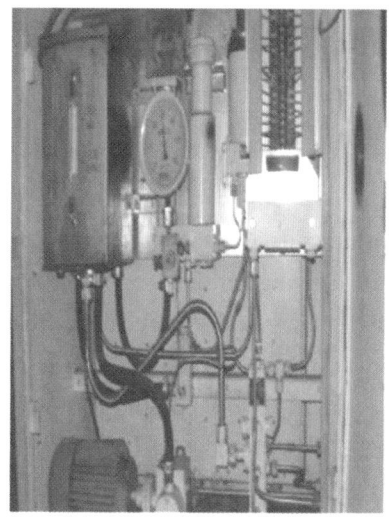

Fig. 3-32 Hydraulic Operating Mechanism

(1) Energy storage component Parts such as pressure accumulators, oil filters, and oil pumps are included. When the electric motor drives the oil pump, the oil is extracted from the oil tank through the oil filter and compressed into the pressure accumulator, compressing N2 to store energy. At the time of operating, the gas expands and does external work, which is transmitted to the working cylinder through hydraulic oil. The mechanical energy is transformed to achieve circuit breaker opening and closing operations.

(2) Control component As the intermediate connection between the energy storage component and the operating component, the control component emits hydraulic pulse signals for opening and closing actions to control the operating component. Control component include opening and closing electromagnets, opening and closing starting valves, and secondary valves.

(3) Operating components. Parts such as working cylinder, pressure switch, safety valve, and oil drain valve. The working cylinder is connected to the circuit breaker body through the connecting piece, and is controlled by the control component to drive the circuit breaker to achieve opening and closing actions. The pressure switch is used to control the starting and stopping of the electric oil pump, as well as opening and closing locking. The safety valve is used to release excessive pressure caused by fault conditions to avoid damage to hydraulic components. The oil drain valve is used to release oil pressure during debugging and maintenance, and is divided into high-pressure oil drain valve and low-pressure oil drain valve.

(4) Auxiliary component. Parts such as signal cylinder, fuel tank, exhaust valves, pressure detector, and auxiliary oil chamber are included. The signal cylinder is used to drive auxiliary switches to switch electrical control circuits, and some cylinders also drive opening and closing indicators and counters. The fuel tank is the oil storage container that is usually open to the atmosphere. During the operation, the discharge of oil from the working cylinder will cause an instantaneous increase in its internal pressure. The exhaust valve is used to exhaust the gas in the working cylinder and pipeline before the hydraulic system pressure is established, in order to avoid

affecting the action time and speed characteristics. The pressure detector is used to measure the pressure value of the hydraulic system. The auxiliary oil chamber is used to fully utilize hydraulic energy, reduce the resistance of the working cylinder when opening and discharging oil, and improve the opening speed.

(5) Electronic component. Parts such as opening and closing coils, heaters, and microswitches are included. The opening and closing coils are respectively used to operate the opening and closing solenoid valves (primary valves). The heater is used to maintain the temperature inside the mechanism box during low external temperatures, prevent oil from freezing and dissipate moisture inside the box. The heater is divided into manual and automatic modes. The microswitch serves as the opening and closing locking contacts, as well as the starting and stopping contacts for the oil pump. Meanwhile the microswitch also converts signals to the main control room for monitoring purposes.

The main advantages of hydraulic operating mechanisms are smooth operation, no noise, and low energy required for control. In case of a small mechanism size, powerful operating force can be obtained. The hydraulic component is lightweight and responds fast, making it easy to achieve automatic control and various protections. When the power supply is temporarily lost, the mechanism can still be operated multiple times, with high action reliability and convenient maintenance. It is particularly suitable for high-voltage, ultra-high voltage, and ultra-high voltage circuit breakers of 126 kV and above. Its main disadvantage is the high requirements for processing technology. If the manufacturing or assembly is poor, it is easy to leak oil, and the speed characteristics are easily affected by the environmental temperature.

2. Action principle

For the hydraulic operating mechanism, the differential principle is adopted. High pressure oil at the same working pressure is used to test the difference in force on different cross-sections on both sides of the piston, thereby causing the piston to move to drive the circuit breaker for opening and closing operations. The working cylinder piston and secondary valve core are designed according to the differential pressure principle. Generally, high-pressure oil is filled on the opening side, while on the closing side, it is controlled by the valve system. High-pressure oil is only filled during the closing operation and in the closed position. When in the open position, it is connected to the low-pressure oil tank.

The working principle of the hydraulic operating mechanism is as shown in Fig. 3-33. The right opening chamber of the working cylinder piston is directly connected to the pressure accumulator, so it is in a constant high-pressure state whether in the closed or open position. The left closing chamber of the piston is controlled by the valve. At the time of closing, the electromagnetic coil is energized to generate magnetic force. The closing solenoid valve (primary valve) is opened to allow high-pressure oil to enter the closing chamber of the control piston of the secondary valve. The operating piston pushes the valve core of the secondary valve to move, thus closing the oil circuit from the working cylinder to the low-pressure cylinder, opening the high-pressure valve port, and allowing the oil in the opening chamber of the control piston to be discharged from the oil drainage hole. The high-pressure oil in the accumulator enters the closing side of the working cylinder piston. Due to the fact that the pressure area on the closing chamber side of the piston is greater than the pressure area on the opening chamber side, the piston quickly moves to the right to achieve the closing. At the time of opening, the high-pressure oil in the closing chamber leaks into the low-pressure oil tank. At the same time, the piston moves to the left to achieve the rapid opening under the action of high-pressure oil in the opening chamber.

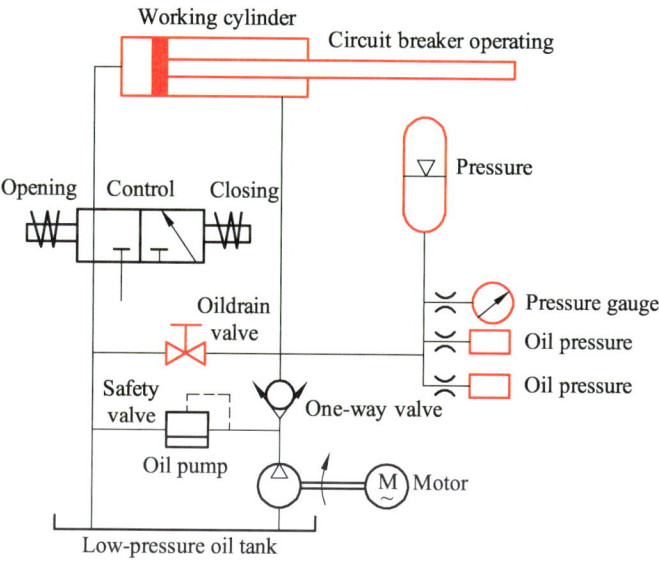

Fig. 3-33　Working Principle Diagram of Hydraulic Operating Mechanism

The oil is used as the medium for transmitting mechanical energy for the hydraulic operating mechanism, and the mechanical energy is stored in the pressure storage barrel. At present, there are two main ways to store energy in the pressure storage barrel: ① Nitrogen gas is used to store energy, which means filling the upper part of the pressure storage barrel piston with nitrogen gas at the specified pre charging pressure. Nitrogen stores energy when it is compressed. ② The method of spring energy storage is to structurally connect the pressure storage barrel piston with the dedicated disc spring. When the oil pump is pressurized, the compressed hydraulic oil pushes the pressure storage barrel to compress the disc spring for energy storage.

3. Locking protection function

(1) Locking function for oil pump timeout operation. Generally, when the oil pump runs for more than 3 to 5 minutes, the normally closed contact of the time relay will delay opening, cut off the power supply of the motor, and stop pressing.

(2) Locking function for prevention of slow opening.

① Electrical lockout. When the circuit breaker and the disconnector are in the closed position, the power supply of the oil pump motor will be cut off and the motor will not start to pressurize if the oil pressure is very low or drops to zero.

② Slow opening valve. There are three methods as follows: First, locking of the secondary valve piston or installation of the device for prevention of slow opening. Second, setting of a manual valve at the third level valve. When the oil pressure drops to zero, the manual valve is tightened to keep the oil pressure system in the closed position. This manual valve is released when the oil pressure is reestablished. Third, setting of a tubular differential cone valve. Regardless of whether the valve is in the open or closed position, the tubular cone valve generates a self-sustaining force to maintain in the open or closed position once the system establishes the pressure, no matter how high the pressure is.

③ Mechanical method. With the mechanical method, the working cylinder piston rod is maintained in the closed position. After the mechanical failure is resolved, the mechanical support can be removed.

(3) Locking function for low oil pressure. When the oil pressure is reduced to a point where it is insufficient to keep the circuit breaker closed or open, the microswitch is used to excite the relevant relay to disconnect the starting circuit for closing and starting circuit for opening, thereby achieving the locking function for opening, closing or reclosing.

4. CY type hydraulic operating mechanism

(1) Brief introduction to the structure.

The common CY series hydraulic operating mechanisms are as shown in Fig. 3-34. This mechanism mainly consists of mechanism box, pressure storage barrel, valve system, working cylinder, oil pump, and control board. The structure of CY3 type hydraulic operating mechanism is as shown in Fig. 3-35, and the structure of CYA8 type hydraulic operating mechanism is as shown in Fig. 3-36.

① Mechanism box. The mechanism box is the supporting base and protective shell of the entire mechanism. There are three valves on the left and right sides and above the box for maintenance, and the upper cover can be opened to 45°. The lower part is equipped with the heater, which can be put into operation when the ambient temperature is below 0°C. There is a hole on the left door for monitoring the pressure gauge.

(a) CYA5 Series   (b) CYA6 Series   (c) CYA8 Series

Fig. 3-34   CY Type Hydraulic Operating Mechanism

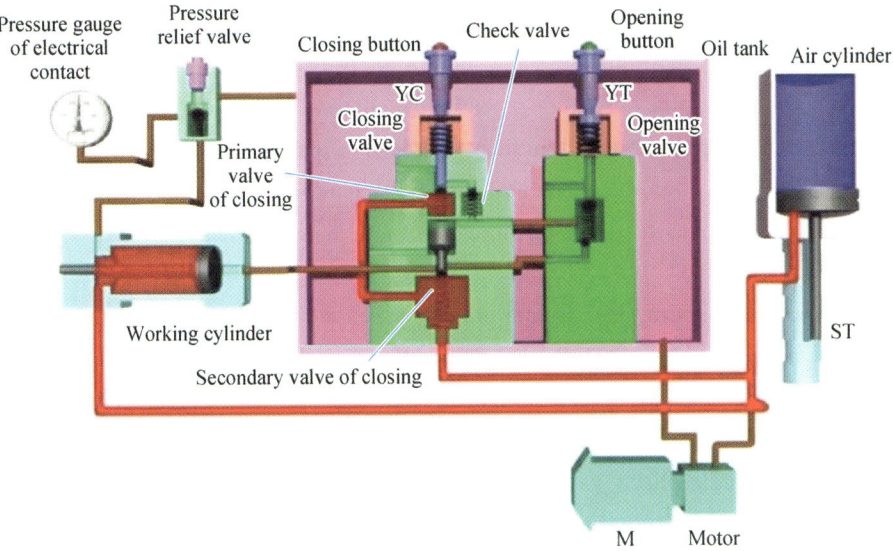

Fig. 3-35　CY3 Type Hydraulic Operating Mechanism

Fig. 3-36　CY8A Type Hydraulic Operating Mechanism

② Pressure storage barrel The upper inner chamber of the pressure storage barrel (above the piston) is pre filled with nitrogen gas at a certain pressure. There are microswitches installed at the piston rod of the pressure storage barrel, which respectively play the roles of starting and stopping the oil pump, closing locking, opening locking, reclosing locking, and sending alarm signals in case of pressure reduction.

③ Valve system. The valve system consists of the oil filter, the oil drain valve, and the operating valve. The operating valve is placed in the oil tank, which contains No. 10 aviation hydraulic fluid.

④ Working cylinder. The working cylinder and piston are high strength and wear-resistant components. The left end of the piston is connected to high-pressure oil. During operation, the left end of the piston always maintains the constant high pressure. The right end of the piston is connected to the operating pipe, and high-pressure oil enters the right end of the piston through the closing circuit. According to the differential principle, the piston moves to the left and drives the circuit breaker to close through the horizontal tie rod. When the pressure in the right closing circuit is released, the piston moves to the right under the action of constant high pressure, driving the circuit breaker to open.

⑤ Oil pump. The oil pump is of a dual plunger type, and the plunger is a pressure component. The plunger and valve seat are in a sliding fit. When the electric motor drives the oil pump to work, low-pressure oil enters the valve seat cavity through the one-way valve. When the crankshaft rotates, the ball bearing pushes the plunger towards the valve seat, causing the oil in the cavity to be squeezed. After the single-phase valve on the other side (HV side) opens, the oil enters the high-pressure oil pipe through the valve port. The single-phase valve opens, and the hydraulic oil enters the pressure storage barrel through the pipeline for storage. The crankshaft rotates one revolution, and both plungers work once each.

⑥ Control board. The control board is equipped with auxiliary switches, contactors, intermediate relays, pressure gauges of electrical contact, and wiring terminal blocks.

(2) Action principle.

① Mechanism energy storage.

After the circuit of the oil pump motor is connected, the oil pump begins to operate. The low-pressure oil in the oil tank enters the oil pump through the oil filter, and the high-pressure oil enters the pressure storage barrel. When the piston rod of the pressure storage cylinder enables the microswitch of the pump stopping position, the oil pump shall stop running to finish the energy storage process. During the running, when the piston rod of the pressure storage cylinder moves down to the pump starting position to enable the microswitch, the oil pump shall run automatically for pressure compensation.

Since the oil pump exhibits no zero pressure locking in the electric circuit, it cannot automatically initiate the pressure at zero gauge pressure, and the locking circuit needs to be temporarily short-connected (or artificially pressed).

② Closing and closing holding.

When the closing electromagnet receives the closing command or the core button is pressed by hand, the movable core of the closing electromagnet moves downward, thus pushing the stem movement of the closing primary starting valve. The oil drain hole under the valve seat is firstly blocked, followed by opening the primary ball valve. The high pressure oil from the closing control tubing passes through the ball valve port pushed open, and propels the movable core of AC contactor via the internal channel. Only when the oil pressure is higher than the low-pressure abnormal locking pressure, automatic start and stop of oil pump can be realized. Opening-closing holding check valve. The high-pressure oil enters the upper part of the cone valve core of the secondary valve, propelling this cone valve core moving downward at high speed. It firstly pre-seals the access to the oil drain hole in the valve seat, and then pushes open the lower ball valve core. With its conical surface, the upper cone valve core seals the upper valve port of the secondary valve and blocks the oil drain circuit. The high-pressure oil from the closing inlet tubing enters the closing chamber of the working cylinder (the side without piston rod) by passing through the lower valve port of the secondary valve and the pipeline. At this time, both ends of the working cylinder piston are subjected to the high-pressure oil with same pressure. Since there is much larger pressure area on the closing side, the piston moves towards the closing direction till to the end of closing. By now, the auxiliary switch also completes the switching. The closing electromagnet loses power, the closing primary valve returns under the force of its reset spring, and the high-pressure oil in the closing primary valve chamber is discharged through the oil drain hole. Simultaneously, due to the function of the high-pressure oil holding circuit, the operating mechanism can be kept in the closed status.

The position of cone valve core of the closing secondary valve determines the closing or open status of the circuit breaker. To maintain the closed status, the cone valve core of the secondary valve must be in the closed position. That is, there shall always be the function of high-pressure oil in the upper part of the cone valve core after the closing operation, thus ensuring the lower valve port of the secondary valve opening and the upper valve port closing.

To prevent the oil pressure in the upper part of the cone valve core from falling to zero due to chronic leakage, the mechanism is equipped with a self-holding high-pressure oil replenishment circuit. It is composed of tubing, connector with 0.5 mm throttle hole, holding valve, tubing and other components, thereby realizing the closing holding.

③ Opening.

When the opening electromagnet receives the opening command or the opening button is manually pressed, the movable core of the opening electromagnet pushes the stem of the opening primary valve moving downward, thus opening the valve port of the opening valve steel ball. The high-pressure oil of the closing holding circuit passes through the tubing and the opening primary valve port, and then is discharged into the low-pressure tank via the oil drain hole in the valve seat. The secondary valve core immediately returns upward under the action of high-pressure oil in the lower part. The upper valve port is opened firstly, and the high-pressure oil in the closing-side cavity of the working cylinder is discharged into the low-pressure tank by passing through the tubing, the upper valve port and the oil drain hole in the valve seat. The working cylinder piston moves towards the opening direction under the action of high-pressure oil on the opening side, finally completing the opening operation. Meanwhile, the auxiliary switch also completes the switching, disconnecting the circuit of opening electromagnet. When the upper valve port of the secondary valve is opened, the secondary valve steel ball quickly closes the lower valve port under the action of its reset spring, preventing excessive discharge of high-pressure oil.

④ Closing locking.

When the oil pressure of the energy storage device is insufficient for closing, the pressure cylinder piston shall move down. Once declining to the closing locking microswitch, the microswitch contact shall close, followed by action of closing locking relay, thereby cutting the closing control circuit. In addition, when the pressure cylinder shows nitrogen leakage, nitrogen pressure abnormally reduces, or oil pressure increases unusually for some reason, the pressure gauge contact shall close, and the relevant relay shall be started, thereby cutting the closing circuit and the oil pump motor power supply circuit.

⑤ Opening locking.

When the oil pressure of energy storage device reduces and does not satisfy opening requirements any more, the pressure cylinder piston shall move down. Once declining to the opening locking microswitch, the microswitch contact shall close, followed by action of opening locking relay, thereby cutting the opening control circuit. As mentioned above, in case of the abnormal increase or decrease of oil pressure and the nitrogen leakage, the electric contact pressure gauge shall act, with the opening circuit and the oil pump motor power supply being cut as well.

5. Integrated hydraulic operating mechanism

Because the circuit breaker needs the operating mechanism to provide the smaller operating power, the integrated hydraulic operating mechanism is designed. Despite its reduced volume, the modularization of functional components is realized. There is no tubing outside the mechanism, the components are miniaturized, and the main components are integrated, thus reducing leakage points, realizing high product quality, avoiding external leakage of mechanism, achieving low operating noise, and facilitating maintenance.

The action principle of the integrated hydraulic operating mechanism is shown in Fig. 3-37.

(1) Opening action.

When receiving the opening command, the opening coil is electrically excited, and the reversing valve moves right. Due to the movement of the reversing valve, the high-pressure oil on the upper side of the piston is discharged. The accumulator fills the high-pressure oil to the lower side of the piston. Under the action of oil pressure, the piston moves towards the opening direction with high speed for the opening action. Due to the pressure difference between the two sides of the piston, the piston stays at the open position reliably.

(2) Closing action.

When receiving the closing command, the closing coil is electrically excited, and the reversing valve moves left. Due to the movement of the reversing valve, the high-pressure oil of the accumulator flows into the upper side of the piston for closing action. Although there are filled with high-pressure oil in the upper and lower sides of the piston, the piston moves down quickly due to the pressure difference caused by the cross-sectional area difference between the upper and lower sides of the piston rod, thereby completing closing action and staying at the closed position reliably (with the inherent pressure-loss anti-slow division function).

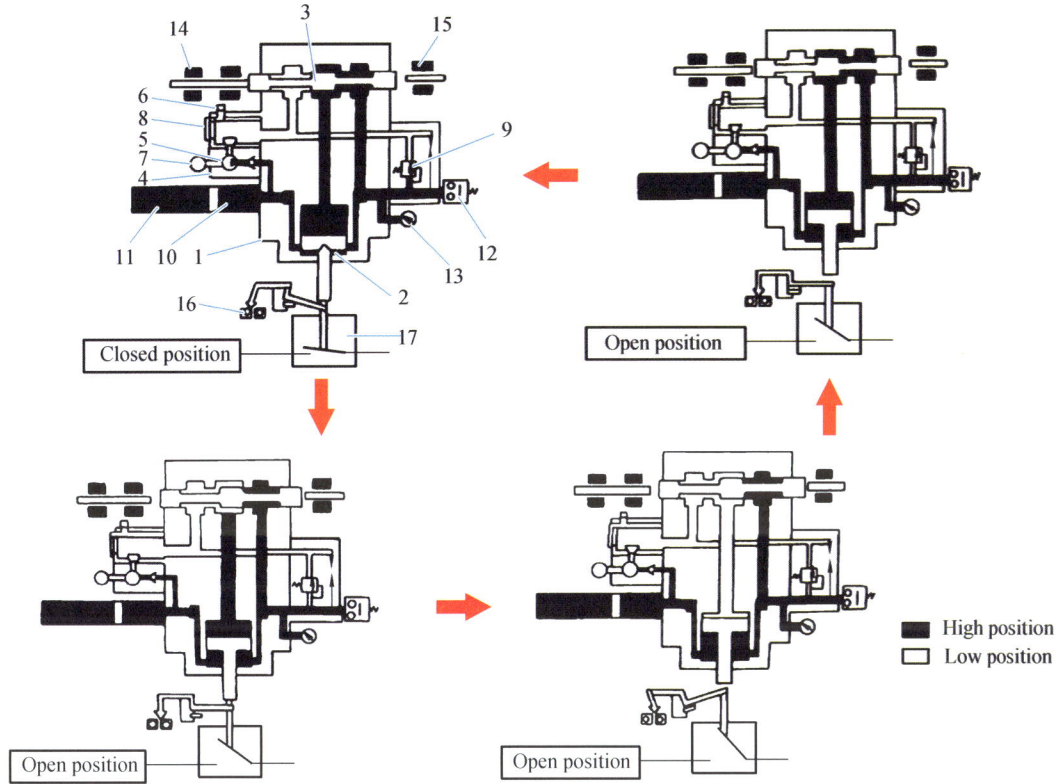

1-Working cylinder; 2-Piston; 3-Reversing valve; 4-Hydraulic pump group; 5-Hydraulic pump; 6-Tank; 7-Motor; 8-Oil level indicator; 9-Safety valve; 10-Accumulator; 11-High-pressure nitrogen; 12-Oil pressure switch; 13-Oil pressure gauge; 14-Opening coil; 15-Closing coil; 16-Opening-closing indicator; 17-Circuit breaker.

Fig. 3-37　Action Principle of the Integrated Hydraulic Operating Mechanism

3.4.4　Pneumatic operating mechanism

Pneumatic operating mechanism is the operating mechanism that uses compressed air as energy to generate thrust. Since compressed air is used as energy, the pneumatic operating mechanism does not need a high-power DC power supply, and the independent air tank can ensure multiple operations of the pneumatic operating mechanism. The opening and closing of the circuit breaker all rely on the compressed air, and the thrust of compressed air is needed by the maintenance of circuit breaker in the open or closed position. Pneumatic operating mechanism is mainly used in 220 kV and 500 kV high-voltage circuit breakers in China.

1. Structure and characteristics

The pneumatic operating mechanism using compressed air as the single power has been eliminated. Currently, the circuit breaker employs the pneumatic operating mechanism, which adopts compressed air as the power for opening operation, supplemented by the closing energy storage element of closing spring. Energy storage of compressed air can be realized by relying on the compressor provided by the operating mechanism. During the opening process, the closing spring stores energy by virtue of the cylinder piston. Simultaneously, the mechanical transmission unit enables the contact terminal to complete the opening operation, and the latching system makes the closing spring being maintained in the energy storage status. When closing, latch tripping can be realized under the magnetic force. The energy released by spring passes through the mechanical transmission unit, enabling the contact terminal completing closing operation.

Pneumatic operating mechanism has the disadvantages of the larger volume and the higher component processing accuracy compared with the electromagnetic operating mechanism. Simultaneously, air compressor device and compressed air tank need to be equipped. Hence, it has very high requirements for air tightness, resulting in high piston and cylinder maintenance requirements. Furthermore, since it is the compressed air that transfers energy in the pneumatic operating mechanism, and action delay may occur during the operation, the pneumatic operating mechanism is rarely used in the UHV circuit breaker.

2. Action principle

The principle of pneumatic mechanism of pneumatic closing and spring opening is shown in Fig. 3-38.

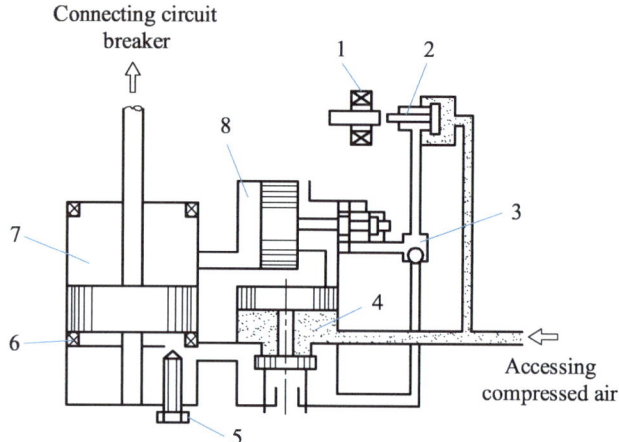

1-electromagnet; 2-starting valve; 3-holding valve; 4-intermediate valve; 5-adjusting screw; 6-buffer gasket; 7-working valve; 8- return valve.

Fig. 3-38　Schematic Diagram of Pneumatic Mechanism

When closing, the electromagnet is energized, the core pushes the ejector rod of the starting valve to open the valve port, and then the compressed air opens the intermediate valve after passing through holding valve and return valve. By entering the part below the working valve piston via the adjusting screw, the compressed air pushes the piston to move upward for closing of circuit breaker. During the closing, if the closing signal is removed in advance, the holding valve ball shall move upward to seal the valve port, thereby ensuring the thorough closing. After completing the closing, the piston of the working valve opens the port accessing to the return valve, making the return valve piston moving right. The compressed air in the upper part of the intermediate valve piston is discharged into the atmosphere via the return valve, and the intermediate valve is reset. From the exhaust port at the lower part of the working valve, the compressed air under the working valve piston is discharged, followed by falling and resetting of the piston.

3. CQ6 pneumatic operating mechanism

CQ6 pneumatic operating mechanism is a driving mechanism composed of piston and cylinder, as well as other components including control valve controlling compressed air supply, closing and opening electromagnet operated by electrical signal, closing spring, buffer, opening holding latch and release, as shown in Fig. 3-39.

Fig. 3-39　Pneumatic Operating Mechanism

CQ6-I pneumatic operating mechanism is used for 252 kV SF_6 circuit breaker or GIS. Each pole circuit breaker is equipped with a mechanism box that contains pneumatic operating mechanism and compressed air tank. 22 mm copper pipe is applied in connecting the compressed air tank between every two pole circuit breakers, thereby maintaining the uniform pressure. Among them, the compressed air is provided by the air compressor unit in the intermediate pole mechanism box. Each circuit breaker can perform not only single-phase operation but also three-phase electrical linkage. CQ6-II pneumatic operating mechanism is used for 126 kV SF_6 circuit breaker or GIS. For each circuit breaker, one mechanism box is equipped in the intermediate phase. The mechanism box is configured with pneumatic operating mechanism, air compressor unit and compressed air tank. For three-phase action of the circuit breaker, the transmission component in the mechanism box is relied to realize three-phase mechanical linkage.

(1) Opening operation.

When the circuit breaker is in the closed position, the cylindrical valve is pressed at the lowest end by the crank arm of cam and locked by the latch 1, thereby sealing the compressed air in the air tank. Due to the action of the spring mounted on the cylindrical valve, the latch 1 is subjected to a counterclockwise torque, while it is also locked by the latch 2, as shown in Fig. 3-40 (a).

The power of opening operation is supplied by the compressed air stored in air tank. By opening the cylindrical valve, the compressed air enters into the cylinder. With downward movement of the piston in the cylinder, the contact terminal is separated.

The opening process is as follows:

The opening coil is energized, and its core rod moves downward, impacting the latch 2. The latch 2 is composed of two connecting rods and three pins, with one pin marked with white dot connecting with two connecting rods, and two pins marked with black dots fixing two connecting rods to the frame respectively. Under the impact of the opening coil core rod, the right connecting rod of the latch 2 rotates clockwise, while the left connecting rod rotates counterclockwise, thus releasing the engagement between latch 1 and latch 2, as shown in Fig. 3-40 (b). After tripping of latch 1, the cylindrical valve shall be free from constraint and be opened by spring force. The compressed air in air tank flows into the cylinder. With the compressed air, the piston moves downward and drives the contact terminal for opening, as shown in Fig. 3-40 (c).

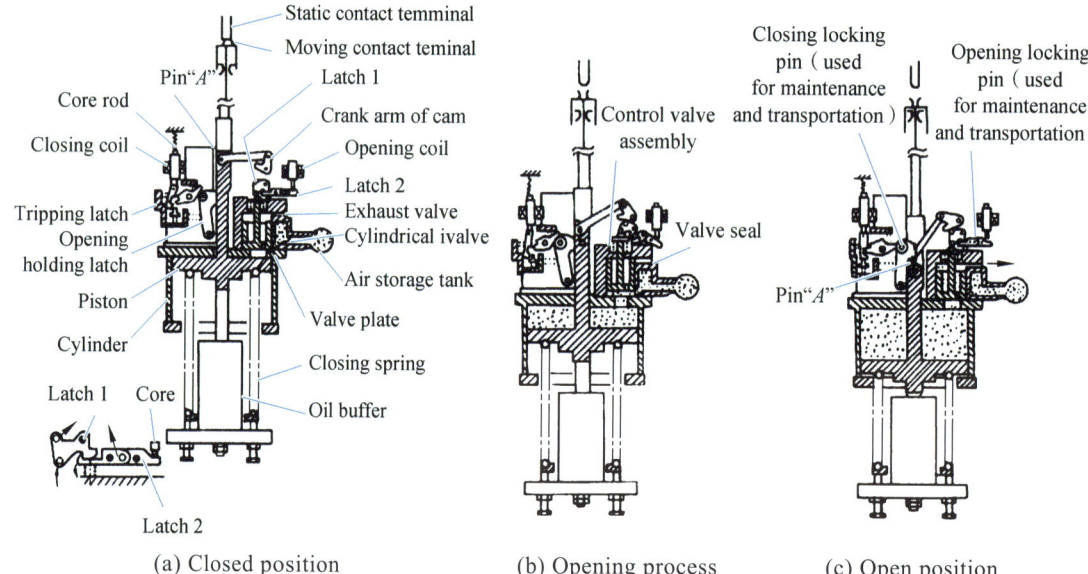

Fig. 3-40 Action Principle of CQ6 (AM) Operating Mechanism

At the end of the opening process, the cylindrical valve is pressed back to the lowest end by the crank arm of cam connected to the piston and is locked by the latch [1]. In this way, the cylindrical valve returns to the closed position, and the compressed air in the cylinder is discharged via the exhaust port. Finally, the pin A is locked by the opening holding latch, and the circuit breaker is kept in the open status.

In the open position, the circuit breaker is locked on the machine by the opening holding latch connected to the frame. The opening holding latch is subjected to the counterclockwise torque generated by the closing spring force. Meanwhile, it forms a "dead point" structure with the tripping latch and the own shaft pin, thus generating clockwise torque and maintaining the open position of the circuit breaker.

In the opening process, the closing spring is compressed for energy storage and for the next closing.

(2) Closing operation.

The power of closing operation is supplied by the closing spring. After tripping of the opening holding latch, the piston moves upward under the action of closing spring, driving the contact terminal for closing.

Closing action process is as follows: the closing coil is energized, and the core rod of the closing coil moves downward, impacting the tripping latch. The engagement of the tripping latch with the opening holding latch is relieved. The opening holding latch rotates counterclockwise, with the pin A releasing from its constraint. The piston and contact terminal move upward under the action of closing spring, thereby completing the closing.

(3) Anti-jump device.

CQ6 operating mechanism is equipped with a mechanical anti-jump device. The principle of the anti-jump device is as shown in Fig. 3-41, and the action composition is as below:

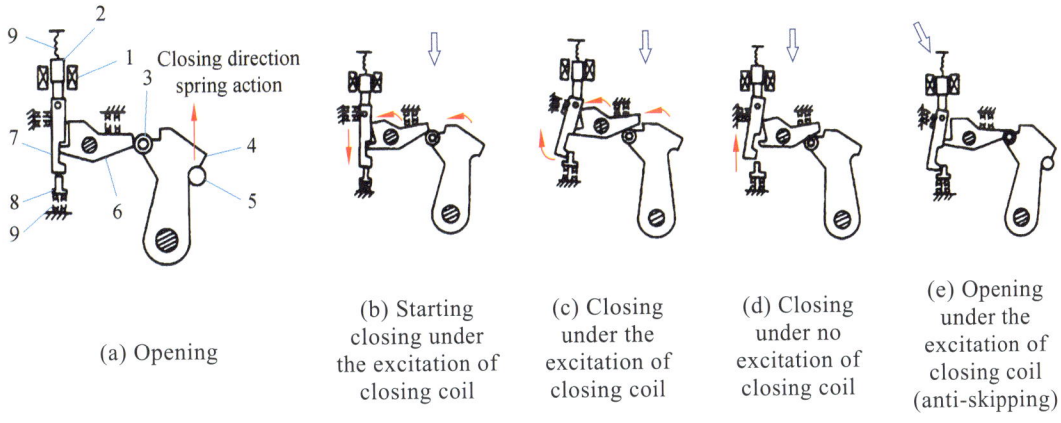

Fig. 3-41　Anti-jump Device

The opening holding latch locks the pin "A", making the circuit breaker staying in the open position. The pin "A" is connected with the operating rod, with the reacting force of closing spring acting on it. Thus, the pin "A" gives a counterclockwise torque to the opening holding latch, while this latch is also locked by the tripping latch via the roller.

When the closing coil is excited by the closing signal, the core rod drives the tripping rod to impact the tripping latch, making it rotating counterclockwise and releasing the constraint to opening holding latch. Under the reacting force of closing spring, the opening holding latch rotates counterclockwise, resulting in tripping of pin "A" and closing of circuit breaker. At the same time, the core rod presses down the anti-jump pin through the tripping rod.

The roller drives the revolution surface of the tripping latch, enabling its further counterclockwise rotation. Hence, the tripping latch makes the tripping rod rotating clockwise [as shown in Fig. 3-41 (b)] and releasing from the anti-jump pin, while the anti-jump pin enables the tripping rod maintaining the inclined status. As shown in Fig. 3-41 (c).

At this time, if the circuit breaker gets an unexpected opening signal and starts opening, the pin "A" will move downward, and the opening holding latch will rotate clockwise under the action of reset spring to lock the pin "*A*". The opening holding latch is then locked by the tripping latch.

In this process, so long as the closing signal is maintained, the tripping rod will be always inclined due to the action of anti-jump pin, so the core rod will not impact the tripping latch. Hence, the circuit breaker cannot repeat the closing operation to achieve the anti-jump function, as shown in Fig. 3-41 (e).

When the closing signal is relieved, the closing electromagnet will lose its excitation, and the core rod will return through the spring in electromagnet. Hence, the core rod and the tripping rod are in the status shown in Fig. 3-41 (a) to prepare for the next closing.

AM and CQ6 pneumatic operating mechanisms are exactly same in structure and action principle.

3.4.5 New operating mechanism

1. Permanent magnetic operating mechanism

Different from electromagnetic mechanism and spring mechanism, permanent magnetic operating mechanism is a brand new operating mechanism for vacuum circuit breaker. It adopts electromagnet for operation, permanent magnet for locking, capacitor for energy storage and electronic device for control, and its structure is shown in Fig. 3-42.

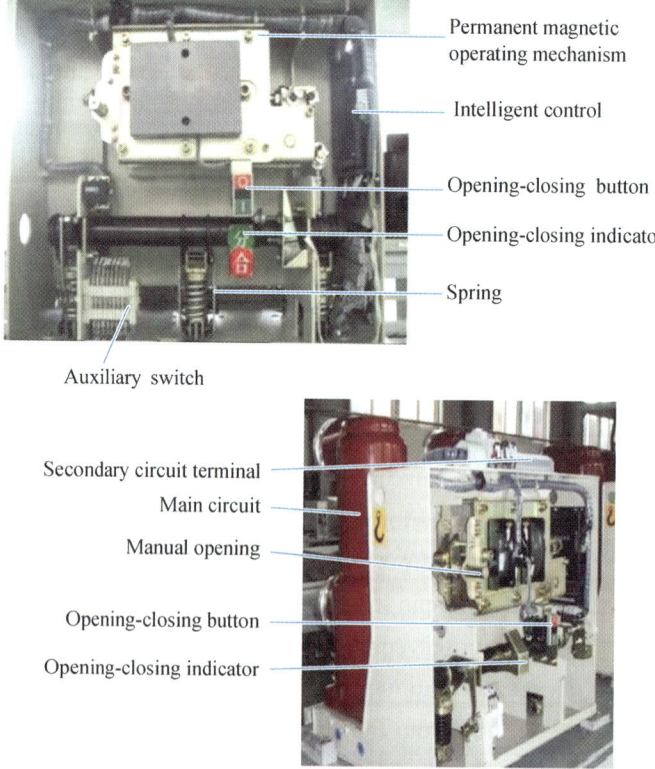

Fig. 3-42 Permanent Magnetic Operating Mechanism

During opening and closing operations, the energy storage capacitor gives electric discharge to the opening or closing coil, making it electrically excited. Thus, electromagnetic attraction is generated to drive the movement of the movable core. The vacuum circuit breaker can be maintained in the open and closed positions merely by the holding power generated by the permanent magnet, while not needing any mechanical energy. The huge pulse energy provided by the capacitor is sufficient for a single reclosing.

For the control of permanent magnetic mechanism, the modern power electronic technology is adopted to constitute the electronic control unit. Generally, the proximity switch is used in detecting the open and closed statuses of the switch.

Permanent magnetic mechanism can be roughly divided into three types, namely double-coil permanent magnetic mechanism, single-coil permanent magnetic mechanism and separate magnetic circuit permanent magnetic mechanism.

(1) Double-coil permanent magnetic mechanism. The double-coil permanent magnetic mechanism has the features that: the permanent magnet is adopted to maintain the vacuum circuit breaker in the limiting positions of opening and closing respectively, the excitation coil is used in propelling the core of mechanism from the open position to the closed position, and another excitation coil is applied to drive the core of mechanism from the closed position to the open position.

Its characteristics are as follows. ① Since it does not need to provide energy for the opening during the closing of mechanism, the smaller closing energy and the thinner closing coil are correspond to the less supply current needed. ② When the mechanism is in the closed position, the permanent magnet only needs to provide the force to overcome the contact terminal spring, excluding the force to overcome the opening spring.

(2) Single-coil permanent magnetic mechanism. The single-coil permanent magnetic mechanism also adopts the permanent magnet to maintain the vacuum circuit breaker in the limiting positions of opening and closing respectively, while the opening and closing share one excitation coil. The closing energy mainly comes from the excitation coil, and the opening energy mainly stems from the opening spring.

Its characteristics are as follows. ① The opening relies on the energy released by the opening spring and the contact terminal spring, and the opening characteristics can be adjusted by adjusting the opening spring. The output characteristics of the opening spring can be consistent with the speed characteristics required by the circuit breaker. ② The opening and closing share one operating coil with simpler structure and smaller volume, which is more suitable for the outdoor enclosed box installation.

(3) Separate magnetic circuit permanent magnetic mechanism. Separate magnetic circuit permanent magnetic mechanism separates the opening, closing and holding magnetic circuits, thereby optimizing the magnetic circuit, with the permanent magnet merely used to maintain the closed position.

This mechanism is characterized by the superior performance in the closing holding force generated by the permanent magnet. Due to the double-magnetic circuit with short working air gap, this mechanism can provide a greater closing holding force with less permanent magnet material.

However, considering its more complicated structure, higher processing precision is required, and there is greater difficulty in processing and assembly.

The aforementioned three different types of permanent magnetic mechanism have their own advantages and disadvantages, needing to be selected according to the requirements of different circuit breakers. The single-coil mechanism has simpler structure, and the opening speed is adjustable. Whereas, it needs to store energy for the opening spring during switch closing, and the closing consumes more energy, so this mechanism is suitable for the switch needing smaller closing power, such as the vacuum circuit breaker on the column; The double-coil mechanism is fit for the switch with greater closing power; If the switch exhibits the greater closing holding force, the separate magnetic circuit permanent magnetic mechanism can be considered.

2. Motor operating mechanism

Motor operating mechanism is completely different from hydraulic, pneumatic, spring and hydraulic spring operating mechanism. By applying advanced digital technology and combining with simple, reliable and mature motor, the motor operating mechanism not only can satisfy all the core requirements of circuit breaker operating mechanism, but also has many new advantages in performance and function.

The basic principle of the motor operating mechanism is that an electronic device-controlled motor is used in directly operating the operating rod of the circuit breaker. This motor operating mechanism is composed of several units, mainly including energy buffer unit, charging unit, converter unit, control unit, motor and solver unit, input/output unit. The motor is powered by energy buffer unit via a converter, and the energy buffer unit is charged by a charging unit (power supply unit). The microprocessor-based control unit is used in speed control and monitoring. The operation of the motor operating mechanism is achieved by virtue of input/output (I/O). The connection among various units is shown in Fig. 3-43.

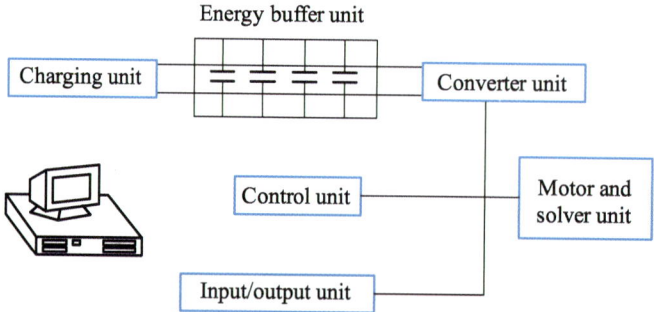

Fig. 3-43　Schematic Diagram of Motor Operating Mechanism

任务一　SN10-10 少油断路器灭弧室检修

一、工作任务

对 SN10-10 少油断路器灭弧室进行检修。掌握油断路器的种类、特点、灭弧原理、灭弧室结构，掌握 SN10-10 少油断路器灭弧室的更换方法流程、操作技巧、检修工艺标准以及作业中的注意事项，积累现场更换少油断路器灭弧室的方法和经验。

二、引用标准

（1）《电气装置安装工程高压电器施工及验收规范》（GB 50147—2010）。
（2）《国家电网公司变电检修管理规定（试行）第 5 分册开关柜检修细则》。
（3）《国家电网公司变电运维管理规定》。
（4）《国家电网公司变电检修管理规定》。
（5）《国家电网公司电力安全工作规程》（变电部分）。
（6）《四川省电力公司高压开关柜现场维护导则》。

三、工作要求

（1）作业为室内停电作业，无天气要求。
（2）被检修设备与其他带电设备均应使用围栏隔离，面向通道处设置唯一出入口。
（3）被检修间隔要求母线与出线电缆均停电并且接地。
（4）作业人员精神状态良好，熟悉工作中安全措施、技术措施以及现场工作危险点。
（5）实训现场要求按生产现场规范布置安全措施，并严格执行标准化作业。
（6）作业人员应规范穿戴劳动保护用品，做好安全防护。

四、工作准备

（一）危险点及预控措施

1. 高压触电

危险点：检修断路器时，接地线装设不当、误入带电间隔或与带电部位安全距离不够造成人身触电；拆接断路器引线时，引线在外力作用下荡向运行设备引起人身触电。

预控措施：工作前工作负责人必须检查确认工作票安全措施已执行完毕，并向工作人员详细交代停电范围和安全注意事项；拆除引线时，不应失去接地保护，并将引线绑扎牢固；在开关柜与相邻带电设备间装设围栏，向内侧悬挂适量"止步，高压危险"标示牌，围栏设置唯一出入口，在出入口处悬挂"从此进出"标示牌；进入断路器间隔处的检修人员不得触动隔离断路器连杆。

2. 低压触电

危险点：控制回路、合闸回路、照明回路空气开关进线侧（上端）均低压电；合上空开后，断路器和开关柜二次回路均视为带电状态，触摸可能造成低压触电。

预控措施：若运行人员在空气开关上悬挂"禁止合闸，有人工作"标示牌时，工作负责人需要向运行人员申请更变安全措施得到许可后，方能合上空气开关；工作负责人反复对工作班成员强调开关空气开关时，切勿将手指伸入空气开关上端头；工作时至少需要两人，一人监护一人操作，听工作负责人指挥。

3. 机械伤害

危险点：断路器分合时未通知其他工作班成员，可能会对工作班成员造成机械伤害；设备储能时进行机械部分操作，有误动风险。

预控措施：断路器分合、接地开关分合时，需要大声呼唱，得到工作班所有人员大声回应之后，方可操作；操作前，首先检查断路器应在分闸位置，并且能量已释放，断开断路器控制和合闸电源，防止机构误动；操作时，工作人员应远离运动部位；工作时至少需要两人，一人监护一人操作，听工作负责人指挥。

4. 设备损坏

危险点：未按照要求分合断路器、分合接地开关、野蛮操作，或者操作时预控措施：禁止野蛮操作，操作过程中，若出现不能操作的现象，应当立即停止，不得使用蛮力，立即汇报负责人；部件复装前要进行全面检查，内部不能遗留任何异物。

（二）工器具及材料选择

SN10-10 少油断路器灭弧室检修工器具及材料见表 3-1。

表 3-1　SN10-10 少油断路器灭弧室检修工器具及材料

类别	名称	规格型号	数量	备注
专用工具	压环专用工具	自制	1 把	
	触指专用工具	自制	1 把	
通用工具	梅花扳手	5.5～32	1 套	
	活动扳手	8 号	1 把	
	开口扳手	6～32	1 套	
	内六角扳手	1.5～10	1 套	
	尖嘴钳	6 号	1 把	
	十字螺丝刀	4～6 号	1 套	
	一字螺丝刀	3、4、6 号	1 套	
	木榔头	中	1 把	
	套筒扳手	32 件	1 套	
	放油桶		1 个	
	加长内六角	10 cm	1 把	
	检修油盘	60 cm × 40 mm	3 个	

续表

类别	名称	规格型号	数量	备注
通用工具	游标卡尺	150 mm	1 把	
	深度游标卡尺	200 mm	1 把	
材料	小毛巾		10 块	
	白布		3 m	
	变压器油		24 kg	
	塑料布		3 m	
	中性凡士林		1 瓶	
	棉纱头		2 kg	
	纱布		6 张	
	密封圈		1 套	
	黄油	1 kg/包	1 包	
	黄绿红黑相色漆		各 0.5 kg	
	洗手液		1 瓶	
	洗涤型汽油	70 号	5 L	
	漆刷	1 号	4 把	
	1032 绝缘清漆		1 kg	
	润滑脂		少量	

（三）作业人员分工

SN10-10 少油断路器灭弧室检修人员分工见表 3-2。

表 3-2　SN10-10 少油断路器灭弧室检修人员分工

序号	工作岗位	数量	职责
1	工作负责人	1	负责本次工作的人员分工、现场查勘、作业方案制定、召开班前会、作业过程中安全监督、工作中突发状况的处理、工作质量的监督、班后会总结
2	操作人员	1	负责 SN10-10 少油断路器灭弧室检修的主要操作
3	辅助操作人员（可选）	1	辅助操作人员进行作业

五、作业程序

SN10-10 少油断路器灭弧室检修作业流程见表 3-3。

表3-3　SN10-10少油断路器灭弧室检修作业流程表

序号	作业内容	作业步骤及标准	安全措施及注意事项
1	开工手续	1. 列队宣读工作票，交代工作内容、安全措施和注意事项； 2. 检查工器具应齐全、合格、摆放位置符合规定； 3. 工作时，检修人员与10 kV带电设备的安全距离必须不得小于0.35 m	1. 相关设备已停电； 2. 禁止无关人员进入现场
2	工作前准备工作	1. 检查工器具是否齐全，检查工器具外观和试验合格； 2. 工作负责人同工作许可人巡视待检修设备，确认工作票所列安全措施已经正确执行，安全措施是否完备，现场是否具备开工条件，必要时进行补充； 3. 执行工作许可手续； 4. 对工作班成员召开班前会； 5. 抄写设备铭牌参数； 6. 布置检修场地，准备废变压器油回收用专用油桶，油桶应能密封和运输	1. 工器具无损伤、变形、失灵现象，需要试验的工器具合格证在有效期内； 2. 巡视现场时禁止无关人员进入现场； 3. 班前会应包含工作地点双重名称；工作时间与内容；工作分工；工作危险点及预控措施；停电范围及工作现场安全措施； 4. 全体工作成员应当正确穿戴安全帽、工作服、工作鞋、劳保手套等劳动保护用品； 5. 在检修现场四周设一留有通道口的封闭式遮拦，并在周围背向带电设备的遮拦上挂适当数量的"止步，高压危险"标示牌，在通道入口处挂"从此进入"标示牌
3	检查设备状态	负责人带领工作班成员检查： 1. 断路器本体、机构及基础构架无变形，无锈蚀；绝缘表面无破损，无脏污，无放电痕迹；导电连接部分金属表面无过热痕迹；相色及标示清楚； 2. 机械指示器指示与实际一致；指示灯、信号与实际一致；机构四连杆机构位于死点位置；分闸弹簧已储能；（如果分闸弹簧储能，说明在合闸位置） 3. 油位位于位于油标1/4至3/4刻度间； 4. 各部密封连接无渗漏；无闪浸；无油滴痕迹；	视工作任务情况可以安排修前测试： 1. 测量各相导电回路电阻； 2. 断路机械特性测试； 3. 测量总行程、超行程及三相不同期性

续表

序号	作业内容	作业步骤及标准	安全措施及注意事项
3	检查设备状态	5. 电动分合闸动作干脆，无异响，无喷油，位置信号变化准确； 6. 操动机构和传动机构动作正常，无异响、无卡涩、死点闭锁可靠； 7. 端子排各接线端子无松动，松脱； 8. 断路器各连接及接地部位螺丝紧固，连接可靠	
4	断路器分解	1. 拧下底部放油螺栓，将油放出； 2. 拧掉上下接线端子引线； 3. 用内六角扳手拧下上帽装配与上接线间的四只内六角螺栓，取下上帽装配，拧下静触座装配的 M8 固定螺栓，取下静触座装配及绝缘套筒； 4. 卸下上接线座，用专用工具旋下上压环，取出灭弧室上边的绝缘环及绝缘衬垫； 5. 用专用工具拧下压环上的四只内六角螺栓，取出绝缘筒装配及下接线座装配； 6. 拆开绝缘拉杆与基座外拐臂的连接，提起导电杆，并卸下与基座内部连接板连接的 10 mm×55 mm 连接销，抽出导电杆装配	1. 不得渗漏； 2. 拆下的零部件应放在清洁干场所，并按相顺序放置，以防丢失； 3. 绝缘部件不得碰伤
5	上帽装配检修	1. 拧下 M8′12 半圆头螺栓，取下排气孔盖，拧下 M12 螺栓，取下油气分离器，拧下回油阀； 2. 清洗、检查各部件；回油阀如密封不严，可用小锤轻敲以下，使其有可靠的密封线； 3. 按拆卸相反顺序装复	1. 上帽无砂眼； 2. 各排气孔道畅通； 3. 回油阀动作灵活，钢球密封可靠
6	静触座装配检修	1. 分解。 （1）用专用工具将触指及弧触指从静触座上卸下，并取出弹簧片； （2）卸下逆止阀； （3）拧下三只螺栓，使静触座与触头架分离。	1. 导电接触面应光滑平整，烧伤面积达 30%且深度大于 1 mm 时，应更换。铜钨合金部分烧伤深度大于 2 mm 时应更换； 2. 触头架与触座间接触应紧密，触座与触指接触面不应有烧伤痕迹；

续表

序号	作业内容	作业步骤及标准	安全措施及注意事项
6	静触座装配检修	2. 清洗，检查。 （1）用合格绝缘油清洗各零件； （2）检查触指及弧触指导电接触面。如有轻微烧伤用细锉及0号砂布修整，烧伤严重时更换； （3）检查触头架与触座的接触面有无烧伤痕迹，若轻微烧伤用0号砂布打磨处理，触指腰部的修整量不允许大于0.5 mm； （4）检查触座的触指尾槽内积垢是否清除干净，隔栅是否完整； （5）检查弹簧片有无变形和损坏； （6）用嘴吹一下，检查逆止阀密封情况，如密封不严，可按上帽回油阀处理方式处理； （7）检查绝缘套筒的漆膜是否完整，有无剥落、起层、起泡现象。 3. 装复。 （1）按分解相反顺序。组装触座及触指时，注意检查弧触指与横吹弧道及定位销间的相对位置，使之符合要求； （2）测量静触指闭合圆直径，应符合要求	3. 触座的隔栅应无裂纹、缺齿现象。固定隔栅的圆柱销无脱落及退出现象； 4. 弹簧片弯曲度不超过0.2 mm，与触指、触座及隔栅不应有烧伤； 5. 逆止阀内不应有铜熔粒及杂质，钢球动作应灵活，挡钢球的圆柱两端应铆好、修平，不得凸出； 6. 内壁不应有严重碳化，烧伤及起层现象，否则应更换； 7. 弧触指必须装在隔栅压有特殊标志处，如隔栅上无特殊标志，则必须将弧触指装于对准横吹弧道的方向； 8. 静触指闭合直径为18.5~20 mm
7	灭弧室装配检修	1. 用合格绝缘油清洗灭弧片1、2、3、4、5； 2. 调整垫片、绝缘衬圈、隔弧壁； 3. 检查灭弧片与绝缘件的烧伤情况，如烧伤轻微时，可用0号砂布轻轻擦拭弧痕，烧伤严重时更换	1. 灭弧片表面光滑平整，无碳化颗粒，无裂纹及损伤； 2. 处理后的灭弧片，第一片长孔径不得超过28 mm，其余灭弧片孔径不得超过26 mm； 3. 绝缘件无烧伤损坏
8	绝缘筒装配检修	1. 分解。 （1）取出绝缘筒内下压环及连接用弹簧； （2）拧下四只螺栓，卸下油位指示计的玻璃罩。 2. 清洗、检查。	1. 上接线座不应有砂眼、裂纹及渗油等现象； 2. 油位指示计座上下孔应畅通； 3. 接触面应平整，手车柜隔离插头触指镀银层应完好，弹簧无变形和损坏；

续表

序号	作业内容	作业步骤及标准	安全措施及注意事项
8	绝缘筒装配检修	（1）检查上接线座内外壁，并用布擦拭干净； （2）清洗油位指示计玻璃及油位指示计座的上下孔； （3）检查接线端子及手车柜隔离插头的接触面有无凹凸不平及过热现象，接触面修整后，涂以中性凡士林油； （4）用布擦拭绝缘筒内外壁，检查外观状况，如漆膜损坏，应重涂1032绝缘清漆； （5）检查下压环与绝缘筒连接用弹簧有无变形。 3. 装复。 检查清擦后，将油位指示计装复	4. 绝缘筒表面漆膜光滑，无掉漆，内壁无放电痕迹，半圆槽无损伤变形； 5. 下压环应完整无损，弹簧应无压扁变形
9	下接线座装配检修	1. 分解。 拧下四只螺栓，卸下导向装置及滚动触头装配。 2. 清洗、检查。 （1）检查导向装置的导电条与接线座的接触是否紧密，两侧导电面是否有烧伤痕迹，如有痕迹须查明原因，加以处理和修整； （2）检查滚动触头装配的滚轮动作情况，轴杆两端铆固情况及弹簧、连板、垫圈是否完整； （3）清洗并检查上下导向绝缘板有无开裂和破损，损坏者将其从导电条上拆下，予以更换； （4）清洗、检查下接线座及手车柜隔离插头导电接触面情况，接触如不良应修复，修复后涂中性凡士林油。 3. 装复。 按相反顺序装复	1. 导电条与滚动触头表面应无烧伤，与接线座的接触应严密； 2. 滚动触头的滚轮转动应灵活，转轴不应弯曲，两端应铆固，各零件齐全，弹簧特性符合要求； 3. 上下导向绝缘板应无破损裂纹，导向口应光滑

续表

序号	作业内容	作业步骤及标准	安全措施及注意事项
10	导电杆的装配检修	1. 分解。 一般动触头可不拆卸，但应检查动触头与导电杆的连接是否松动。如烧损严重需拆下处理时，可用专用工具把动触头拧下，若拆卸困难，须卸下导电杆装配后在虎钳上进行拆卸，以防止基座各连板受力弯曲变形。 2. 检查。 （1）在动触头卸下时，应检查导电杆的螺纹及内部弹簧是否变形； （2）检查导电杆与缓冲器的铆接是否牢固，缓冲器下端口有无严重撞击痕迹。如有严重撞击痕迹，应查出原因予以消除； （3）检查导电杆的弯曲度，不合格时应校直。 3. 装复。 按相反顺序装复	1. 动触头铜钨合金部分烧伤深度大于 2 mm 时应更换，紫铜部分不应有烧伤； 2. 动触头与导电杆的连接应紧密牢固； 3. 导电杆装配各结合处应光滑无凸台； 4. 连接螺纹不应有乱扣现象； 5. 弹簧应无断裂及严重锈蚀； 6. 铆接牢固，铆钉两端应修平； 7. 缓冲器下断口部分不应有严重撞击痕迹； 8. 导电杆的弯曲度应小于 0.15 mm
11	基座装配检修	1. 分解。 （1）拧下正面凸起部分的特殊螺栓，用专用工具打下转轴上的弹性销，旋出转轴，取出基座的内拐臂、连板、拆开轴销，使内拐臂及连板分离； （2）用专用工具拧开转轴密封的螺纹套，取出铜垫圈及骨架密封圈； （3）拧下三只螺栓，卸下油缓冲器活塞杆。 2. 清洗、检查。 （1）清洗检查各部轴销、开口销是否齐全完整，内拐臂，连板是否有变形，铆钉是否牢固，橡胶制动块是否完整； （2）清洗转轴及外摇臂，检查各部焊口情况； （3）更换骨架密封圈时应将内唇翻过来仔细检查有无破损；	1. 连板、内拐臂应无变形损坏； 2. 橡胶制动块不应有裂纹、损坏、铆钉应牢固； 3. 转轴与外摇臂上的轴销不平行度 ≤0.3 mm； 4. 转轴表面不应有机械伤痕； 5. 各部焊口应牢固； 6. 骨架密封圈不应破损，密封面无毛刺、麻纹、气孔、缺损等； 7. 骨架密封圈上的抱簧应完整无损，接头对接良好； 8. 活塞杆端部应无严重撞击现象； 9. 活塞杆与圆盘间的铆接应牢固，但活塞杆应仍能活动； 10. 基座无砂眼、裂纹等。轴孔应光滑无毛刺； 11. 骨架密封圈的唇应向内，弹性销带倒角的一端应向内，弹性销的两端要与拐臂轴套平齐；

续表

序号	作业内容	作业步骤及标准	安全措施及注意事项
11	基座装配检修	（4）清洗、检查油缓冲活塞杆及下部的圆盘； （5）用合格绝缘油清洗基座内部及轴孔，进行内外部检查，轴孔内如有毛刺应磨光。 3. 装复。 按分解相反顺序装复。在装转轴时，要先在骨架密封圈外表面涂以少量钙基润滑脂，然后均匀用力将骨架密封圈压入孔内，再将螺纹套、垫圈套在转轴上，再把转轴对准内拐臂的轴套孔慢慢旋入，并用专用工具将螺纹套适当拧紧，最后将转轴及拐臂旋至分闸位置，使两者的孔对齐，打入弹性销	12. 外摇臂与内拐臂间的相对位置正确，转轴装复后转动应灵活
12	断路器的组装	1. 更换全部密封圈； 2. 将导电杆装配与基座内的，连板装在一起； 3. 清擦下接线座上下密封圈，放正密封圈，然后将接线座放在基座口找正； 4. 将弹簧放入绝缘内壁半圆槽内，在放下压环，并使其内圆弧台均匀压在弹簧上。然后将绝缘筒放在下接线座找正，用专用工具将下压环上的四只内六角螺栓对角均匀拧紧，以保证导电杆上下运动灵活； 5. 依次装入绝缘衬套、灭弧片、调整垫片，并注意最下面的绝缘衬圈的安装方向，以保证横吹弧道与上接线座的接线端子方向相反，装完灭弧片后，可继续装绝缘衬垫及绝缘环； 6. 用专用工具装上上压环，压紧灭弧室，测量 M 尺寸。如不合格，可调整第四、五片灭弧片间的绝缘垫片厚度； 7. 组装完毕后，将传动拉杆与拐臂连接，手动操作几次，检查连接是否正确	1. 密封圈不应有破损、麻纹等，密封面应平整，无堆漆； 2. 上下接线端子中心与基座上装卸弹性销孔的中心必须在一条直线上； 3. 绝缘筒与下压环的连接弹簧须卡入绝缘筒内壁半圆形凹槽内，下压环不许偏斜； 4. 密封圈不许压偏，紧固后的缝隙均不许超过 0.1 mm； 5. 第一横吹弧道与上接线端子夹角 180°； 6. M 尺寸标准：（135±0.5）mm； 7. 导电杆动作应灵活，无卡涩现象

续表

序号	作业内容	作业步骤及标准	安全措施及注意事项
13	整体调试	1. 调整灭弧片上端面至上引线座上端面的距离； 2. 调整动触头合闸位置的高度； 3. 调整导电杆的行程； 4. 调同期性； 5. 调整合闸弹簧缓冲器； 6. 调整动静触头的同心度	1. 燃弧距离要求 SN10-10 型断路器为（135±0.5）mm，SN10-10III 型为（153±0.5）mm，如不合要求，可由调整灭弧片之间的垫片来达到； 2. 动触头上端面至上引线座上端面的距离要求 SN10-10I 型为（130±1.5）mm，SN10-10II 型为（110±1.5）mm、SN10-10III 型为 [122±(1～2)] mm，这样才能满足超行程的要求。它可由调主轴至机构室绝缘联杆的长短来达到，也可调主轴到操动机构的传动杆的长短。即连杆调短就使上述尺寸减少，超行程增大，而连杆调长则使上述尺寸增加，超行程减小； 3. 要求总行程 SN10-10I 型为（145±3）mm、SN10-10II 型为（155±3）mm、SN10-10III 型为（157±3）mm，不合格时可调传动拉杆或连杆的长短来达到，亦可增减分闸限位器的铁片和橡皮垫圈数来达到，调后不影响超行程； 4. 三相分闸不同期性要求不大于 2 mm，不合格时可由改变各相绝缘连杆的长短来达到，调连杆时应注意不能影响动触头端面至上引线座上端面的距离，应为上述中各型号要求的数据，若保证了这个尺寸不超过±2 mm 的误差，也可以不测三相分闸不同期性。注意调同期可和调行程同时进行； 5. 在断路器处于合闸位置时，拐臂的终端滚子打在缓冲器上距极限位置还应留有 2～4 mm 的间隙
14	工作结束	1. 清理工作现场，将工器具全部收拢并清点，废弃物按相关规定处理，材料及备品备件回收清点； 2. 关闭检修电源； 3. 做好检修记录，记录本次检修内容、反措或技改情况，有无遗留问题； 4. 召开班会总结，整理技术文件资料，并存档保管； 5. 接受现场验收，办理工作票终结手续，检修人员全部撤离工作现场	严禁在清理场地的同时负责人前去终结工作票

Task 1   Maintenance of Arc Extinguishing Chamber of SN10-10 Low Oil Circuit Breaker

1.1   Work Tasks

Give maintenance to the arc extinguishing chamber of SN10-10 low oil circuit breaker. Grasp the types, characteristics, arc extinguishing principle and arc extinguishing chamber structure of the oil circuit breaker, master the change method process, operation skills, maintenance technology standard and operation precautions of the arc extinguishing chamber of SN10-10 low oil circuit breaker, and accumulate methods and experience in changing the arc extinguishing chamber of low oil circuit breaker on site.

1.2   References

(1) *Code for Construction and Acceptance of High-voltage Electrical Apparatus of Electric Equipment Installation Engineering* (GB 50147-2010).

(2) *Substation Maintenance Management Regulations of State Grid Corporation of China (Trial Implementation), Volume 5 – Rules for Maintenance of Switch Cabinets*.

(3) *Regulations of State Grid Corporation of China on Management of Substation Operation and Maintenance*.

(4) *Substation Maintenance Management Regulations of State Grid Corporation of China*.

(5) *Electric Power Safety Working Regulations (Power Transformation) of State Grid Corporation*.

(6) *Field Maintenance Guidelines for High-voltage Switch Cabinets of Sichuan Electric Power Company*.

1.3   Work Requirements

(1) The operation belongs to indoor power interruption operation, without weather requirements.

(2) Equipments under maintenance are to be isolated from other electrified equipments by a fence. The only exit is to be provided at the place oriented towards the passage.

(3) For the maintenance bay, the bus and outlet cable are required to be powered off and grounded.

(4) Operators are to be in good mental state, and are aware of safety measures, technical measures and hazards to site operation during operation.

(5) Take safety measures on practical training site as per regulations on production site, and strictly implement standard operation.

(6) Operators are requested to wear labor protection appliances to ensure safety protection.

1.4 Preparation for Work

1.4.1 Hazards and Preventive and Control Measures

1. High-voltage electric shock

Hazard: during the maintenance of circuit breaker, the improper installation of grounding wire, the stray into charging bay, or the insufficient safe distance from the live part can all cause personal electric shock; When disconnecting and connecting the circuit breaker lead, the lead may swing to the operating equipment under the action of external force, thus causing personal electric shock.

Preventive and control measures: before work, the person in charge of work must inspect and confirm the complete implementation of work ticket safety measures, and notify the operators of the power interruption scope and safety precautions in detail; When removing the lead, grounding protection shall be maintained, and the lead shall be bound securely; Fence needs to be installed between the switch cabinet and the adjacent live equipment, with appropriate number of sign boards indicating "Stop! High Voltage, Danger!" hanging on the inner side. The fence is set with the only access, with the sign board indicating "Entrance/Exit" hanging at the access; The maintainer entering the circuit breaker bay shall not touch the connecting rod of isolating circuit breaker.

2. Low-voltage electric shock

Hazard: there is low-voltage current in the incoming side (upper end) of air switch of control circuit, closing circuit and lighting circuit. After closing the air switch, the secondary circuits of circuit breaker and switch cabinet are all considered as live circuits, and touching the circuits may cause low-voltage electric shock.

Preventive and control measures: if an operator hangs the sign board indicating "No Closing, Work in Progress" on an air switch, the person in charge of work must request for safety precautions change from the operator before closing the air switch; When the person in charge of work repeatedly emphasizes the switching of air switch to the members of the work group, it is forbidden to stretch the fingers into the upper end of the air switch; At least two persons are required during work, one for supervision and the other for operation, who both are instructed by the person in charge of work.

3. Mechanical injury

Hazard: during opening and closing of circuit breaker, if notification is not given to other members of the work team, the members may suffer mechanical injury; For the mechanical operation during energy storage of equipment, there shall be the risk of misoperation.

Preventive and control measures: during opening and closing of circuit breaker and grounding switch, it is necessary to shout and sing loudly, and to give operation only after receiving the loud response from all members of the work team. Before operation, it needs to first inspect the open position of circuit breaker and the release of energy, and then to disconnect control and closing power of circuit breaker to avoid misoperation of mechanism; During operating, the operators shall stay away from the moving parts; At least two persons are required during work, one for supervision and the other for operation, who both are instructed by the person in charge of work.

4. Equipment damage

Hazard: the equipment may be damaged to varying degrees if opening and closing of circuit breaker and grounding switch are not in accordance with requirements, or rough handling is adopted, or tools or components are left behind during operation.

Preventive and control measures: rough handling is prohibited. If there is a phenomenon that operation cannot be given during the operation, the operation shall be stopped immediately, no brute force shall be used, and report shall be sent to the person in charge immediately; Before reassembly of the parts, comprehensive inspection shall be performed to prevent foreign matter from being left inside.

1.4.2 Work Tools and Material Selection

Tools, instruments and materials for the maintenance of arc extinguishing chamber of SN10-10 low oil circuit breaker are as shown in Table 3-1.

Table 3-1 Tools, Instruments and Materials for the Maintenance of Arc Extinguishing Chamber of SN10-10 Low Oil Circuit Breaker

Category	Name	Specification and model	Quantity	Remarks
Specialized tools	Special tool for junk ring	Homemade	1	
	Special tool for contact finger	Homemade	1	
General tools	Spline end wrench	5.5–32	1 set	
	Movable spanner	8	1	
	Open-end spanner	6–32	1 set	
	Allen wrench	1.5–10	1 set	
	Long-nose pliers	6	1	
	Cross screwdriver	4–6	1	
	Slotted screwdriver	3, 4, 6	1	
	Wood hammer	Center	1	
	Socket spanner	32 packs	1	
	Oil drum		1 drum	
	Extended allen wrench	10 mm	1	
	Oil tray for maintenance	60 cm × 40 mm	3 drum	
	Vernier caliper	150 mm	1	
	Depth vernier caliper	200 mm	1	
Material	Small towel		10 pieces	
	White cloth		3 m	

Continued

Category	Name	Specification and model	Quantity	Remarks
Material	Transformer oil		24 kg	
	Plastic cloth		3 m	
	Neutral petroleum jelly		1 Bottle	
	Cotton waste		2 kg	
	Gauze		6	
	Seal ring		1 set	
	Grease	1kg/ package	1 package	
	Yellow, green, red and black phase paints		0.5 kg respectively	
	Hand sanitizer		1 Bottle	
	Washing gasoline	70	5 L	
	Paint brush	1	4	
	1032 insulating varnish		1 kg	
	Grease		A little	

1.4.3 Division of Labor Among Operators

Personnel allocation of the maintainers for the maintenance of arc extinguishing chamber of SN10-10 low oil circuit breaker is as shown in Table 3-2.

Table 3-2 Personnel Allocation of the Maintainers for the Maintenance of Arc Extinguishing Chamber of SN10-10 Low Oil Circuit Breaker

S/N	Job	Quantity	Responsibilities
1	Person in charge of work	1	Be responsible for work division of working staffs, site survey, stipulation of operation scheme, pre-shift meeting, safety supervision during operation, handling of emergencies during work, supervision of work quality and summary of post-shift meeting
2	Operator	1	Responsible for the main operation of the maintenance of arc extinguishing chamber of SN10-10 low oil circuit breaker
3	Auxiliary operators (optional)	1	Assist the operators to perform the operation

1.5 Operation Procedures

The operation process of maintenance of arc extinguishing chamber of SN10-10 low oil circuit breaker is as shown in Table 3-3.

Table 3-3　Operation Process of Maintenance of Arc Extinguishing Chamber of SN10-10 Low Oil Circuit Breaker

S/N	Scope of work	Operational steps and standards	Safety measures and precautions
1	Commencement procedures	1. Read the work tickets after falling into procession, and explain the work content, safety measures and precautions; 2. Inspect and determine that tools and instruments should be complete, qualified, and placed in accordance with the provisions; 3. During operation, the safety distance between the maintainer and the 10 kV live equipment must be not less than 0.35 m	1. Relevant equipment has been in power interruption; 2. Uninvolved personnel shall not enter into the site
2	Preparations before work	1. Check if all instruments and tools are complete, and if their appearance and test are acceptable; 2. The person in charge of work and work permitter shall make an tour inspection for equipments under maintenance, and confirm all safety measures as listed by work ticket have been properly implemented. It is also necessary to check if safety measures are complete, and if the site is provided with conditions for commencement of work, and make supplements as required; 3. Implement work permit procedures; 4. Call in toolbox meeting participated by members of work team; 5. Record parameters on the equipment nameplate; 6. Arrange the maintenance site and prepare the special oil drum for waste transformer oil recovery. The oil drum should be able to be sealed and transported	1. Instruments and tools are free of damage, deformation and malfunction. Qualification certificates for instruments and tools to be tested are within the term of validity; 2. Prevent other persons from entering the site during tour inspection; 3. Toolbox meeting shall cover dual designations of work place; Working hours and contents; Work division; Working hazards as well as preventive and control measures; Power-cut scope and safety measures on work site; 4. All work members shall properly wear such labor protection appliances as safety helmet, working clothes, working shoes and protective gloves; 5. An enclosed barrier with access shall be provided around the maintenance site, and appropriate number of sign boards indicating "Stop! High Voltage, Danger!" shall be hanged on the surrounding barrier back facing the live equipment, and the sign board indicating "Entrance/Exit" shall be hanged at the access
3	Equipment status inspection	The person in charge shall lead members of the shift team to inspect and confirm that: 1. The main body, structure and base frame of circuit breaker are free from deformation and corrosion; the insulating surface is free from damage or dirt and shows no discharge mark; the metal surface of conductive connection exhibits no overheating mark; the phase color and identification are clear;	The test before maintenance can be arranged according to the work task: 1. Measure the resistance of each phase conductive circuit; 2. Test mechanical characteristics of circuit breaker; 3. Measure the total stroke, over-stroke and three-phase non-synchronism

Continued

S/N	Scope of work	Operational steps and standards	Safety measures and precautions
3	Equipment status inspection	2. The indication of mechanical indicator is consistent with the actual condition; the indicator light and signal are in line with the actual condition; the four-connecting rod mechanism is located at the dead point position; 3. The oil level is located between 1/4 and 3/4 scales of oil pointer; 4. The various parts are in tight connection without leakage; there is no flash leaching; there is no oil drop mark; 5. The electric opening and closing actions are crisp, there is no abnormal sound or oil injection, and the position signal change is accurate; 6. The operating mechanism and transmission mechanism operate normally and give no noise or jam, and the dead point locking is reliable; 7. Each wiring terminal of terminal strip shows no loosening or slippage; 8. The connection and grounding parts of circuit breaker are fastened with screw, and the connection is reliable	
4	Circuit breaker disassembly	1. Unscrew the bottom oil drain bolt and release the oil; 2. Unscrew the upper and lower terminal leads; 3. Use Allen wrench to unscrew the four hex bolts between the upper cap assembly and the upper wiring, remove the upper cap assembly, unscrew the M8 fixed bolt of the static contact base assembly, and take down the static contact base assembly and insulating sleeve; 4. Remove the upper wiring base, unscrew the upper pressure ring with special tool, and take out the insulating ring and insulating gasket above the arc extinguishing chamber; 5. Use special tool to unscrew the four hex bolts on the pressure ring, and take out the insulating cylinder assembly and the lower wiring base assembly;	1. No leakage is permitted; 2. The removed parts and components should be placed in a clean and dry place and put in phase sequence to prevent loss; 3. Insulating components should not be damaged

Continued

S/N	Scope of work	Operational steps and standards	Safety measures and precautions
4	Circuit breaker disassembly	6. Disassemble the connection between the insulating tie rod and the external crank arm of the base, lift the conductive rod, remove the 10 mm × 55 mm connecting pin connected with the connecting plate inside the base, and draw out the conductive rod assembly	
5	Upper cap assembly maintenance	1. Unscrew the M8′12 snap-head bolt, remove the exhaust vent cover, screw out the M12 bolt, take down the oil-gas separator, and screw off the oil return valve; 2. Clean and inspect the various parts; If the oil return valve is not tightly sealed, small hammer can be used in tapping to ensure the reliable sealing line; 3. Reinstall in the reverse order of removal	1. The upper cap shall show no sand hole; 2. Each exhaust duct shall be smooth; 3. The oil return valve shall act flexibly and the steel ball shall be sealed reliably
6	Static contact base assembly maintenance	1. Disassembly. (1) Remove the contact finger and arc contact finger from the static contact base with special tool, and take out the spring leaf; (2) Remove the check valve; (3) Unscrew the three bolts to separate the static contact base from the contact terminal holder. 2. Cleaning and inspection. (1) Clean the various parts with qualified insulating oil; (2) Inspect the conductive contact surface of contact finger and arc contact finger. If there are minor burns, fine file and no.0 abrasive cloth can be used in finishing. Change is needed for serious burns; (3) Inspect and confirm whether there is burn mark on the contact surface between the contact terminal frame and the contact base. For minor burns, no.0 abrasive cloth shall be used in polishing, and the finishing at the waist of contact finger shall be not greater than 0.5 mm; (4) Inspect whether the dirt in the contact finger tail groove of the contact base is completely removed, and whether the barrier is intact; (5) Inspect whether the spring leaf suffers deformation and damage;	1. The conductive contact surface shall be smooth and flat. When the burn has area reaching 30% and depth greater than 1 mm, change is needed. If the burn depth is greater than 2 mm in the part of copper tungsten alloy, change is needed; 2. The contact between the contact terminal frame and the contact base shall be tight, and there shall be no burn mark on the contact surface between the contact base and the contact finger; 3. The barrier of the contact base shall be free from cracks and teeth missing. The cylindrical pin fixing the barrier shall show no phenomenon of peeling or exiting; 4. The bending degree of the spring leaf shall not exceed 0.2 mm, and the contact finger, contact base and barrier shall not be burned; 5. There shall be no copper molten particles and impurities in the check valve, the steel ball shall act flexibly, and the two ends of the cylinder resisting steel ball shall be riveted, smooth, and free from protruding; 6. The inner wall shall be free from serious carbonization, burn and flaking, or change shall be needed;

Continued

S/N	Scope of work	Operational steps and standards	Safety measures and precautions
6	Static contact base assembly maintenance	(6) Blow with the mouth to check the sealing condition of the check valve. In case of untight sealing, treatment can be given with the same treatment method for the oil return valve of the upper cap; (7) Inspect whether the paint film of the insulating sleeve is complete, and whether there are phenomena of peeling, flaking and foaming. 3. Reassembly. (1) Reassemble in the reverse order of disassembly. When assembling the contact base and the contact finger, it needs to pay attention to the relative position between the arc contact finger and the cross-blown arc path and positioning pin for the conformance with requirements; (2) Measure the diameter of the closed circle of static contact finger for the conformance with requirements	7. Arc contact finger must be installed in the position of barrier with special mark. If there is no special mark on the barrier, the arc contact finger shall be installed aligning the direction of the cross-blown arc path; 8. The closing diameter of the static contact finger is 18.5−20 mm
7	Arc extinguishing chamber assembly maintenance	1. Clean the arc extinguishing plates 1, 2, 3, 4 and 5 with qualified insulating oil; 2. Adjust the gasket, insulating ring and arc separation wall; 3. Check the burn condition of the arc extinguishing plate and insulating part. For minor burns, No. 0 abrasive cloth can be used in gently wiping the arc mark. In case of serious burns, change shall be needed	1. The surface of the arc extinguishing plate shall be smooth, flat and free from carbonization particles, cracks or damage; 2. After treatment, the length aperture of the first arc extinguishing plate shall not exceed 28 mm, and the aperture of the remaining parts shall not exceed 26 mm; 3. The insulating part shall be free from burn damage
8	Insulating cylinder assembly maintenance	1. Disassembly. (1) Take out the lower pressure ring in the insulating cylinder and the connecting spring; (2) Unscrew the four bolts, and remove the glass cover of the oil level indicator. 2. Cleaning and inspection. (1) Check the inner and outer walls of the upper wiring base, and wipe them clean with cloth; (2) Clean the glass of oil level indicator, as well as the upper and lower holes in the base of oil level indicator; (3) Check whether the contact surface of the wiring terminal and the isolating plug of handcart cabinet exhibits roughness and overheating. After finishing the contact surface, neutral petroleum jelly shall be applied;	1. The upper wiring base shall show no phenomena of sand holes, cracks or oil seepage; 2. The upper and lower holes in the base of the oil level indicator shall be unblocked; 3. The contact surface shall be smooth, the silver-plated layer of contact finger of isolating plug of handcart cabinet shall be intact, and the spring shall be free from deformation or damage; 4. The paint film on the surface of insulating cylinder shall be smooth and free from paint-shedding, the inner wall shall show no discharge mark, and the semicircular groove shall be free from damage or deformation; 5. The lower pressure ring shall be intact, and the spring shall show no flattening deformation

Continued

S/N	Scope of work	Operational steps and standards	Safety measures and precautions
8	Insulating cylinder assembly maintenance	(4) Wipe the inner and outer walls of the insulating cylinder with a cloth, and check the state of appearance. If the paint film is damaged, 1032 insulating varnish shall be recoated; (5) Check whether the spring used to connect the lower pressure ring with the insulating cylinder shows deformation. 3. Reassembly. Reinstall oil level indicator after inspection and cleaning	
9	Lower wiring base assembly maintenance	1. Disassembly. Unscrew the four bolts, and remove the guide device and rolling contact terminal assembly. 2. Cleaning and inspection. (1) Check whether the conductive strip of guide device is in close contact with the wiring base, and whether there are burn marks on the conductive surface on both sides. If there are marks, the cause shall be ascertained, and treatment and finishing shall be given; (2) Check the roller action of rolling contact terminal assembly, the riveting at the both ends of shaft rod, and the completeness of spring, connecting plate and washer; (3) Clean and check whether the upper and lower guide insulating plates are cracked or damaged. Remove the damaged ones from the conductive strip, and change them. (4) Clean and check the conductive contact surface of the lower wiring base and the isolating plug of handcart cabinet. If the contact is bad, repair shall be needed, and neutral petroleum jelly shall be applied after repair. 3. Reassembly. Reinstall in the reverse order	1. The surface of conductive strip and rolling contact terminal shall not be burned, and the contact with the wiring base shall be tight; 2. The roller rotation of rolling contact terminal shall be flexible, the rotating shaft shall not be bent, the both ends shall be riveted, the various parts shall be complete, and the spring characteristics shall meet the requirements; 3. The upper and lower guide insulating plates shall be free from damage or crack, and the guide port shall be smooth
10	Conductive rod assembly maintenance	1. Disassembly. The moving contact terminal usually cannot be disassembled, while the tightness of the connection between the moving contact terminal and the conductive rod shall be checked. If the burning damage is serious, and disassembly is required for treatment, special tools shall be used to unscrew the moving contact terminal. If the disassembly is difficult,	1. For the moving contact terminal, if the burn depth is greater than 2 mm in the part of copper tungsten alloy, change shall be needed. The copper part shall not be burned; 2. The connection between the moving contact terminal and the conductive rod shall be tight and firm;

Continued

S/N	Scope of work	Operational steps and standards	Safety measures and precautions
10	Conductive rod assembly maintenance	conductive rod assembly shall be removed, and then disassembly can be performed on the jaw vice, thereby preventing various connecting plates of the base from bending and deformation under the force. 2. Inspection. (1) When removing the moving contact terminal, the thread of conductive rod and the internal spring shall be checked for deformation; (2) Check whether the riveting of the conductive rod and the buffer is firm, and whether there are serious impact marks at the lower port of buffer. If there are serious impact marks, the cause shall be found and eliminated; (3) Check the bending degree of the conductive rod, and straighten the unqualified one. 3. Reassembly. Reinstall in the reverse order	3. Various joints of the conductive rod assembly shall be smooth and free from convex; 4. The connecting thread shall show no disorder phenomenon; 5. The spring shall be free from breakage and serious corrosion; 6. The riveting shall be firm, and the both ends of the rivet shall be smooth; 7. There shall be no serious impact marks in the part of lower break of buffer; 8. The bending degree of conductive rod shall be less than 0.15 mm
11	Base assembly maintenance	1. Disassembly. (1) Unscrew the special bolt in the front convex part, use special tools to remove the elastic pin on the rotating shaft, screw out the rotating shaft, take out the internal crank arm and connecting plate of the base, and disassemble the shaft pin to separate the internal crank arm and the connecting plate; (2) Unscrew the thread bushing sealed by revolving shaft with special tool, and take out the copper washer and the skeleton seal ring; (3) Unscrew the three bolts, and remove the oil buffer piston rod. 2. Cleaning and inspection. (1) Clean and check whether the various shaft pins and split pins are complete, whether the internal crank arm and the connecting plate are deformed, whether the rivets are firm, and whether the rubber brake blocks are intact; (2) Clean the rotating shaft and the external rocker arm, and check the welding condition of each part; (3) When replacing the skeleton seal ring, it needs to turn over the inner lip for careful damage inspection; (4) Clean and check the oil buffer piston rod and the lower disc;	1. The connecting plate and the internal crank arm shall not be deformed or damaged; 2. The rubber brake block shall not be cracked or damaged, and the riveting shall be firm; 3. The non-parallel degree between the rotating shaft and the shaft pin on the external rocker arm is not greater than 0.3 mm; 4. There shall be no mechanical scars on the surface of the rotating shaft; 5. The welding joints of each part shall be firm; 6. The skeleton seal ring shall not be damaged, and the sealing surface shall be free from burrs, cracks, pores, defects, etc; 7. The spring on the skeleton seal ring shall be intact and undamaged, and the joint shall be well butted; 8. There shall be no serious impact phenomenon at the end of the piston rod; 9. The riveting between the piston rod and the disc shall be firm, while the piston rod shall still be able to move; 10. The base shall be free from sand holes, cracks, etc. The shaft holes shall be smooth and free from burrs;

Continued

S/N	Scope of work	Operational steps and standards	Safety measures and precautions
11	Base assembly maintenance	(5) Clean the base interior and shaft holes with qualified insulating oil, perform internal and external inspection, and polish any burrs in the shaft holes. 3. Reassembly. Reinstall in the reverse order of disassembly. When installing the rotating shaft, it needs to first apply a small amount of calcium-based grease on the outer surface of the skeleton seal ring, and then press the skeleton seal ring into the hole with uniform force, put the thread sleeve and washer on the rotating shaft, slowly screw in the rotating shaft by aligning the shaft sleeve hole of the internal crank arm, properly tighten the thread bushing with special tool, and finally rotate the rotating shaft and the crank arm to the open position to ensure alignment of the two holes and drive the elastic pin	11. The lip of the skeleton seal ring shall be inward, the chamfer end of the elastic pin shall be inward, and the two ends of the elastic pin shall be flush with the bushing of the crank arm; 12. The relative position between the external rocker arm and the internal crank arm shall be correct, and the rotation of the rotating shaft shall be flexible after reinstallation
12	Circuit breaker assembly	1. Replace all seal rings; 2. The conductive rod assembly is assembled with the connecting plate in the base; 3. Clean the upper and lower seal rings of the lower wiring base, place the seal ring in the proper position, and then put the wiring base in the base port for alignment; 4. Put the spring into the semicircular groove on the inner wall of insulation, lay down the pressure ring, and make its inner circular arc table evenly pressing on the spring. Then place the insulating cylinder on the lower wiring base for alignment, and evenly and diagonally tighten the four hex bolts on the lower pressure ring with special tools, to ensure that the conductive rod moves up and down flexibly; 5. Install the insulating bushing, arc extinguishing plate and adjusting gasket in turn, and pay attention to the installation direction of the bottom insulating ring, to ensure that the direction of the cross-blown arc path is opposite to that of the wiring terminal of the upper wiring base. After completing the installation of arc extinguishing plate, the insulating gasket and insulating ring can continue to be installed;	1. Seal ring shall be free from damage and crack, etc. The sealing surface shall be flat and free from piled paint; 2. The center of the upper and lower wiring terminals must be in a straight line with the center in the base for elastic pin loading and unloading; 3. The connecting spring of the insulating cylinder and the lower pressure ring shall be stuck into the semi-circular groove on the inner wall of the insulating cylinder, and the lower pressure ring shall not be deflected; 4. The seal ring shall not be inclined, and the gap after tightening shall not exceed 0.1 mm; 5. The angle between the first cross-blown arc path and the upper wiring terminal is 180°; 6. M dimension standard: (135 ± 0.5) mm; 7. The conductive rod shall act flexibly and be free from jam phenomenon

S/N	Scope of work	Operational steps and standards	Safety measures and precautions
12	Circuit breaker assembly	6. Use special tools to install the upper pressure ring, press the arc extinguishing chamber, and measure the M dimension. In case of disqualification, the thickness of the insulating gasket between the fourth and fifth arc extinguishing plates can be adjusted; 7. After completing the assembly, connect the transmission tie rod with the crank arm. After several times of manual operation, check the correctness of the connection	
13	Overall commissioning	1. Adjust the distance between the upper end of the arc extinguishing plate and that of the upper lead base; 2. Adjust the height of the closed position of the moving contact terminal; 3. Adjust the stroke of the conductive rod; 4. Adjust the synchronism; 5. Adjust the closing spring buffer; 6. Adjust the concentricity of the moving and static contact terminals	1. For the arcing distance, it is (135 ± 0.5) mm for SN10-10 circuit breaker and (153 ± 0.5) mm for SN10-10III circuit breaker. If the requirements cannot be satisfied, the gasket between the arc extinguishing plates can be adjusted; 2. For the distance between the upper end face of the moving contact terminal and that of the lead base, it is (130 ± 1.5) mm for SN10-10I, (110 ± 1.5) mm for SN10-10II, and $[122\pm(1\sim2)]$ mm for SN10-10III, thereby meeting the requirements of overstroke. It can be achieved by adjusting the length of the insulating link from the main shaft to the mechanism chamber, or by adjusting the length of the transmission rod from the main shaft to the operating mechanism. That is, when the connecting rod is shortened, the aforementioned dimension shall decrease and the overstroke shall increase, otherwise the aforementioned dimension shall increase and the overstroke shall decrease. 3. For the total stroke, it is (145 ± 3) mm for SN10-10I, (155 ± 3) mm for SN10-10II, and (157 ± 3) mm for SN10-10III. In case of disqualification, the length of the transmission tie rod or connecting rod can be adjusted, or the number of iron plates and rubber washers of the opening stopper can be increased or decreased to satisfy the requirements. The overstroke shall not be affected after adjustment.

Continued

S/N	Scope of work	Operational steps and standards	Safety measures and precautions
13	Overall commissioning		4. Three-phase opening non-synchronism is required to be not greater than 2 mm. In case of disqualification, the length of each phase insulating connecting rod can be adjusted to achieve the requirements. During connecting rod adjustment, the distance between the end face of moving contact terminal and the upper end face of upper lead base shall not be affected, and the data required by aforementioned various types shall be maintained. If the error of ±2 mm of this dimension can be guaranteed, it shall not need to measure the three-phase opening non-synchronism. Note that the synchronism adjustment can be conducted at the same time of stroke adjustment; 5. When the circuit breaker is in the closed position, the end roller of the crank arm shall be hit on the buffer with a spacing of 2~4 mm away from the limit position
14	End of work	1. Clean up the work site, collect and count all tools and instruments, dispose the waste according to relevant regulations, and recycle and count materials and spare parts; 2. Turn off the maintenance power supply; 3. Complete maintenance records, record the maintenance content, counter-measures or technical changes, and determine whether there are any remaining problems; 4. Hold shift meetings to summarize, sort out technical documents and materials, and archive them for maintenance; 5. Receive the field acceptance inspection, transact the work ticket termination procedures, and evacuate all maintainers from the work site	It is strictly prohibited for the person in charge to terminate the work ticket while cleaning the site

任务二　断路器二次回路故障处理

一、工作任务

按照标准化作业流程，观察现场设备故障现象，查找 KYN28-12 型开关柜（配 VS1-12 型断路器）二次回路故障，分析故障原因并处理。

二、引用标准

（1）《电气装置安装工程高压电器施工及验收规范》(GB 50147—2010)。
（2）《国家电网公司变电检修管理规定（试行）第 5 分册开关柜检修细则》。
（3）《3.6 kV~40.5 kV 交流金属封闭开关设备和控制设备》(GB 3906—2006)。
（4）《国家电网公司电力安全工作规程（变电部分）》。
（5）《四川省电力公司高压开关柜现场维护导则》。

三、工作要求

本工作内容为室内作业，不受天气影响。基本工作规范及要求如下：
（1）作业前应做好现场查勘。
（2）着装规范，工具、仪表、材料准备充分，检修技术资料（包括运行缺陷记录、标准化作业书）齐全。
（3）现场工作票已办理，工作开始前须履行工作许可手续。开工前应向工作班成员交代现场安全措施、危险点及控制措施。
（4）作业全过程必须严格遵守《国家电网公司电力安全工作规程》（以下简称《安规》），工作过程中严格按标准化作业书进行作业。

四、工作准备

（一）危险点及预控措施

1. 高压触电

危险点：误入、误登带电间隔导致高压触电。

预控措施：工作前向作业人员交代清楚临近带电设备，并加强监护；检修设备与相邻运行设备必须用围栏明显隔离，并悬挂"止步，高压危险"标示牌，标示牌应面对检修设备；工作人员应走指定通道，在遮栏内工作，严禁擅自移动和跨越遮栏。

2. 低压触电

危险点：控制回路、合闸回路、照明回路低压断路器进线侧（上端）均低压电；合上空开后，断路器和开关柜二次回路均视为带电状态，触摸可能造成低压触电。

预控措施：应由两人进行，一人操作，另一人监护；检修电源应有漏电保护器，移动电具金属外壳均应可靠接地；检修前应断开交、直流操作电源及储能电机电源，严禁带电拆、接操作回路电源接头。

3. 机械伤害

危险点：断路器分合、接地刀闸分合时未通知其他工作班成员，可能会对工作班成员造成机械伤害；将断路器从"试验"位置转移至"检修"位置或将断路器从"检修"位置转移至"试验"位置时，手车可能发生倾倒，对操作者造成机械伤害。

预控措施：应统一指挥，做好协调配合；检修前应将机构放在分闸位置；处理故障前，应断开储能电机电源，将能量全部释放；保护传动必须得到现场一次负责人许可，传动时本体及机构上禁止任何工作，人员撤离；弹簧机构操动时，不得靠近电动机和弹簧。

4. 设备损坏

危险点：未按照要求分合断路器、分合接地刀闸、野蛮操作，或者操作时遗留工具在柜体内，均会对设备造成不同程度的损坏。

预控措施：禁止野蛮操作，操作过程中，禁止遗留任何工具于柜内，若出现不能操作的现象，应当立即停止，不得使用蛮力，立即汇报负责人；工作时至少需要两人，一人监护一人操作，听工作负责指挥。

（二）工器具及材料选择

断路器二次回路故障处理工具及材料见表3-4。

表3-4 断路器二次回路故障处理工具及材料表

类别	名称	规格型号	数量	备注
专用工具	储能把手		1把	
	手车摇柄		1个	
通用工具	一字螺丝刀	ST3×100 mm	1把	
	十字螺丝刀	ST3×100 mm	1把	
	尖嘴钳	6号	1把	
	平口钳	6号	1把	
	万用表		1只	
	信号指示灯		3个	红、绿、白各1
	按钮		2个	红、绿各1
	微动开关		1个	
	辅助开关		1个	
材料	分、合闸线圈		1个	
	闭锁电磁铁		1个	
	整流元件板		1个	

（三）作业人员分工

断路器二次回路故障处理人员分工见表3-5。

表 3-5　断路器二次回路故障处理人员分工

序号	工作岗位	数量	职责
1	工作负责人	1	负责本次工作的人员分工、现场查勘、作业方案制定、召开班前会、作业过程中安全监督、工作中突发状况的处理、工作质量的监督、班后会总结
2	操作人员	1	主要负责断路器二次回路故障的排查、处理
3	辅助操作人员（可无）	1	辅助操作人员进行二次回路故障的处理

五、作业程序

断路器二次回路故障处理作业流程见表 3-6。

表 3-6　断路器二次回路故障处理作业流程

序号	作业内容	作业步骤及标准	安全措施及注意事项
1	开工手续	1. 确认工作地点正确，设备名称和编号标识清楚； 2. 得到许可工作命令	1. 核对设备名称和设备编号，确认工作地点； 2. 得到许可后才能进入工作场地
2	工作前准备工作	1. 检查工器具是否齐全，检查工器具外观和试验合格； 2. 工作负责人同工作许可人巡视待检修设备，确认工作票所列安全措施已经正确执行，安全措施是否完备，现场是否具备开工条件，必要时进行补充； 3. 执行工作许可手续； 4. 对工作班成员召开班前会； 5. 抄写设备铭牌参数	1. 工器具无损伤、变形、失灵现象，需要试验的工器具合格证在有效期内； 2. 巡视现场时禁止无关人员进入现场； 3. 班前会应包含工作地点双重名称；工作时间与内容；工作分工；工作危险点及预控措施；停电范围及工作现场安全措施； 4. 全体工作成员应当正确穿戴安全帽、工作服、工作鞋、劳保手套等劳动保护用品
3	确认设备状态	负责人带领工作班成员确认设备状态，观察故障现象，要求手动操作，观测仔细，记录清晰	观测和记录，检查故障记录情况
4	分析故障原因	正确分析造成故障的原因，得出正确结论，确定故障点	认真阅读电气原理图，通过故障现象分析动作原理，找出故障点
5	故障处理	接线正确，连接点牢固可靠	加强监护，发现违章立即制止，故障点处理后检查确认
6	断路器动作试验	电动合、分正常，机械、电气位置指示信号正确	1. 通电前必须呼唱； 2. 加强监护，发现违章立即制止
7	工作终结	1. 自检并清理打扫现场，所检设备完好，无遗留问题，现场清扫做到工完、料尽、场地清； 2. 办理工作终结手续，所有工作人员确已全部撤离现场，向工作许可人交代检修的问题，做检修记录	1. 清点人数确认所有人员确已撤离现场，向许可人交代清楚并做检修记录； 2. 严禁负责人前去终结工作票的同时清理场地

Task 2 Fault Treatment of the Secondary Circuit of Circuit Breaker

2.1 Work Tasks

In accordance with the standardized operation process, observe the fault phenomenon of field equipment, seek for the fault of the secondary circuit of KYN28-12 switch cabinet (equipped with VS1-12 circuit breaker), analyze the cause of the fault, and give treatment.

2.2 References

(1) *Code for Construction and Acceptance of High-voltage Electrical Apparatus of Electric Equipment Installation Engineering* (GB 50147-2010).

(2) *Substation Maintenance Management Regulations of State Grid Corporation of China (Trial Implementation), Volume 5 – Rules for Maintenance of Switch Cabinets*.

(3) *Alternating-current Metal-Enclosed Switchgear and Controlgear for Rated Voltages Above 3.6 kV and up to and Including 40.5 kV* (GB 3906-2006).

(4) *Electric Power Safety Working Regulations (Power Transformation) of State Grid Corporation*.

(5) *Field Maintenance Guidelines for High-voltage Switch Cabinets of Sichuan Electric Power Company*.

2.3 Work Requirements

This operation belongs to indoor operation and will not be affected by weather. The basic operating specifications and requirements are as follows:

(1) Site survey shall be done before operation.

(2) The dressing shall satisfy requirements, tools, instruments and materials shall be fully prepared, and maintenance technical documents (including operating defect records and standardized operation books) shall be complete.

(3) The on-site work ticket has been processed, and the work permit procedures must be completed before the operation. Field safety measures, hazards and control measures should be explained to the members of work team before the commencement.

(4) The whole process of operation must strictly observe the *Electric Power Safety Working Regulations of State Grid Corporation of China* (hereinafter referred to as *Safety Regulations*), and operation shall be performed in strict accordance with the standardized operation book.

2.4 Preparation for Work

2.4.1 Hazards and Preventive and Control Measures

1. High-voltage electric shock

Hazard: entering or ascending the charging bay by mistake may cause high-voltage electric shock.

Preventive and control measures: inform operators of nearby live equipment before start of work and strengthen the supervision; Maintenance equipment and adjacent operating equipment must be clearly separated by fence, and the sign board indicating "Stop! High Voltage, Danger!" needs to be hanged by facing the maintenance equipment; The operators shall take the designated channel and work in the barrier, and they are strictly prevented from moving and crossing the barrier without authorization.

2. Low-voltage electric shock

Hazard: there is low-voltage current in the incoming side (upper end) of low-voltage circuit breaker of control circuit, closing circuit and lighting circuit. After closing the air switch, the secondary circuits of circuit breaker and switch cabinet are all considered as live circuits, and touching the circuits may cause low-voltage electric shock.

Preventive and control measures: implement by two persons, one for operation and one for supervision; The maintenance power supply shall have leakage protector, and the metal shell of the mobile electric appliance shall be reliably grounded; Before maintenance, the AC and DC operating power supply and the energy storage motor power supply shall be disconnected, and live disconnection and connection of the power supply connector of operating circuit with power is strictly forbidden.

3. Mechanical injury

Hazard: during opening and closing of circuit breaker and grounding knife switch, if notification is not given to other members of the work team, the members may suffer mechanical injury; When the circuit breaker is moved from the "Test" position to the "Maintenance" position or the circuit breaker is moved from the "Maintenance" position to the "Test" position, the handcart may be tripped, causing mechanical injury to the operator.

Preventive and control measures: unified command, coordination and cooperation are needed; The mechanism shall be placed in the open position before maintenance; Before fault treatment, the energy storage motor power supply shall be disconnected to release the energy thoroughly; The protective transmission must be approved by the site primary supervisor, it is forbidden to conduct any work on the body and mechanism during transmission, and the personnel shall be evacuated; When the spring mechanism is operating, it is forbidden to approach the motor and the spring.

4. Equipment damage

Hazard: the equipment may be damaged to varying degrees if opening and closing of circuit breaker and grounding knife switch are not in accordance with requirements, or rough handling is adopted, or tools are left in the cabinet during operation.

Preventive and control measures: rough handling is prohibited. It is forbidden to leave any tool in the cabinet during operation. If there is a phenomenon that operation cannot be given during the operation, the operation shall be stopped immediately, no brute force shall be used, and report shall be sent to the person in charge immediately; At least two persons are required during work, one for supervision and the other for operation under instructions of person in charge of work.

2.4.2 Work tools and material selection

Tools and materials for the fault treatment of the secondary circuit of circuit breaker are shown in Table 3-4.

Table 3-4　Tools and Materials for the Fault Treatment of the Secondary Circuit of Circuit Breaker

Category	Name	Specification and model	Quantity	Remarks
Specialized tools	Energy storage handle		1	
	Rocking handle of handcart		1 drum	
General tools	Slotted screwdriver	ST3 × 100 mm	1	
	Cross screwdriver	ST3 × 100 mm	1	
	Long-nose pliers	6	1	
	Flat-nose pliers	6	1	
	Multimeter		1	
	Signal indicator		3 drum	1 for red, green and white respectively
	Button		2 drum	1 for red and green respectively
	Microswitch		1 drum	
	Auxiliary switch		1 drum	
Material	Opening and closing coil		1 drum	
	Latching electromagnet		1 drum	
	Rectifier element board		1 drum	

2.4.3 Division of labor among operators

Allocation of the personnel for fault treatment of the secondary circuit of circuit breaker is shown in Table 3-5.

Table 3-5　Allocation of the Personnel for Fault Treatment of the Secondary Circuit of Circuit Breaker

S/N	Job	Quantity	Responsibilities
1	Person in charge of work	1	Be responsible for work division of working staffs, site survey, stipulation of operation scheme, pre-shift meeting, safety supervision during operation, handling of emergencies during work, supervision of work quality and summary of post-shift meeting

Continued

S/N	Job	Quantity	Responsibilities
2	Operator	1	Mainly responsible for fault investigation and handling of the secondary circuit of circuit breaker
3	Auxiliary operators (optional)	1	Assist the operator in fault treatment of the secondary circuit

2.5 Operation Procedures

The operation process of fault treatment of the secondary circuit of circuit breaker is as shown in Table 3-6.

Table 3-6　Operation Process of Fault Treatment of the Secondary Circuit of Circuit Breaker

S/N	Scope of work	Operational steps and standards	Safety measures and precautions
1	Commencement procedures	1. Make sure that the work place is correct and the equipment name and number are clearly marked; 2. Obtain the work permit command	1. Check the equipment name and number, and confirm the work place; 2. Enter the workplace after obtaining permission
2	Preparations before work	1. Check if all instruments and tools are complete, and if their appearance and test are acceptable; 2. The person in charge of work and work permitter shall make an tour inspection for equipments under maintenance, and confirm all safety measures as listed by work ticket have been properly implemented. It is also necessary to check if safety measures are complete, and if the site is provided with conditions for commencement of work, and make supplements as required; 3. Implement work permit procedures; 4. Call in toolbox meeting participated by members of work team; 5. Record parameters on the equipment nameplate	1. Instruments and tools are free of damage, deformation and malfunction. Qualification certificates for instruments and tools to be tested are within the term of validity; 2. Prevent other persons from entering the site during tour inspection; 3. Toolbox meeting shall cover dual designations of work place; Working hours and contents; Work division; Working hazards as well as preventive and control measures Power-cut scope and safety measures on work site; 4. All work members shall properly wear such labor protection appliances as safety helmet, working clothes, working shoes and protective gloves
3	Confirm equipment status	The person in charge leads the members of the work team to confirm the equipment status and observe the fault phenomenon. Manual operation, careful observation and clear record are required	Give observation and record, and check the fault record
4	Fault cause analysis	Analyze the cause of fault correctly, draw the correct conclusion, and determine the fault point	Read the electrical schematic diagram carefully, analyze the action principle through the fault phenomenon, and find the fault point
5	Fault treatment	The wiring shall be correct, and the connection point shall be firm and reliable	Strengthen the monitoring, immediately stop once discovering violation, and check and confirm after fault treatment
6	Circuit breaker operation test	The electric closing and opening are normal, and the mechanical and electrical position indicating signals are correct	1. Loudly shouting is required before energization; 2. Strengthen the monitoring, and immediately stop once discovering violation

Continued

S/N	Scope of work	Operational steps and standards	Safety measures and precautions
7	End of work	1. Conduct self-inspection and clean up the site. The inspected equipment shall be in good condition, there shall be no remaining problem, and the site cleaning shall be completed after completing operation and exhausting materials; 2. Handle end-of-work procedures. All the operators have actually evacuated from the site, explained the maintenance problems to the work permitter, and made maintenance records	1. Count the number of persons to confirm that all personnel have evacuated from the site, explain clearly to the permitter, and make maintenance records; 2. It is strictly forbidden for the person in charge to clean up the site while terminating the work ticket

任务三 断路器机械特性测试

一、工作任务

断路器机械特性测试。

二、引用的规程规范

(1)《高压电气设备试验方法》(第二版)。
(2)《国家电网公司电力安全工作规程》(变电部分)。
(3)《国家电网公司变电检修管理规定(试行)第 2 分册断路器检修细则》。
(4)《电力设备预防性试验规程》(DL/T 596—2005)。
(5)《电气设备交接试验标准》(GB 50150—2006)。

三、作业条件

(1) 机械特性测试应在良好、干燥天气下进行,在测量过程中,遇到 6 级以上大风以及雷暴雨、冰雹、大雾、沙尘暴等恶劣天气时应停止工作。
(2) 机械特性测试仪的输出电源严禁短路。
(3) 机械特性测试仪尽可能使用外接电源作为测试电源,防止因为内部电源电力不足而影响测试结果。
(4) 进行断路器低电压特性测试时,加在分合闸线圈上的操作电压时间不宜过长,防止烧损线圈。
(5) 测量仪器必须有校验合格证。
(6) 测试仪器的使用环境:-10 ~ +50 °C,相对湿度:≤80%。

四、作业前准备

1. 现场勘察的基本要求及条件

(1) 查阅被试断路器运行情况、了解试验场地等条件。查阅该断路器历年试验报告、相关交接预试规程、断路器运行记录和缺陷情况记录。
(2) 查勘现场设备停电范围及安全措施。
(3) 查勘场地检修电源箱,核实电源容量是否满足要求。
(4) 现场工器具、作业车辆、机具、材料等定置摆放位置。

2. 工器具及材料选择(见表 3-7)

表 3-7 断路器机械特性测试操工器具及材料清单

序号	名称		单位	数量	备注
1	高压断路器机械特性测试仪		台	1	
2	测试仪附件箱	时间通道接线	组	1	
3		控制电源通道接线	组	1	

续表

序号	名　称		单　位	数　量	备注
4	测试仪附件箱	传感器信号通道接线	组	1	
5		储能电源接线	组	1	
6		传感器及固定夹件	组	1	
7		电源线	组	1	
8		外壳接线地	组	1	
9		线夹及插件针脚附件		若干	
10		150 mm 钢角尺	把	1	
11		8～10 呆扳手	把	1	
12		17～19 呆扳手	把	1	
13		ST3-150 mm 十字螺丝刀	把	1	
14		白布	张	1	
15		万用表	个	1	
16		内六角 1.5 mm	把	1	
17		电源盘	个	1	
18		插线板	个	1	
19		油盘	个	2	

3. 危险点及预防措施

（1）高压触电。

危险点：误入带电间隔可能导致高压触电。

预控措施：用围栏将被检修间隔与相邻带电设备（间隔）隔离，并且向作业现场内悬挂"止步，高压危险"标示牌，在靠近道路侧设置唯一出入口，悬挂"从此进出"标示牌；被测试断路器的两侧，各挂装一组三相短路接地线；工作时至少需要两人，一人监护一人操作，听工作负责指挥。

（2）低压触电。

危险点：断路器机构箱内控制回路、合闸回路、照明回路空气开关进线侧（上端）均低压电；合上空开后，断路器二次回路均视为带电状态，触摸可能造成低压触电。外搭接电源及测试过程中，触碰到带电部分可能会造成低压触电。

预控措施：若运行人员在空气开关上悬挂了"禁止合闸，有人工作"标示牌时，工作负责人需要向运行人员申请更变安全措施得到许可后，方能合上空开；工作负责人反复对工作班成员强调开关空气开关时，切勿将手指伸入空气开关上端头；测试过程中，遇到掉线的情况，要断电后才能接线，重新工作；外搭接电源时，工作时至少需要两人，一人监护一人操作，听工作负责指挥。

（3）机械伤害。

危险点：断路器分合、接地开关分合时未通知其他工作班成员，可能会对工作班成员造成机械伤害；作业人员在操作过程中，手误碰到带有能量的机构上，造成夹持伤害。

预控措施：断路器分合、接地开关分合时，需要大声呼唱，得到工作班所有人员大声回应之后，方可操作；工作时至少需要两人，一人监护一人操作，听工作负责指挥。

（4）设备、仪器损坏。

危险点：未按照要求进行测试、野蛮操作，对仪器和被测设备损坏。

预控措施：禁止野蛮操作，操作过程中，按步骤和要求进行测试，防止仪器损坏。若出现不能操作的现象，应当立即停止，不得使用蛮力，立即汇报负责人；工作时至少需要两人，一人监护一人操作，听工作负责指挥。

（5）高处坠落。

危险点：工作班成员爬上断路器时，使用爬梯无人扶持，也未固定，爬梯滑落造成坠落受伤；工作班成员在断路器上接线时，未正确使用安全带可能坠落受伤。

预控措施：在工作人员上下的梯子上，应悬挂"从此上下！"的标示牌；梯子应坚固完整，有防滑措施。梯子的支柱应能承受作业人员及所携带的工具、材料攀登时的总重量；工作人员在上下爬梯时，应有人扶持爬梯，或将爬梯固定在断路器构架上；在没有脚手架或者在没有栏杆的脚手架上工作，高度超过 1.5 m 时，应使用安全带；安全带的挂钩或绳子应挂在结实牢固的构件上，或专为挂安全带用的钢丝绳上，并应采用高挂低用的方式；工作时至少需要两人，一人监护一人操作，听工作负责指挥。

4. 作业人员分工

作业人员分工如表 3-8 所示。

表 3-8　断路器机械特性测试人员分工

序号	工作岗位	数量（人）	工作性质
1	操作人员	1	专门负责操作
2	监护人员	1	专职监护
3	辅助人员	1	辅助接线、工器具传递、记录等工作

五、工作程序

1. 操作流程

操作流程如表 3-9 所示。

表 3-9　断路器机械特性测试操作流程

序号	作业内容	作业标准	安全注意事项	责任人
1	前期工作准备	（1）履行工作票手续。 （2）核对现场的设备位置和情况。 （3）装设安全围栏、悬挂标示牌。 （4）按规程要求，正确使用劳动防护用品，工作服穿戴整齐	（1）按照安规要求办理工作票的相关手续。 （2）进入作业现场要正确戴好安全帽，穿工作服，着软底鞋	

续表

序号	作业内容	作业标准	安全注意事项	责任人
2	工作环境确认	（1）检查被检设备初始状态，通过机械位置指示，确认断路器在分闸位置；通过储能指示，确认断路器未储能，记录设备铭牌参数。询问过往检修记录、试验记录。 （2）核查围栏、标示牌 （3）检查需要使用的工器具、材料、机具与仪表，检查方式正确 （4）工作负责人与辅工召开班前会，按照"四清楚"原则，向辅助工进行工前交底，明确工作任务、安全措施、危险点、工作流程并履行确认手续	（1）设备初始状态应该是分闸未储能状态。 （2）围栏和标示牌设置正确。 （3）安全用具、工器具外观检查合格、无损伤、变形、破损等情况。 （4）开工前，对辅助工进行安全交底，正确做到"四清楚"，正确履行确认手续	
3	测试前检查	（1）拆除面板、防尘盖，同时可以再次根据拐臂和弹簧的状态确认设备是在分闸未储能状态。 （2）清洁断路器外壳 （3）检查控制回路电压及功能完整性：储能完成后，检查合闸回路的完成性；合闸后检查分闸回路的完整性	（1）再次确认设备状态。 （2）设备外壳清洁、无尘。 （3）设备储能后，避免触碰到已储能的传动部件 （4）储能、分合闸断路器前均应该呼唱	
4	测试仪接线	（1）正确进行接线，首先将测试仪外壳和被测试设备都可靠接地。 （2）连接时间通道接线：将断路器的上出线座的A、B、C、三相分别与测试仪器的A1、B1、C1（或其他设置的连接通道)通道相连接，下出线座短接后同测试仪器的公共端连接。 （3）连接控制通道接线：从断路器的合闸控制回路两端引出线，连接到测试仪器的合闸控制端；将分闸控制回路两端引出线连接到测试仪器的分闸控制端。	（1）接线前，若需要高处作业，需采取正确的安全措施，正确使用安全带。 （2）接地前，应注意接线顺序：先接接地端，再接设备和测试仪器，拆除的顺序与之相反。 （3）所有接线完成后，检查接线是否牢靠，有无松动；确保接线避免缠绕、交叉、混乱；确保所有接线正确无误。 （4）通电测试过程中若出现测试线掉落的情况，一定要断电后，接线。	

续表

序号	作业内容	作业标准	安全注意事项	责任人
4	测试仪接线	（4）连接储能电源接线。可选择储能回路同测试仪器的储能通道连接，由测试仪器驱动储能；也可根据实际情况，直接在设备上完成储能。 （5）安装角位移传感器：正确选择角位移传感器；组装传感件（传感器、延长杆、联轴器）并检查水平并紧固；在主传动轴顶端水平安装传感件，安装好后应对中主传动轴中心并紧固；组装传感器磁力固定底座；旋转底座手柄将底座吸附于断路器本体外壳（尽可能水平安装），爪臂固定传感件，爪臂保留适当调节裕度并稍加紧固；检查传感器旋转无卡涩、初始角度符合厂家要求；检查传感件与主传动轴保持水平，满足要求后紧固爪臂。 （6）连接通道线仪器端	（5）传感器安装正确，水平，且传感器的初始角度要避开盲区。 （6）发生安全违章行为，听从制止的	
5	开机前检查	（1）测试仪电源总开关、控制脉冲触发回路电源开关、储能电源回路电源开关均应置于断开位置。 （2）检查各回路保险丝有无熔断（针对有保险丝熔断的测试仪器，检查）。 （3）检查打印纸余量足够。 （4）检查测试仪测试电源切换开关。 （5）连接电源、开机	（1）测试前，用万用表检查电压是否在额定电压。 （2）电源在未连接的状态下，进行开机前检查。 （3）开机通电前，要注意呼唱	
6	测试参数设置	（1）设置（输入）被测开关型号。 （2）选择测试传感器类型（旋转/直线/激光）。 （3）选择传感器行程测量参数（类型/长度）。 （4）选择速度测量方式（合前分后 10 ms 或 6 mm）。 （5）选择弧直转换参数（起始角度/标准角度/标准长度）。 （6）保存参数	（1）根据被测试设备的型号，在仪器中设置（输入）被测开关型号。 （2）选择测试传感器类型或选择与实际一致（旋转/直线/激光）。 （3）设置传感器行程测量参数或选择与实际一致。 （4）设置速度测量方式或选择与实际一致（合前分后 10 ms 或 6 mm）。 （5）选择弧直转换参数（起始角度/标准角度/标准长度）	

续表

序号	作业内容	作业标准	安全注意事项	责任人
7	测试低电压动作特性	（1）正确操作测试仪，调整电压前断开测试仪输出。 （2）合闸测试，输出控制电压为$(80\%\sim110\%)U_e$时，断路器应均能可靠动作。 （3）分闸测试，输出控制电压为$30\%U_e$时，应可靠拒动；输出控制电压为$65\%U_e$时，应可靠动作。 （4）循环操作三次（口述后两次）。 （5）操作前相互呼唱	（1）测试过程中，设备储能需在额定电压下完成，且储能完成后，应立即断掉储能电源。 （2）分别按照$80\%U_e$、$110\%U_e$、$30\%U_e$、$65\%U_e$几个边界值进行测试，检查断路器的动作情况。断路器应在$80\%\sim110\%U_e$时，可靠合闸；在小于$30\%U_e$时，可靠拒分，大于$65\%U_e$时，可靠分闸。 （3）正确掌握低压动作特性测试的电压值和动作情况。 （4）设备在分合操作前应相互呼唱。 （5）不得触碰已储能的传动部件。 （6）不得随意在未断电前变更接线	
8	测试机械动作特性	（1）正确操作测试仪或软件平台。 （2）控制输出电压为$100\%U_e$。 （3）合闸测试操作： ①打印测试数据，并根据厂家标准判断数据； ②按规定保存电子文档（设备硬盘或外接U盘）； ③循环操作三次（口述后两次）。 （4）操作前相互呼唱	（1）按照测试要求，循环正确操作3次动作特性测试。 （2）分合闸操作前应相互呼唱。 （3）发生安全违章行为，听从制止的。 （4）不得触碰已储能的传动部件。 （5）不得随意在未断电前变更接线	
9	文明施工	操作结束后，整理工具、清理工作现场	（1）完成测试后，工器具按规定放置，不得出现掉落、踩踏等情况。 （2）清理场地、办理工作终结手续	

续表

序号	作业内容	作业标准	安全注意事项	责任人
10	检修记录填写	（1）按照流程、准确执行检查维护，及时履行确认。 （2）准确记录各项检查参数。 （3）规范完成维护环境信息记录及人员签字确认。 （4）提供以实际一致的明确结论	（1）按照规范，整理测试数据，并做好记录。 （2）根据测试数据，对测试设备的情况进行分析总结	

2. 操作示例图

（1）工器具示例图见图 3-44、3-45。

图 3-44　测试工具材料

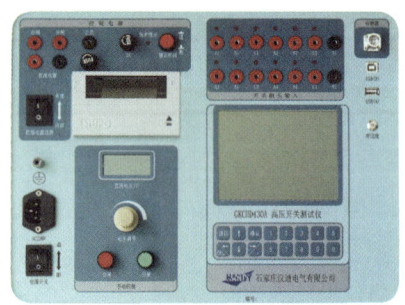

图 3-45　机械特性测试仪器

（2）测试接线图。

以 VS1-12 真空断路器的测试为例，说明开展断路器机械特性测试接线的一般方法（见图 3-46）。

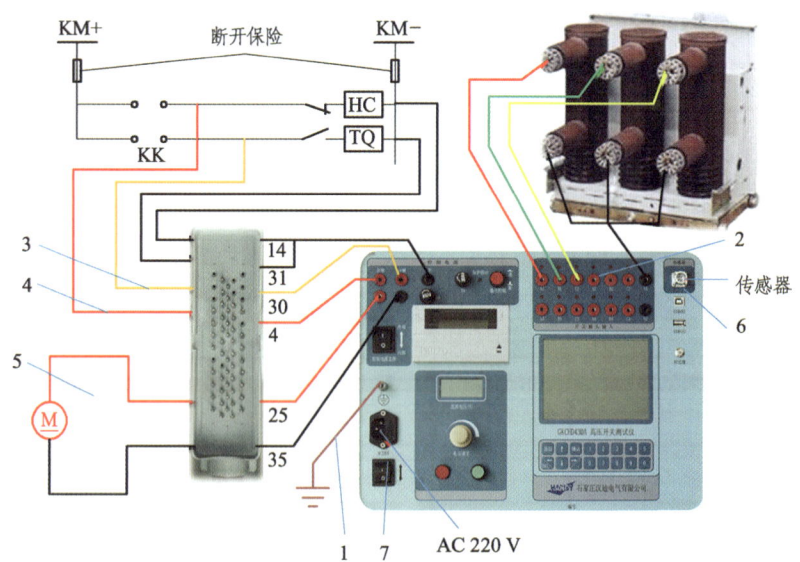

1—接地；2—时间通道；3—分闸控制通道；4—合闸控制通道；
5—储能控制；6—传感器连接通道；7—电源。

图 3-46　测试接线原理图

（3）图 3-47 为被测设备和测试仪器接地示例图。见图 3-48 至图 3-54。

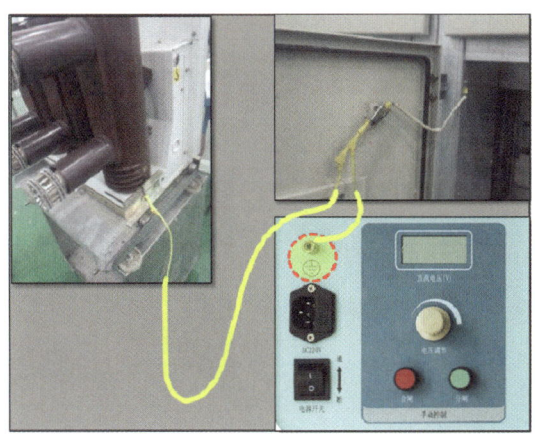

图 3-47　测设备和测试仪器接地

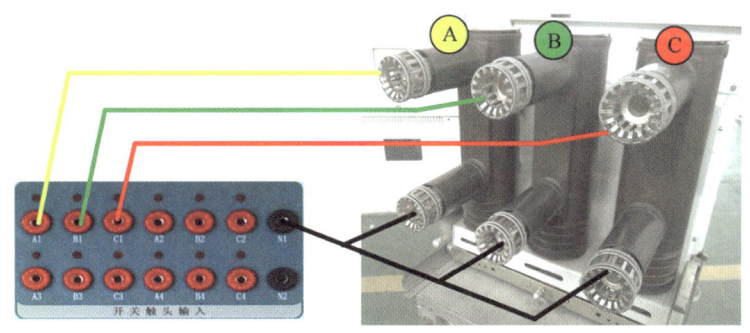

图 3-48　时间通道接线

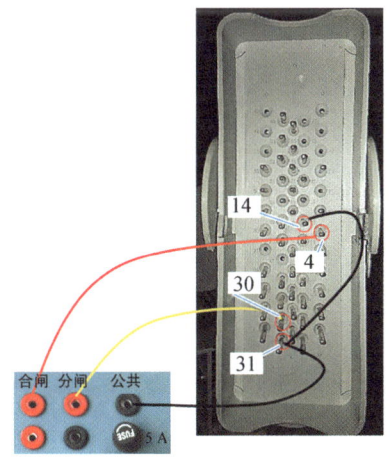

图 3-49　分合闸控制接线连接

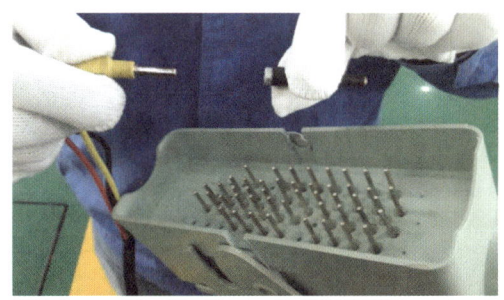

图 3-50　专用接线插件针脚

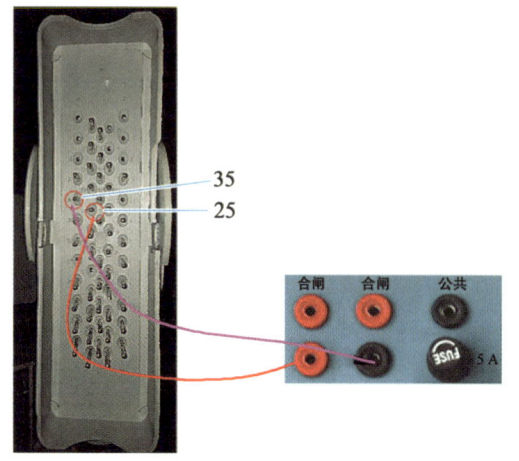

图 3-51　储能通道接线

图 3-52　传感器连接

图 3-53　输出电压调节

图 3-54　输出控制按钮

六、相关知识

（一）机械特性相关参数

（1）合闸时间：从合闸回路带电到所有断口都接触的时间间隔（见图 3-55）。

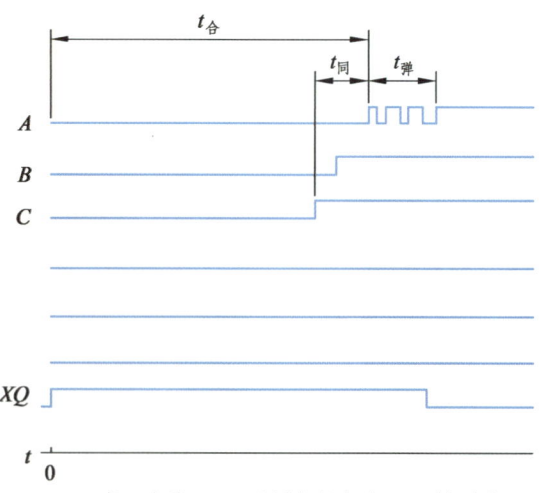

A、B、C—断口波形；XQ—线圈电压波形；t—时间坐标。

图 3-55　合闸时间量示意图

（2）不同期时间：各相关断口间动作时间之差。

（3）弹跳时间：动、静触头间在稳定接通之前的过渡过程时间。

（4）分闸时间：从分闸回路带电到所有断口都分开的时间间隔（见图 3-56）。

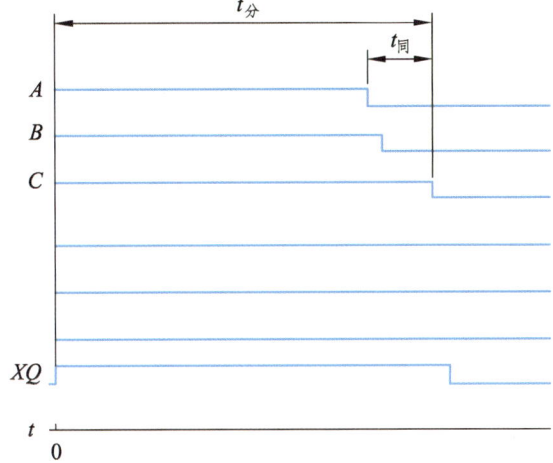

A、B、C—各断口波形；XQ—线圈电压波形；t—时间坐标。

图 3-56　分闸时间量示意图

（5）合闸行程相关参数。

① 合闸速度：开关合闸时，触头接通前某一区段，动触头的平均速度。常是指刚合前 6 mm 内的平均速度，合闸速度 $=\dfrac{\Delta L}{\Delta t}$。

② 开距：分闸状态时，动、静触头间的距离。

③ 接触行程（超程）：开关在合闸后，碟簧压缩的距离。

④ 总行程：为开距与超程之和。

合闸行程曲线如图 3-57 所示。

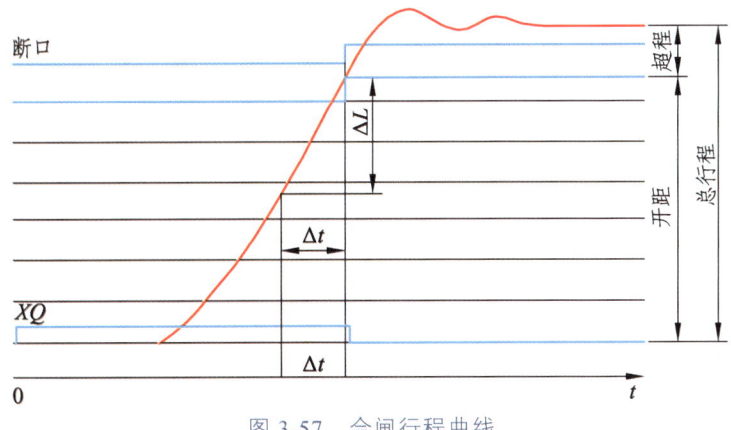

图 3-57　合闸行程曲线

（6）分闸行程相关参数。

分闸速度：指开关分闸时，触头分离后动触头在某一区段的平均速度，通常是指刚分后 6 mm 内的平均速度，分闸速度 $=\dfrac{\Delta L}{\Delta t}$。

分闸行程曲线如图 3-58 所示。

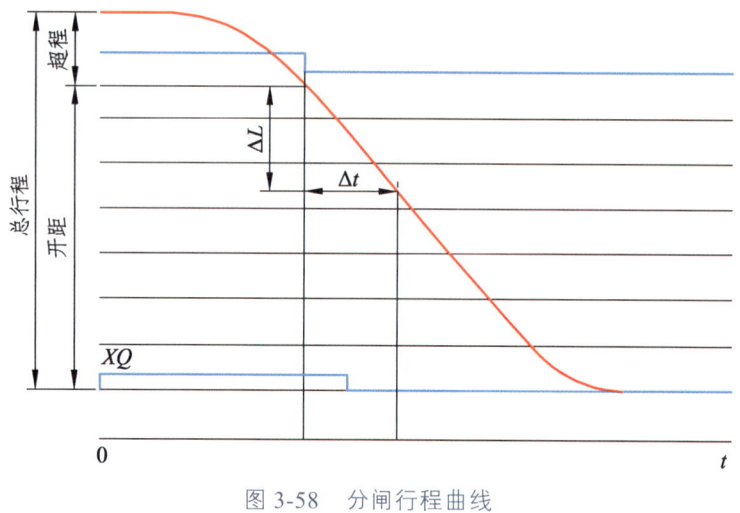

图 3-58　分闸行程曲线

（二）相关参数分析

1. 速度

速度是保证断路器政策工作和系统安全运行的主要参数，断路器分、合闸时，触头运动速度是断路器的重要特性参数，分、合闸速度不足将会引起触头合闸振颤，预击穿时间过长。分闸速度过慢会加长灭弧时间，切除故障时易导致加重设备损坏和影响电力系统稳定。同时会致使断路器内存压力增大，轻者烧坏触头，使断路器不能继续工作，重者易造成越级跳闸，扩大停电范围，引起断路器爆炸。

当触头运动速度过高时，造成运动机构受过度机械应力使个别零部件损坏或寿命缩短。同时由于强烈的机械冲击和振动还将使触头弹跳时间加长。由于合闸弹跳过程中，触头断开距离小，电弧不会熄灭，导致触头电磨损加重，从而影响灭弧室的电寿命。

2. 行程曲线

行程曲线主要反映了开关动触头的运动过程，了解此运动过程是否正常，有无卡涩，缓冲器工作是否正常，有无机械缺陷等。必要时，可以根据速度的定义来计算某些区段的平均速度值，并与仪器显示的速度值进行验证。通常情况下，一台性能良好的高压开关，其整个行程曲线在各处的过渡都是比较平滑的，不会有异常的拐点，或在某处出现强烈的反方向跳动。

3. 断开波形

反映了开关动、静触头的接通情况。可了解各断口的动作时间，各断口动作的先后情况，即同期差；各断口的"弹跳"情况。需注意是异常的弹跳要排除在测试过程中，由于接线松动造成的某种假象。

Task 3　Mechanical Characteristic Test of Circuit Breaker

3.1　Work Tasks

Mechanical characteristic test of circuit breaker.

3.2　Referenced Code of Practice

(1) *Test methods for high voltage electrical equipment* (second edition).

(2) *State Grid Corporation of China Electric Power Safety Work Regulations* (substation part).

(3) *State Grid Corporation Substation Maintenance Management Regulations (Trial) Volume 2 Breaker Maintenance Rules*.

(4) *Preventive Test Code for Power Equipment* (DL/T 596-2005).

(5) *electrical equipment handover test standard* (GB 50150-2006).

3.3　Operating Conditions

(1) Mechanical property test shall be carried out in good and dry weather. During the measurement, work shall be stopped in case of severe weather such as gale above Grade 6, thunderstorm, hail, fog and sandstorm.

(2) The output power supply of the mechanical property tester is strictly prohibited from short circuit.

(3) The mechanical property tester uses an external power supply as the test power supply as much as possible to prevent the test result from being affected by insufficient internal power supply.

(4) When testing the low voltage characteristic of circuit breaker, the operating voltage applied to the opening and closing coil should not be too long to prevent the coil from burning.

(5) Measuring instruments must have calibration certificates.

(6) The operating environment of the testing instrument: $-10 \sim +50$ °C, relative humidity: $\leqslant 80\%$.

3.4　Preparation before Operation

1. Basic requirements and conditions of site investigation

(1) Read the operation of the circuit breaker under test, and understand the test site and other conditions. Read the test reports of the circuit breaker over the years, relevant handover pre-test procedures, circuit breaker operation records and defect records.

(2) Survey the power outage range and safety measures of field equipment.

(3) Investigate the site to repair the power box, and verify whether the power capacity meets the requirements.

(4) On-site instruments, working vehicles, machines and tools, materials, etc.

2. Selection of instruments and materials (see Table 3-7)

Table 3-7 List of Instruments and Materials for Testing Mechanical Characteristics of Circuit Breakers

No.	Name		Unit	Quantity	Remarks
1		Mechanical characteristic tester of high voltage circuit breaker	Set	1	
2	Tester accessory box	Time channel wiring	Group	1	
3		Control power channel wiring	Group	1	
4		Sensor signal channel wiring	Group	1	
5		Energy storage power supply wiring	Group	1	
6		Sensor and fixing clamp	Group	1	
7		Power cord	Group	1	
8		Enclosure wiring ground	Group	1	
9		Wire clamp and plug-in pin accessories		A number of	
10		150 mm steel square	Piece	1	
11		8-10 wrench	Piece	1	
12		17-19 dummy wrench	Piece	1	
13		ST3-150 mm Phillips screwdriver	Piece	1	
14		Calico	Sheet	1	
15		Multimeter	Piece	1	
16		Inner hexagon 1.5 mm	Piece	1	
17		Power plate	Piece	1	
18		Wiring board	Piece	1	
19		Food tray	Piece	2	

3. Dangerous points and preventive measures

(1) High voltage electric shock.

Dangerous point: straying into live interval may lead to high voltage electric shock.

Pre-control measures: Separate the maintenance interval from the adjacent live equipment (interval) with a fence, and hang the sign of "Stop, High Voltage Danger" in the operation site, set up the only entrance and exit near the road side, and hang the sign of "Enter and Exit from here"; Two sides of the circuit breaker to be tested are respectively hung with a set of three-phase short-circuit grounding wires; At work, at least two people are required, one for supervision and one for operation, and one for listening to the work.

(2) Low voltage electric shock.

Dangerous points: The inlet side (upper end) of the air switch of the control circuit, closing circuit and lighting circuit in the circuit breaker mechanism box is low voltage;After closing and opening, the secondary circuit of the circuit breaker is regarded as charged, and touching may cause low voltage electric shock. Touching the live part may cause low voltage electric shock during the external lap power supply and test.

Pre-control measures: If the operator hangs the sign "No switching on, someone works" on the air switch, the person in charge of the work needs to apply to the operator for permission to change the safety measures before closing the air switch; The person in charge of the work repeatedly emphasizes to the members of the work class that when switching the air switch, do not put your fingers into the upper end of the air switch; The person in charge of the work repeatedly emphasizes to the members of the work class that when switching the air switch, do not put your fingers into the upper end of the air switch.

(3) Mechanical damage.

Dangerous points: Failure to notify other members of the working class when the circuit breaker is opened or closed and the grounding switch is opened or closed may cause mechanical damage to the members of the working class; during the operation, the operator accidentally touches the mechanism with energy, causing clamping injury.

Pre-control measures: When the circuit breaker is opened and closed and the grounding switch is opened and closed, it is necessary to sing loudly, and it can only be operated after all personnel in the working class respond loudly; At work, at least two people are required, one for supervision and one for operation, and one for listening to the work.

(4) Equipment and instrument damage.

Dangerous points: Failure to carry out testing and rough operation as required, causing damage to instruments and tested equipment.

Pre-control measures: Rough operation is prohibited. During operation, test according to steps and requirements to prevent instrument damage. In case of inoperability, stop immediately, do not use brute force, and immediately report to the person in charge; At work, at least two people are required, one for supervision and one for operation, and one for listening to the work.

(5) Falling from a height.

Dangerous points: When the members of the working class climb the circuit breaker, the ladder used is unsupported and not fixed, and the ladder slips and causes falling injury; Members of the shift may fall and get injured if they don't use the safety belt correctly when wiring the circuit breaker.

Pre-control measures: On the ladder where the staff get up and down, hang "Up and down from now on!" The signboard; Ladder shall be firm and complete with anti-skid measures. The pillars of the ladder should be able to bear the total weight of the operators and the tools and materials they carry when climbing; When the staff climbs the ladder up and down, someone should support the ladder, or fix the ladder on the circuit breaker frame; Safety belts should be used when working on scaffolding without scaffolding or railings, and the height exceeds 1.5 m: The hook or rope of the safety belt should be hung on a strong component or a steel wire rope specially used for hanging the safety belt, and the method of hanging high and using low should be adopted; At work, at least two people are required, one for supervision and one for operation, and one for listening to the work.

4. Division of labor among operators

The division of labor of operators is shown in Table 3-8.

Table 3-8　Division of Labor of Circuit Breaker Mechanical Characteristics Tester

No.	Work post	Quantity (person)	Nature of work
1	Operation personnel	1	Be responsible for operation
2	Guardianship personnel	1	Full time guardianship
3	Auxiliary personnel	1	Auxiliary wiring, instrument transmission, recording, etc.

3.5　Working Routine

1. Operating process

Operation process is shown in table 3-9.

Table 3-9　Operation Flow of Mechanical Characteristic Test of Circuit Breaker

No.	Job content	Performance standard	Safety precautions	Person liable
1	Preparatory work preparation	(1) Perform working ticket procedures. (2) Check the location and situation of equipment on site. (3) Install safety fences and hang signs. (4) According to the requirements of regulations, use labor protection articles correctly, and wear work clothes neatly	(1) Go through relevant formalities of working ticket according to safety requirements. (2) Enter the job site to wear safety helmet, overalls and soft-soled shoes correctly	
2	Work environment confirmation	(1) Check the initial state of the tested equipment, and confirm that the circuit breaker is in the opening position through mechanical position indication; Confirm that the circuit breaker does not store energy through energy storage indication, and record the nameplate parameters of the equipment. Ask about past maintenance records and test records. (2) Check fences and signs. (3) Check the instruments, materials, machines and instruments that need to be used, and check them correctly. (4) The person in charge of the work shall hold a pre-job meeting with the auxiliary workers, and make pre-job disclosure to the auxiliary workers according to the principle of "four clear points", so as to clarify the work tasks, safety measures, dangerous points and work flow and perform confirmation procedures	(1)The initial state of the equipment should be the state of no energy storage after switching off. (2) Fences and signs are set correctly. (3) Safety appliances and instruments are qualified in appearance inspection without damage, deformation and breakage, etc. (4) Before starting work, make safety disclosure to the auxiliary workers, correctly achieve the "Four Clearances" and correctly perform the confirmation procedures	
3	Pre-test inspection	(1) Remove the panel and dust cover. At the same time, it can be confirmed again that the equipment is in the open state without energy storage according to the state of crank arm and spring. (2) Clean the breaker housing.	(1) Confirm the equipment status again. (2) Equipment shell shall be clean and dust-free. (3) After the equipment stores energy, avoid touching the stored transmission parts.	

Continued

No.	Job content	Performance standard	Safety precautions	Person liable
3	Pre-test inspection	(3) Check the voltage and functional integrity of the control circuit: after the energy storage is completed, check the completion of the closing circuit; Check the integrity of the opening circuit after closing	(4) Energy storage, before opening and closing circuit breaker should be singing	
4	Tester wiring	(1) For correct wiring, the tester shell and the tested equipment should be grounded reliably at first. (2) Connection time channel wiring: connect the A, B, C and three phases of the upper outlet seat of the circuit breaker with the A1, B1 and C1 (or other set connection channels) channels of the test instrument respectively, and connect the lower outlet seat with the public end of the test instrument after being shorted. (3) Connection control channel wiring: lead wires from both ends of the closing control loop of the circuit breaker and connect to the closing control end of the test instrument; Connect the outgoing lines at both ends of the break-off control loop to the break-off control end of the test instrument. (4) Connect the energy storage power supply wiring. The optional energy storage loop is connected with the energy storage channel of the test instrument, and the test instrument drives the energy storage; Energy storage can also be completed directly on the equipment according to the actual situation. (5) Install angular displacement sensor: select angular displacement sensor correctly; Assemble sensing parts (sensor, extension rod and coupling), check the level and tighten them;Install the sensor horizontally at the top of the main drive shaft, and center and fasten the main drive shaft after installation; Assemble the magnetic fixing base of the sensor;Rotate the handle of the base to absorb the base to the shell of the breaker body (install it horizontally as far as possible), fix the sensor with the claw arm, and keep proper adjustment margin and tighten it slightly;Check that the rotation of the sensor is free from jamming and the initial angle meets the requirements of the manufacturer; Check that the sensor is level with the main transmission shaft, and fasten the claw arm after meeting the requirements. (6) Connect the instrument end of channel line	(1) Before wiring, if high-altitude operation is required, correct safety measures should be taken and safety belts should be used correctly. (2) Before grounding, pay attention to the wiring sequence: connect the grounding terminal first, then connect the equipment and test instruments, and the removal sequence is opposite to it. (3) After all wiring is completed, check whether the wiring is firm and loose; Ensure the wiring to avoid winding, crossing and confusion; Make sure all wiring is correct. (4) If the test line drops during the power-on test, it must be connected after the power is cut off. (5) The sensor is installed correctly and horizontally, and the initial angle of the sensor should avoid the blind area. (6) Safety violations occur and are stopped	

Continued

No.	Job content	Performance standard	Safety precautions	Person liable
5	Check before starting up	(1) The main power switch of the tester, the power switch of the control pulse trigger circuit and the power switch of the energy storage power circuit shall be placed in the off position. (2) Check whether the fuses in each circuit are blown (for test instruments with blown fuses, check). (3) Check that the printing paper margin is sufficient. (4) Check the tester to test the power switch. (5) Connect the power supply and power on	(1) Before testing, use a multimeter to check whether the voltage is at the rated voltage. (2) Check before starting the power supply when it is not connected. (3) Pay attention to singing before powering on	
6	Test parameter setting	(1) Set (input) the model of the tested switch. (2) Select the test sensor type (rotary/linear/laser). (3) Select the sensor travel measurement parameter (type/length). (4) Select the speed measurement method (10 ms or 6 mm after closing). (5) Select arc-to-straight conversion parameters (starting angle/standard angle/standard length). (6) Save parameters	(1) According to the model of the tested equipment, set (input) the model of the tested switch in the instrument. (2) Select the type of test sensor or choose to match the actual (rotary/linear/laser). (3) Set the sensor travel measurement parameters or select them in accordance with the actual conditions. (4) Set the speed measurement mode or choose to be consistent with the actual situation (10 ms or 6 mm after closing). (5) Select arc-to-straight conversion parameters (starting angle/standard angle/standard length)	
7	Test low voltage operation characteristics	(1) Operate the tester correctly and disconnect the tester output before adjusting the voltage. (2) During the closing test, when the output control voltage is 80%–110%U_e, the circuit breakers should all operate reliably. (3) Break-brake test, when the output control voltage is 30%U_e, it should reliably refuse to operate; When the output control voltage is 65%U_e, it should act reliably. (4) Loop operation three times (twice after dictation). (5) Sing to each other before operation	(1) During the test, the energy storage of the equipment should be completed at the rated voltage, and the energy storage power should be cut off immediately after the energy storage is completed. (2) Test according to the boundary values of 80%U_e, 110%U_e, 30%U_e and 65%U_e, and check the action of circuit breaker.The circuit breaker should be reliably closed when 80%–110%U_e; In case of less than 30%U_e, it can reliably reject points, and in case of more than 65%U_e, it can reliably switch off. (3) Correctly grasp the voltage value and action situation of low-voltage action characteristic test.	

Continued

No.	Job content	Performance standard	Safety precautions	Person liable
7	Test low voltage operation characteristics		(4) Equipment should sing to each other before opening and closing operation. (5) Do not touch the stored energy transmission parts. (6) Do not arbitrarily change the wiring before power failure	
8	Test mechanical action characteristics	(1) Operate the tester or software platform correctly. (2) Control the output voltage to 100%U_e. (3) Closing test operation: (a) Print the test data and judge the data according to the manufacturer's standard; (b) Save electronic documents (equipment hard disk or external USB flash drive) as required; (c) Loop operation three times (twice after dictation). (4) Sing to each other before operation	(1) According to the test requirements, cycle and correctly operate the action characteristic test for 3 times. (2) Before opening and closing operation, they should sing to each other. (3) Safety violations occur and are stopped. (4) Do not touch the stored energy transmission parts. (5) Do not arbitrarily change the wiring before power failure	
9	Civilization construction	After the operation, arrange tools and clean up the work site	(1) After the test is completed, the tools and instruments shall be placed according to the regulations, and no dropping or trampling shall occur. (2) Clean up the site and go through the formalities for ending the work	
10	Fill in maintenance records	(1) According to the test data, the situation of the test equipment is analyzed and summarized. (2) Accurately record all inspection parameters. (3) Standardize and complete the maintenance of environmental information records and personnel signature confirmation. (4) Provide clear conclusions that are practically consistent	(1) According to the specifications, arrange the test data and make records. (2) According to the test data, the situation of the test equipment is analyzed and summarized	

2. Operation example diagram

(1) Example drawing of tools and instruments (see Fig.3-44, 3-45).

(2) Test wiring diagram.

Taking the test of VS1-12 vacuum circuit breaker as an example, this paper explains the general method of testing and wiring the mechanical characteristics of circuit breaker (see Fig. 3-46).

(3) Fig. 3-47 is an example diagram of grounding of the tested equipment and test instrument. see Fig. 3-48 to Fig. 3-54.

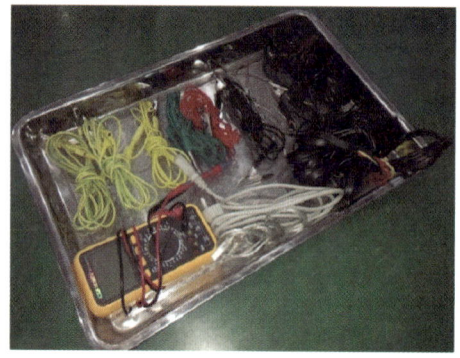
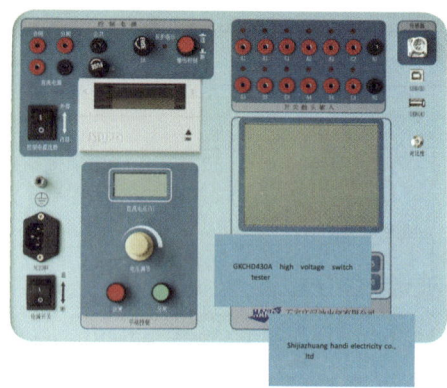

Fig. 3-44　Test Tool Material 　　Fig. 3-45　Mechanical Characteristic Testing Instrument

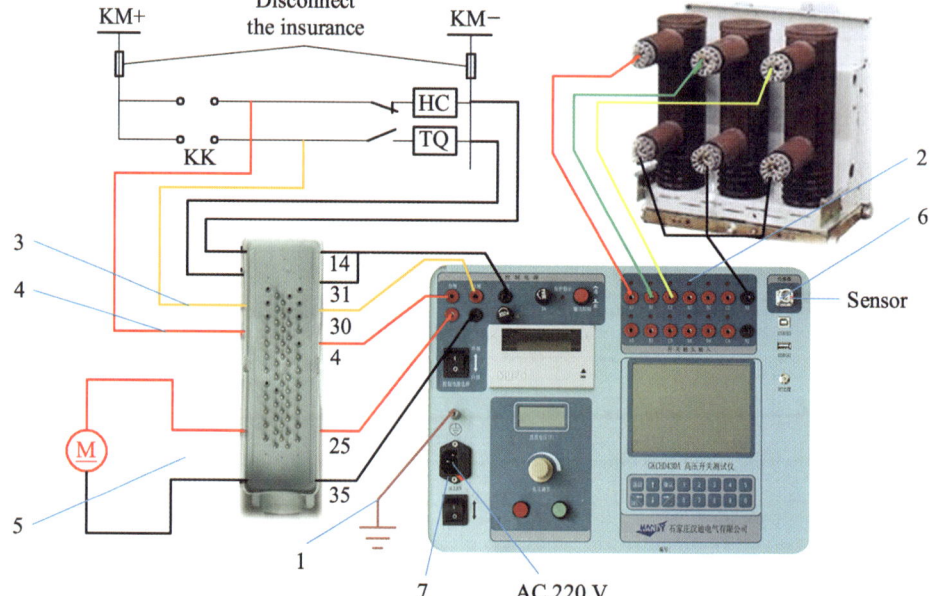

1-Grounding; 2-Time channel; 3-Opening control channel; 4-Closing control channel; 5-Energy storage control; 6-Sensor connection channel; 7-Power Supply.

Fig. 3-46　Test Wiring Schematic Diagram

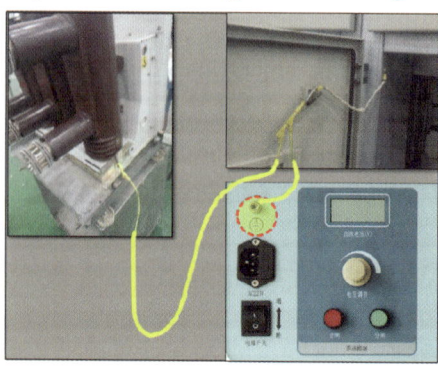

Fig. 3-47　Grounding of Testing Equipment and Instruments

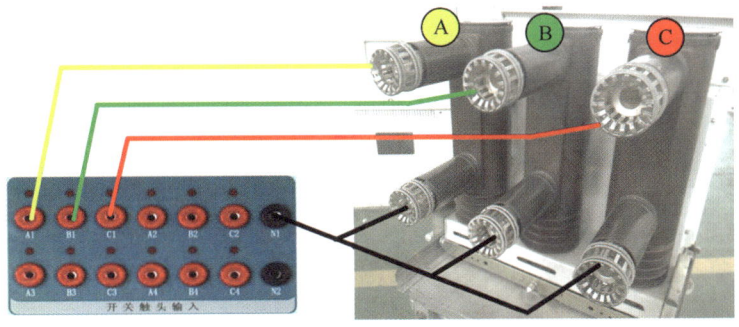

Fig. 3-48　Time Channel Wiring

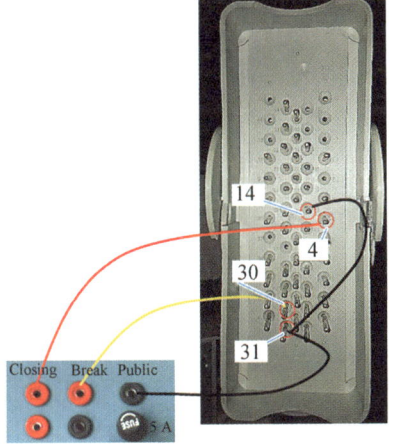

Fig. 3-49　Opening and Closing Control Wiring Connection

Fig. 3-50　Pin of Special Wiring Plug-in

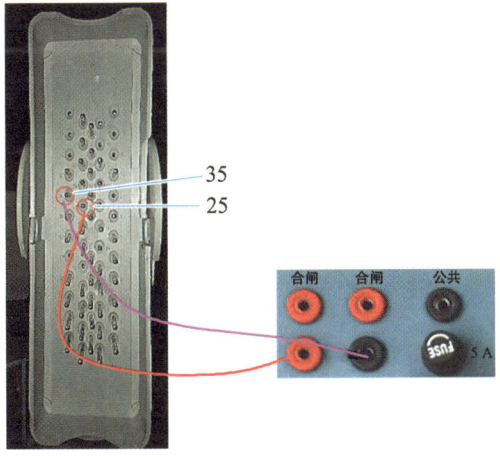

Fig. 3-51　Energy Storage Channel Wiring

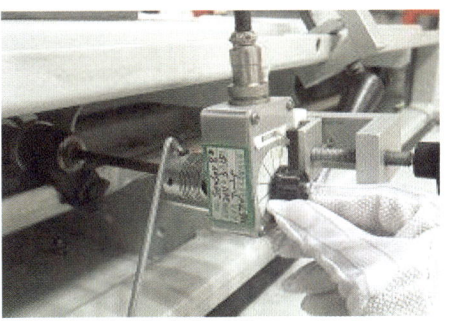

Fig. 3-52　Sensor Connection

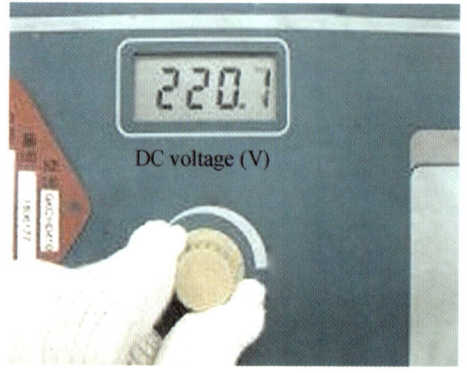

Fig. 3-53　Output Voltage Regulation

Fig. 3-54　Output Control Button

3.6　Relevant Knowledge

1. Parameters related to mechanical properties

(1) Closing time: The time interval from the electrification of the closing circuit to the contact of all fractures (see Fig. 3-55).

(2) Different period time: The difference of action time between related fractures.

(3) Bouncing time: The transition time between moving and stationary contacts before stable connection.

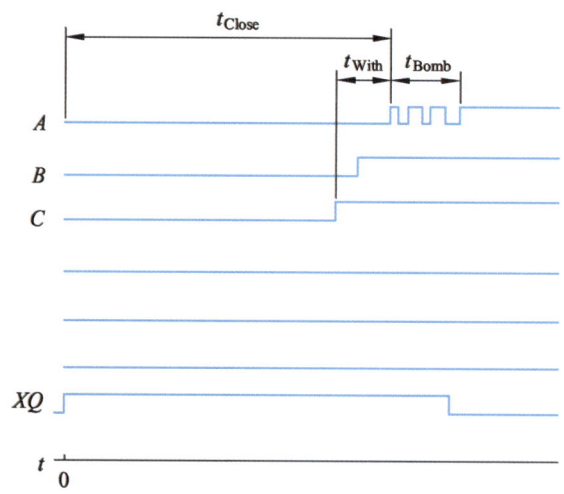

A, B, C-Fracture waveform; XQ-Coil voltage waveform; t-Time coordinate.

Fig. 3-55　Schematic Diagram of Closing Time

(4) Opening time: the time interval from the electrification of the opening circuit to the separation of all fractures (see Fig. 3-56).

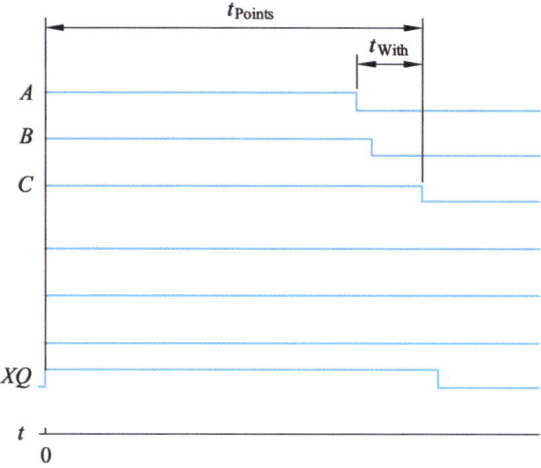

A, *B*, *C*-The waveform of each fracture; *XQ*-Coil voltage waveform; *t*-Time coordinate.

Fig. 3-56　Schematic Diagram of Opening Time

(5) Closing stroke related parameters.

① Closing speed: when the switch is closed, the average speed of the moving contact in a certain section before the contact is connected. Often refers to the average speed within 6 mm before closing, closing speed $= \dfrac{\Delta L}{\Delta t}$.

② Open distance: the distance between the moving and static contacts when the brake is opened.

③ Contact stroke (overtravel): the distance that the disc spring compresses after the switch is closed.

④ Total travel: it is the sum of distance and over-travel.

Closing stroke curve is shown in Fig. 3-57.

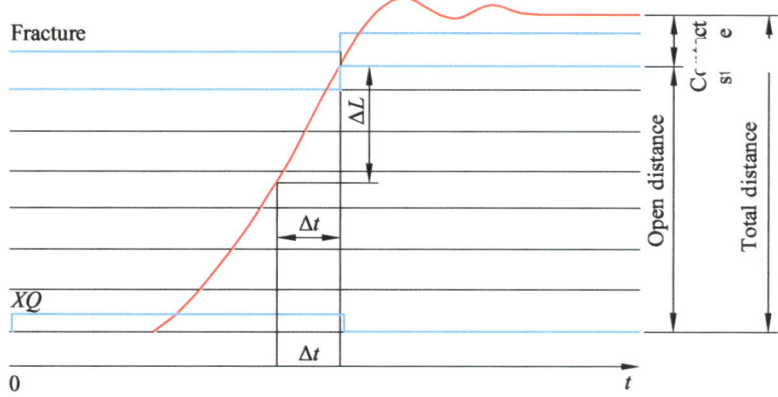

Fig. 3-57　Closing Stroke Curve

(6) Parameters related to tripping stroke.

Opening speed: Refers to the average speed of moving contact in a certain section after contact separation when the switch is opened, which usually refers to the average speed within 6 mm after opening and opening speed $= \dfrac{\Delta L}{\Delta t}$.

Opening stroke curve is shown in Fig. 3-58.

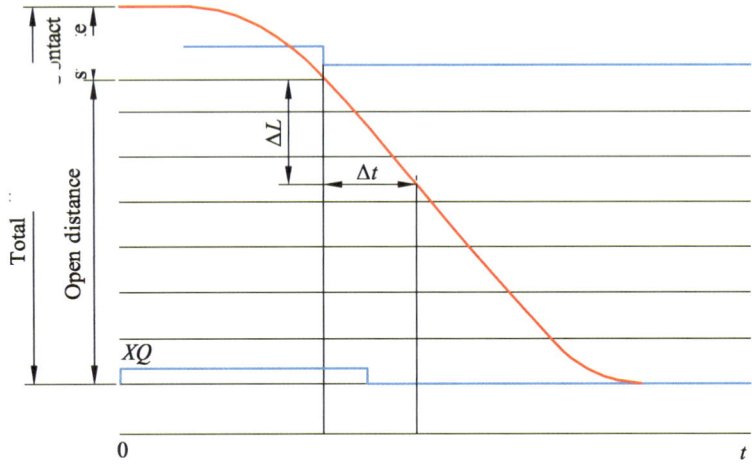

Fig. 3-58   Opening Stroke Curve

2. Analysis of related parameters

(1) Speed.

Speed is the main parameter to ensure the policy work and safe operation of the circuit breaker. When the circuit breaker opens and closes, the contact movement speed is an important characteristic parameter of the circuit breaker. The insufficient opening and closing speed will cause the contact closing vibration and the pre-breakdown time is too long. The slow opening speed will prolong the arc extinguishing time, which will easily lead to the damage of heavy equipment and affect the stability of power system. At the same time, the memory pressure of the circuit breaker will increase, the lighter one will burn out the contacts, which will make the circuit breaker unable to continue working, and the heavier one will easily cause leapfrog tripping, expand the power outage range and cause the circuit breaker to explode.

When the moving speed of the contact is too high, the moving mechanism is subjected to excessive mechanical stress, which damages individual parts or shortens the service life. At the same time, due to strong mechanical shock and vibration, the contact bounce time will be lengthened. In the process of closing and bouncing, the contact disconnection distance is small, and the arc will not be extinguished, which leads to increased electrical wear of the contact, thus affecting the electrical life of the arc extinguishing chamber.

(2) Travel curve.

The travel curve mainly reflects the moving process of the switch moving contact, so as to know whether the moving process is normal, whether there is jamming, whether the buffer works normally, whether there is mechanical defect, etc. If necessary, the average speed value of some sections can be calculated according to the definition of speed, and verified with the speed value displayed by the instrument. Usually, the transition of the whole stroke curve of a high-voltage switch with good performance is smooth everywhere, and there will be no abnormal inflection point, or there will be strong jumping in the opposite direction somewhere.

(3) Disconnect waveform.

It reflects the on-state of the moving and stationary contacts of the switch. You can know the action time of each fracture and the sequence of action of each fracture, that is, the difference of the same period; the "bounce" of each fracture. It should be noted that abnormal bounce should be excluded from some illusion caused by loose wiring during the test.

项目 四 高压开关柜检修

模块一 配电装置介绍

配电装置是指根据电气主接线的要求，由开关设备、母线、保护和测量电气以及必要的辅助设备等一、二次设备，按照一定的技术要求建造而成的电工建筑物。配电装置在电力系统正常运行中起着接受、传输和分配电能的作用，当电力系统发生故障时能够迅速切断故障部分，以及对系统运行方式的改变、对线路和设备的操作都在其中进行，所以配电装置是维持电力系统正常运行的重要环节，是发电厂和变电站的重要组成部分。

一、配电装置基本要求

配电装置是维持电力系统正常运行的重要环节，是发电厂和变电站的重要组成部分，为此，对于配电装置的设置，应当有以下要求。

1. 安全

为保证检修、操作时工作人员的安全，设计配电装置时应当采取一系列必要的措施，例如用隔墙把相邻回路的设备隔开、为遮拦设置足够的安全距离、设置安全出口、设备配置五防联锁功能、设备外壳保护接地等措施。建筑方面还要考虑防火防爆等措施。

2. 可靠

配电装置出现事故的主要原因有绝缘电介质由于污秽闪络、五防系统不全导致发生误操作、断路器开端能力不足导致爆炸等。因此，配电装置的设计当考虑足够的安全净距、考虑防火防爆、考虑防风抗震等因素，按照系统要求和自然条件以及相关规程合理选择相应的设备，使其有正确的技术参数，保证配电装置的可靠运行。

3. 方便

电力系统运行方式的改变、设备、线路的切换等，都需要对变电站、发电厂配电装置中的设备进行操作，如果对设备的操作过于复杂、繁琐，如进行一个回路的操作需要来回走几层楼或者几个不同的走廊，会使得发生误操作的可能性更高。所以配电装置的结构布置应当力求整齐、清晰，便于操作和检修。

4. 经济

在满足上述三个要求的前提下，应当对配电装置的设备进行合理布置，节约占地面积，节约钢材等原材料，降低造价，节省投资。

5. 发展

根据变电站、发电厂所处位置以及当地未来发展规划，分析将来是否有扩建的必要，若有，应当预留备用间隔、备用容量，便于将来安装和扩建。

二、配电装置的相关术语

1. 安全净距

为了满足配电装置运行和检修的需要，各个带电设备之间，应当保证一定的距离。配电装置各部分之间，为确保人身、电网、设备的安全，所必须保持的最小电气距离，称为配电装置的安全净距。

按照我国《高压配电装置设计技术规程》（DL/T 5352—2018）规定，屋内外配电装置各部分最小安全净距可分为 A、B、C、D、E 五类，如表 4-1、表 4-2 所示。其中最基本的是带电部分与地之间的距离，称 A1 值，和带电部分与不同相带电部分之间的距离，称 A2 值。

表 4-1　部分屋内配电装置安全净距

符号	解释	安全净距/mm			
		10 kV	35 kV	110 kV	220 kV
A_1	1. 带电部分与接地部分； 2. 网状、板状遮栏向上延伸距地 2.3 m 处与遮拦上方带电部分之间	125	300	850	1800
A_2	1. 不同带电部分距离； 2. 断路器和隔离开关断口两侧引线带电部分之间	125	300	900	2000
B_1	1. 带电部分至栅状遮拦间； 2. 交叉的不同时停电的无遮拦带电部分之间	$B_1=A_1+750$ mm			
B_2	网状遮拦至带电部分之间	$B_2=A_1+30$ mm$+70$ mm			
C	无遮拦裸导体至地面（楼面）	2500	2600	3150	4100
D	平行的不同时停电检修的无遮拦裸导线之间	$D=A_1+1800$ mm			
E	通向屋外出线套管至屋外通道的路面	4000	4000	5000	5500
		此项考虑了人站在汽车车厢中举手的高度			
备注		110 kV、220 kV 默认为中性点直接接地			

表 4-2　部分屋外配电装置安全净距

符号	解释	安全净距/mm			
		10 kV	35 kV	110 kV	220 kV
A_1	1. 带电部分与接地部分； 2. 网状遮栏向上延伸距地 2.5 m 处与遮拦上方带电部分之间	200	400	900	1800
A_2	1. 不同带电部分距离； 2. 断路器和隔离开关断口两侧引线带电部分之间	200	400	1000	2000
B_1	1. 设备运输时，外廓至无遮拦带电部分； 2. 交叉的不同时停电检修的无遮拦带电部分； 3. 栅状遮拦至绝缘体和带电部分之间	$B_1=A_1+750$ mm 750 mm 为人员手臂误入遮拦的长度，以及设备移动时的摆动长度			
B_2	网状遮拦至带电部分之间	$B_2=A_1+30$ mm$+70$ mm 30 mm 为水平方向施工误差，70 mm 为人员手指误入网状遮栏长度			
C	1. 无遮拦裸导体至地面； 2. 无遮拦裸导体至建筑物顶部之间	$C=A_1+2300$ mm$+200$ mm 2300 为人员站立举手后高度，200 mm 为垂直方向施工误差			
D	1. 平行的不同时停电检修的无遮拦裸导线之间； 2. 带电部分与建筑物边沿部分之间	$D=A_1+1800$ mm$+200$ mm 1800 mm 为考虑检修人员和工具允许的活动范围，200 mm 为垂直方向施工误差			
备注		110 kV、220 kV 默认为中性点直接接地			

2. 间隔

间隔是配电装置中最小的组成部分，大体上对应电气主接线中一个接线单元，一主设备为主，加上附属设备的一整套电气设备。

在发电厂或变电站内，一个完整的电气连接部分，包括断路器、隔离开关、电流互感器、电压互感器、端子箱等，称为一个间隔。根据间隔中主设备的不同作用，间隔的功能也各不相同，包括变压器间隔、母线间隔、出线间隔、母联间隔，等等。

例如出线间隔，以出线断路器为主设备，此间隔还包括出线侧隔离开关、母线侧隔离开关、接地刀闸、电流互感器、避雷器、端子箱等附属设备。又如变压器间隔，以变压器为主设备，此间隔还包括变压器各侧隔离开关、避雷器、中性点接地等附属设备。

3. 层

层是指设备布置的层次，有单层布置、双层布置、三层布置等。

4. 列

一个间隔断路器的排列次序称为列，有单列布置、双列布置、三列布置等。

5. 通道

为便于操作、检修和搬运,配电装置在设计时,不同间隔之间均应设置维护通道、操作通道、防爆通道。

三、配电装置的类型

配电装置的结构形式,与电气主接线、电压等级和采用的电气设备形式有密切的关系,其分类方式很多。

1. 按设备地点

按照设备地点可以分为屋内式配电装置与屋外式配电装置。

(1) 屋内式:布置在建筑内的配电装置。由于其允许安全净距较小,可以分层布置,占地面积较小;维修、巡视、操作均在室内进行,可以减轻工作量,不受气候影响;外界污秽空气与气候对设备影响较小,减少维护工作量;房屋建筑投资较大,建设周期长,但户内设备价格较户外设备普遍较低。

(2) 屋外式:布置在建筑外的配电装置。土建工作量和费用较小,建设周期短;扩建更方便;相邻电气设备之间安全净距较大,方便带电作业;占地面积较大;受外界环境影响,设备运行条件较差,绝缘要求更高;不良天气条件对设备检修、维护、操作有一定影响。

在发电厂或变电站中,35 kV 及以下配电装置通常采用屋内式,110 kV 及以上配电装置通常采用屋外配电装置;也有农村、偏远地区 35 kV 配电装置布置在户外,有的 110 kV、220 kV 配电装置受环境及地理条件所限采用户内布置。

2. 按组装方式

屋内配电装置按照组装方式可分为装配式配电装置和成套式配电装置。

(1) 装配式:在现场进行电气设备的组装而成的配电装置。目前需要安装重型设备的屋内配电装置通常采用装配式。

(2) 成套式:制造厂商按要求预先将开关电器、互感器、保护电器等组装成套后,运输至现场进行安装的配电装置。这种配电装置布置在封闭或半封闭的金属外壳中,结构紧凑,占地面积非常小;现场安装工作量小,建设周期短,便于搬迁和扩建;运行可靠性高,维护方便;消耗材料较多,造价较高。

3~10 kV 等级配电装置大多采用屋内成套配电装置,对 110~220 kV 配电装置有特殊要求时,例如室外空气长期处于严重污秽状态,也可采用成套配电装置。

四、屋内配电装置

屋内配电装置是将各种电气一、二设备安装在屋内,避开大气污染和恶劣气候影响,它的特点有:

(1) 允许安全净距较小,可以分层布置,占地面积小。

(2) 维修、巡视和操作在室内进行,不受天气影响。

(3) 外界污秽的空气对电气设备影响较小。可以减少维护工作量。

(4) 房屋建筑的投资较大。

（一）屋内配电装置的类型

屋内式配电装置按照设备布置形式可以分为单层式、二层式和三层式。

（1）单层式：一般用于出线不带电抗器的配电装置，所有的电气设备布置在单层房屋内。这种布置形式占地面积较大，通常可采用成套开关柜，主要用于单母线接线、单母线分段接线、中、小容量发电厂和变电站。

（2）二层式：一般用于出线带电抗器的情况，所有电气设备按照重量分别布置在两层建筑中，较重的设备如断路器、电抗器、互感器等布置在第一层，较轻的设备如母线、隔离开关布置在第二层。这种布置形式结构较简单，占地面积较少，运行和检修较为方便，综合造价较低。

（3）三层式：所有电气设备按照重量分别布置在三层建筑中，具有安全、可靠、占地面积小等特点，但是结构复杂，施工时间长，造价高，运行和检修不方便，因此我国很少采用三层式布置的屋内配电装置。

（二）屋内配电装置布置原则

1. 总体布置原则

（1）尽量将电源布置在每段母线中部，使母线截面通过较小的电流，但有时为了连接的方便，根据主厂房或变电站的布置而将发电机或变压器间隔设置在每段母线的端部。

（2）同一回路的电气设备和导体应布置在一个间隔内，以保证检修的方便和安全，并且限制故障的范围。

（3）较重的设备布置在下层，以减轻楼板的荷重，并便于安装。

（4）满足安全净距的前提下，充分利用间隔的位置，少留无用间隙以降低占地面积。

（5）设备布置力求简单清晰，结构对称，便于操作，利于扩建。间隔内设备尺寸除满足4-1最小安全净距外，还应考虑设备的安装和检修条件，进而确定间隔的确切高度和宽度，设计时可以参考典型方案。

2. 设备的布置

（1）母线及隔离开关。

母线通常装设在配电装置的上部，呈水平、垂直或直角三角形布置。

水平布置相比垂直布置不便于观察，但建筑部分简单，可以降低建筑高度，安装容易，在中、小型容量发电厂变电站内采用较多。

垂直布置时，可以利用建筑物的高度，让相间距可以做得比较大，就无需增加间隔深度，节约占地面积，而且便于观察。但是垂直布置结构复杂，建筑高度较高。垂直布置可用于 20 kV 及以下、短路电流较大的装置中。

直角布置结构紧凑，可以充分利用户内建筑的高度与深度，但三相为非对称布置，外部短路时各相导体和绝缘子的动稳定性以及机械强度都各不一样，这种方式可以用于 6~35 kV 大、中容量的配电装置中。

母线相间距离除了考虑最小安全净距外，还需要考虑短路时母线和绝缘子的机械强度与安装条件等。在 6~10 kV 小容量装置中，母线水平布置时，相间距取 250~350 mm；母线垂直布置时，相间距取 700~800 mm；35 kV 配电装置中母线水平布置时，相间距约 500 mm。

若主接线为双母线布置，则两组母线之间应该由隔墙或隔板分开，若母线分段时，两段母线间也应当有隔墙或隔板分开，这样任意一组母线故障时，不会影响另一组母线，并且母线检修时也更加安全。

母线侧隔离开关通常在母线下方，为了防止带负荷拉合隔离开关引起飞弧短路，在双母线布置的屋内配电装置中，母线与隔离开关之间宜装设耐火隔板，两层以上的配电装置中，母线隔离开关应当单独布置在一个小室内。为确保运行人员和检修人员的安全，屋内配电装置应设置完备的"五防"系统。

（2）断路器及其操动机构。

断路器通常设在单独的小室内。油断路器小室的形式，按照油量和防火防爆的要求，可分为敞开式、防爆式及封闭式。敞开式小室完全或者部分使用非实体的隔板或遮拦；封闭式小室四壁采用实体墙壁、顶盖和无网眼的门全封闭；若封闭式小室出口直接通向屋外或者专设有防爆通道，则为防爆式小室。

一般 35 kV 及以下的屋内断路器和油浸式互感器安装在开关柜内或者用混泥土墙、砖墙隔开的单独小间内；35~220 kV 屋内断路器与油浸式互感器则应安装在防爆隔板隔开的单独小间内，当单台设备总油量在 100 kg 以上时，应设储油或挡油设施，以防止事故漏油后燃烧范围过大。

断路器本体与其操动机构应当使用隔板隔开，操动机构应当位于通道内并且面向通道。

（3）互感器和避雷器。

电流互感器无论干式还是油浸式，都可以和断路器放在同一小室内，穿墙式电流互感器应尽可能作为穿墙套管使用，以减少配电装置的体积和造价。

电压互感器经隔离开关接至母线上，它需要占用专门的间隔，但在同一间隔可以装设几个不同用途的电压互感器。

当母线接有架空线路时，母线上应当装设避雷器，可以和电压互感器装设在同一间隔，可共用一个隔离开关，两者之间应采用隔板隔开。

（4）电抗器。

电抗器较重，大多布置在封闭小室的第一层。电抗器按其容量不同有三种布置方式：三相垂直，品字形和三相水平布置。

通常线路电抗器采用垂直布置或垂直布置；当电抗器额定电流超过 1000 A，电抗值超过 5%~6%时，由于重量尺寸过大，垂直布置会有困难，故采用品字形布置；当额定电流超过 1500 A 的母线分段电抗器或变压器低压侧的电抗器，宜采用水平布置。

（5）电容器。

高压电容器室的大小主要由电容器容量和对通道的要求决定。电容器室的建筑面积可按照每 100 kvar 约需 4.5 m²。电容器室应当有良好的通风，若不能保证室内温度不超过 40 ℃，应增设机械通风装置。若电容器容量不大，则可以考虑设置在高压配电装置或无人值班的高低压配电室内。

（6）变压器室。

变压器室的最小尺寸应根据变压器外形尺寸和变压器外廓至变压器室四壁的最小安全净距而定，按规程不应小于表 4-3 所列的数值。

表 4-3 变压器外廓与变压器室四壁的最小安全净距

变压器容量/（kV·A）	320 及以下	400~1000	1250 及以上
至墙壁净距（mm）	600	600	800
至大门净距（mm）	600	800	1000

变压器室的高度与变压器高度、运行方式、通风条件等因素有关。根据通风要求，变压器室的地坪有抬高和不抬高两种情况。变压器室的地坪是否抬高由其通风条件以及通风面积确定，当变压器室的进风窗和出风窗面积不能满足通风条件时，就应抬高变压器室的地坪。地坪不抬高时，变压器放置在混凝土地面上，变压器室的高度一般为 3.5~4.8 m；地坪抬高时，变压器放置在抬高的地坪上，下面是进风洞，地坪抬高高度一般有 0.8 m、1.0 m、1.2 m 三种，变压器室的高度一般为 4.8~5.7 m。

变压器室的进风窗位置较低，须加铁丝网防止小动物进入，出风窗位置高于变压器，要考虑金属百叶窗挡雨。

当变电站有两台变压器时，一般各自单独安装在变压器室内，以防止一台变压器发生火灾，影响另一台变压器正常运行。

（7）电缆构筑物。

电缆隧道或电缆沟道是用来放置电缆的。电缆隧道为封闭狭长的构筑物，高度 1.8 m 以上，两侧设有数层敷设电缆的支架，可放置较多的电缆，人在隧道内能够方便地进行电缆的敷设和维修工作。电缆隧道其造价高，一般用于大型发电厂主厂房内。电缆沟则为有盖板的沟道，沟宽 1 m 左右，敷设和维修电缆必须揭开盖板。雨水较多时，电缆沟内容易出现积水，给维修带来一定困难，电缆沟造价较低，施工简单，多用于中、小型发电厂和变电站。

（8）通道和出口。

配电装置的布置应便于设备的操作、检修和搬运，故需设置必要的通道。凡是专门用来维护和搬运电气设备的通道，称为维护通道；如通道内设有开关类设备的操动机构、就地控制屏等，则称为操作通道；仅和防爆小室想通的通道，称为防爆通道。通道的最小宽度应满足室内最小安全净距要求。一般情况下，维护通道最小宽度应比最大搬运设备大 0.4~0.5 m，操作通道最小宽度为 1.5~2 m，防爆通道最小宽度为 1.2 m。

为保证工作人员的安全以及出入方便，不通长度的屋内配电装置室都应设有一定数目的出入口。长度小于 7 m 的小室可以仅设一个出入口，长度大于 7 m 的小室应当设置两个出入口，当长度大于 60 m 时，中部应当增设一个出入口。为了便于工作人员的逃生，配电装置室的门应当向外开启，并装设弹簧锁，相邻配电装置室之间有门，则门应当能够向两个方向开启。

五、屋外配电装置

目前，我国 110 kV 及以上电压等级一般多采用屋外配电装置。屋外配电装置是将电气设备安装在露天场地的基础、支架或者构架上，它的特点有：

（1）土建工作量和费用较小，建设周期短。

（2）扩建比较方便。

（3）相邻设备之间距离较大，便于带电作业。

（4）占地面积大。

（5）受外界环境影响，设备运行条件较差，需加强绝缘。

（6）不良天气对设备的维修和操作有一定影响。

（一）屋外配电装置的类型

屋外配电装置根据电气设备和母线布置的高度，可以分为低型、中型、半高型和高型。

（1）低型配电装置：电气设备直接放置在地面基础上，母线布置高度比较低，为了保证安全距离，设备周围设有常设围栏。低型配电装置造价低，占地面积大，设备高度过低，安全性较差。

（2）中型配电装置：所有的电气设备都安装在同一水平面内，并且装在一定高度（2~2.5 m）的基础上，使带电部分对地保持比较高的高度，以便工作人员能在地面安全地活动。母线所在的水平面要稍高于其他设备所在的水平面。

中型配电装置按照隔离开关的布置方式可以分为普通中型和分相中型，所谓分相中型是指隔离开关采用垂直伸缩式，直接布置在母线正下方，其余设备与普通中型配电装置相同。

中型配电装置布置清晰，不易误操作，运行可靠，施工维护方便，所使用钢材较少，造价较低，占地面积大。

（3）高型配电装置：将一组母线及其隔离开关与另一组母线及其隔离开关上下重叠布置，相比中型配电装置，节省占地面积 50%左右，但是消耗钢材非常多，造价昂贵，操作和维护困难，检修上层设备十分不便。

（4）半高型配电装置：将母线置于高一层的水平面上，与断路器、互感器、隔离开关等其他设备上下重叠布置，其占地面积比普通中型配电装置少 30%左右。其特点是介于高型配电装置与中型配电装置，具有两者的优缺点，除母线侧隔离开关外，其余设备维护还是较为方便。

（二）屋外配电装置的选型

屋外配电装置的选型除了与电气主接线有很大关系之外，还与场地位置、面积、地质、地形等因素以及总体布置有关，并且受到设备材料的供应、施工、运行和检修要求等因素的影响和限制，故应通过技术经济比较来选择最佳方案。

（1）低型配电装置：虽然其建设成本低廉，但是由于占地面积过大，设备高低过低导致安全性较差等缺点，一般情况都不采用低型配电装置。

（2）中型配电装置：普通的中型配电装置施工、检修和运行都比较方便，抗震能力好，造价比较低，缺点是占地面积较大。此种型式一般用在非高产农田地区，以及不占良田土石方工程量不大的地方，并宜在地震烈度高的地区采用。

分相中型配电装置采用硬母线配单柱式合垂直伸缩隔离开关，布置简单清晰，可以省去大量构架，比普通中型配电装置更节约用地。但单柱式隔离开关抗震能力较差，不宜在地震烈度较高的地区采用。

（3）高型配电装置：高型配电装置最大的优点就是占地面积非常少，比普通中型配电装置节约占地 50% 作用，但是耗用钢材较多，检修运行不方便，一般在以下情形中使用高型配电装置：①配电装置设置在高产农田或者是地少人多的城市；②由于地形条件限制，场地狭窄或者需要大量开挖、回填土石方等土石方工程量较大的地方；③原有配电装置需要改建或者扩建，而场地受到限制。高型配电装置适用于220 kV及以上电压等级，在地震烈度高的地区，不宜采用高型配电装置。

（4）半高型配电装置：半高型配电装置结缘占地不如高型显著，但是其运行、维护和检修更方便，施工难度更小，耗用钢材也较少。半高型适宜于110 kV配电装置。

（三）屋外配电装置布置原则

1. 母线及构架

（1）屋外配电装置的母线有软母线和硬母线两种。

软母线为钢芯铝绞线、软管母线和分裂导线，三相水平布置，用悬式绝缘子悬挂于母线构架上。

硬母线有矩形和管型的。矩形母线用于35 kV及以下配电装置中；管型母线用于110 kV及以上配电装置中。管型母线一般安装在支柱绝缘子上，母线不会摇摆，相间距可以缩小，与单柱式垂直伸缩隔离开关配合可以节约占地面积；管型母线直径大，表面光滑无毛刺，更不容易产生电晕；但是管型母线抗震能力差，档距小，一般不能上人检修。

（2）屋外配电装置构架一般由型钢或钢筋混凝土制成。

钢构架经久耐用，机械强度大，可以按任何负荷和尺寸制造，便于固定设备，抗震能力强，运输方便，但金属消耗巨大，为防锈需要经常维护。

钢筋混凝土可以节约大量钢材，也能够满足各种强度和尺寸要求，经久耐用，维护简单。

2. 电力变压器

电力变压器外壳为金属，并且不带电，因此采用落地布置，安装在变压器基础上。

变压器基础一般做成双梁并铺铁轨，铁轨间距等于变压器滚轮中心距。为了防止变压器发生事故时，变压器油流动造成事故扩大，单个邮箱油量超过1000 kg以上的变压器，按照防火要求，在设备下方应当设置储油池或挡油墙，其尺寸应当比设备外廓大1 m，储油池内一般铺设厚度不小于0.25 m的鹅卵石。

建筑物与户外油浸变压器的外廓间距不宜小于10 m，当其间距小于10 m，且在5 m以内时，在变压器外轮廓投影范围外侧各3 m内的屋内配电装置楼、主控制楼及网络控制楼面向油浸变压器的外墙不应开设门、窗和通风孔；当其间距在5~10 m时，在上述外墙上可设甲级防火门。

3. 其他电气设备

（1）断路器：按照断路器在配电装置中所占的位置，可以分为单列、双列和三列布置。若在进出线方向均呈三列布置，则称为三列布置；当断路器布置在主母线两侧时，则称为双列布置；如将断路器集中布置在主母线一侧，则称为单列布置。断路器的布置方式，须根据主接线、场地条件、总体布置和出线方向等多种因素综合考虑。

按断路器基础高度分，低型和高型两种布置方式。低型断路器放置在0.5~1 m混泥土基础上，检修方便，抗震能力强，但是由于基础低，安全性差，四周必须常设遮拦，占地面积大，影响通道畅通，此方式用于低型配电装置。高型布置断路器放置在高约2 m的基础上，断路器的操动机构须装在相应的基础上,用于中型配电装置中。

(2)隔离开关和互感器:均采用高型布置,要求与断路器相同,隔离开关操动机构装设在其边相基础的一定高度上。为保证安全,每段母线应装设 1~2 组接地隔离开关。

(3)避雷器:有高型和低型两种布置方式。110 kV 及以上的阀式避雷器由于器身细长,如采用高型布置,安装在 2 m 基础上,总高度达到接近 6 m,检修、试验和维护都不方便,并且抗震能力差,因此多采用落地式低型布置,安装在 0.4 m 基础上,四周常设遮拦。

(4)电缆沟:屋外配电装置中电缆沟的布置,应使电缆所走的路径最短。电缆沟按其布置方向,可分为纵向电缆沟和横向电缆沟。

六、成套配电装置

成套配电装置:制造厂商按照主接线要求,预先将每一回路的设备(如断路器、隔离开关、避雷器、互感器、测量仪器等)装配在封闭或者封闭的金属壳(柜)内,形成标准模块,各模块运输至现场后,进行组合装配,形成完整的配电装置称为成套式配电装置,它的特点有:

(1)成套配电装置有金属外壳(柜体)的保护,各电气设备和载流导体不容易积灰,便于维护,特别是污秽地区优势更为突出。

(2)成套配电装置易于实现系列化、标准化。

(3)装配质量好、运行可靠性高。

(4)结构紧凑、布置合理,减少占地面积。

(5)工程建设时间短,建设工作量小。

(一)成套配电装置的类型

(1)成套配电装置按柜体特点,可以分为开启式和封闭式。

① 开启式成套配电装置母线外露,柜内各单元之间未隔开,结构简单,造价低。

② 封闭式成套配电装置也叫做封闭式开关柜,所有设备均封闭于全金属封闭外壳中,其中母线、电缆头、断路器、测量仪表均被金属隔间隔开,运行安全,可防止事故扩大,适用于工作环境差,要求较高的地方。

(2)成套式配电装置按断路器的固定特点,可分为固定式和可移动式。

① 固定式成套配电装置中的全部电气设备均固定于柜内构架上,若要移动则必须停电后使用工具拆卸。

② 可移动式成套配电装置也称为手车式开关柜,其断路器和断路器的操动机构等部件均装设在可以推进拉出的底盘车上,便于检修、试验和维护,隔离开关采用插入式触头,代替传统的闸刀式触头。

(3)成套配电装置按母线数,可分为单母线和双母线两种。

(4)成套配电装置按电压等级可分为高压配电装置和低压配电装置两种。高压配电装置也称为高压开关柜,低压配电装置也称为低压配电屏。

(二)高压成套配电装置

高压成套配电装置也称为高压开关柜,以断路器为主体,将检测仪表、保护设备和辅助设备按一定主接线要求都装设在封闭或半封闭金属柜体中,通常以一个柜体构成一个间隔。

柜内各电气设备相互绝缘，绝缘材料大多采用支柱式绝缘子和空气，安全净距较小，结构可以做得比较紧凑，从而节省占地面积和材料。根据运行经验，高压开关柜可靠性高，维护方便，安装简单，在 3~35 kV 系统中大量采用。

（三）箱式变电站

在配电系统中，由于以变电站为中心的供电半径过大，线路损耗随用电负荷和线路长度增大而增大，将导致末端电压较低，大大降低供电质量。为减少线路损耗，保证供电质量，就必须提高供电电压。为此，国家在城乡供电网络建设中，要求高压电直接进入负荷中心，供电电压从 0.4 kV 提高到 10 kV 甚至 35 kV，线路损耗大大减少，也减少了总用铜量和总投资，经济效益非常可观。要实现高压深入负荷中心，箱式变电站是最经济、最有效的成套配电装置。

箱式变电站（见图 4-1）是一种将高压开关设备、变压器和低压配电装置按一定接线方式组成一体而成。箱式变电站具有成套性强，体积小，占地面积小，提高供电质量，减少线路损耗，缩短送电周期安装方便，环境适应性强，运行可靠，投资少等一系列优点，在电力系统配电网络中大量使用。

图 4-1　箱式变电站

（四）气体全封闭组合电器（GIS）

其全称为 GasInsultedSwitchgear，简称 GIS。它是由断路器、隔离开关、接地刀闸、互感器、避雷器、母线、出线套管等电气设备按接线要求组合成的一个整体，并且全封闭在接地的金属外壳中，壳内充装 SF_6 气体，作为绝缘介质和灭弧介质。

Program 4 High-voltage Switch Cabinet Maintenance

Module 1 Power Distribution Unit Introduction

The power distribution unit refers to the electrical building constructed by the primary and secondary equipment such as switchgear, bus, protective and measuring electrical equipment and necessary auxiliary equipment according to the main electrical wiring requirements and certain technical requirements. The power distribution unit plays the role of accepting, transmitting and distributing electric energy in the normal operation of the power system. When the power system involves fault, the power distribution unit can quickly cut off the fault part. The change of system operation mode and the operation of line and equipment both are carried out in the power distribution unit. Hence, the power distribution unit is an important link to maintain the normal operation of the power system, and is an important part of the power plant and substation.

4.1.1 Basic Requirements of Power Distribution Unit

The power distribution unit is an important link to maintain the normal operation of the power system, and is an important component of the power plant and substation. Therefore, for the setting of the power distribution unit, there shall be the following requirements.

1. Safety

To ensure the safety of operators during maintenance and operation, a series of necessary measures shall be taken when designing the power distribution unit, such as separating the equipment in the adjacent circuits with partition wall, setting a sufficient safety distance for the barrier, reserving a safety exit, equipping the equipment with five-prevention interlocking function, and protective grounding of the equipment shell. For buildings, fire and explosion prevention measures also need to be considered.

2. Reliability

The main causes of the accidents of power distribution unit include insulating dielectric flashover for filth, misoperation for incomplete five-prevention system, and explosion for insufficient opening capacity of circuit breaker. Hence, the design of the power distribution unit shall consider sufficient net safety distance, fire and explosion prevention, wind and earthquake resistance and other factors. According to system requirements, natural conditions and relevant regulations, corresponding equipment shall be selected reasonably, thereby ensuring correct technical parameters and reliable operation of power distribution unit.

3. Convenience

For the change of operation mode of power system and the switching of equipment and line, the equipment in the power distribution unit of substation and power plant needs to be operated. If the equipment operation is too complex and tedious, such as that it needs to walk back and forth several floors or several different corridors to operate one circuit, there shall be a higher possibility of misoperation. Therefore, the structural arrangement of the power distribution unit shall be neat, clear, and easy for operation and maintenance.

4. Economy

Under the premise of meeting the above three requirements, the equipment of the power distribution unit shall be rationally arranged, thereby saving floor area, raw materials (such as steel), cost and investment.

5. Development

According to the location of substation and power plant and the local future development plan, the necessity of future expansion shall be analyzed. If there is necessity, spare bay and spare capacity shall be reserved for future installation and expansion.

4.1.2 Relevant Terms of Power Distribution Unit

1. Net safety distance

To meet the needs of the operation and maintenance of the power distribution unit, a certain distance shall be guaranteed among live equipment. To ensure the safety of human body, power grid and equipment, the minimum electrical distance that must be maintained among various parts of the power distribution unit is called the net safety distance of power distribution unit.

According to the *Code for Design of High Voltage Electrical Switchgear* (DL/T 5352-2018) of our country, the minimum net safety distance for the various parts of the indoor and outdoor power distribution units can be divided into the five categories of A, B, C, D and E, as shown in Tables 4-1 and 4-2. Among them, the most basic distance is the distance between the live part and the ground (called A1 value), as well as the distance between the live part and the different phase live part (called A2 value).

Table 4-1　Net Safety Distance of Partial Indoor Power Distribution Unit

Symbol	Explanation	Net safety distance /mm			
		10 kV	35 kV	110 kV	220 kV
A_1	1. Between live part and grounding part; 2. Between the position at 2.3 m away from the ground after the mesh and plate barrier extending upward and the live part above the barrier	125	300	850	1800

Continued

Symbol	Explanation	Net safety distance /mm			
		10 kV	35 kV	110 kV	220 kV
A_2	1. Distance between different live parts; 2. Between the live parts of the leads on both sides of the break of circuit breaker and disconnector	125	300	900	2000
B_1	1. Between live part and grid barrier; 2. Between unblocked live parts with intersecting power interruption at different times	$B_1 = A_1 + 750$ mm			
B_2	Between mesh barrier and live part	$B_2 = A_1 + 30$ mm $+ 70$ mm			
C	Between unblocked bare conductor and ground (floor);	2500	2600	3150	4100
D	Between parallel unblocked bare conductors for power interruption maintenance at different times	$D = A_1 + 1800$ mm			
E	Between the outlet bushing to the outside and the pavement of the outside access road	4000	4000	5000	5500
		This item takes into account the height at which person raises their hands while standing in the car compartment			
Remarks		By default, 110 kV and 220 kV are directly grounded at neutral points			

Table 4-2 Net Safety Distance of Partial Outdoor Power Distribution Unit

Symbol	Explanation	Net safety distance /mm			
		10 kV	35 kV	110 kV	220 kV
A_1	1. Between live part and grounding part; 2. Between the position at 2.5 m away from the ground after the mesh barrier extending upward and the live part above the barrier	200	400	900	1800
A_2	1. Distance between different live parts; 2. Between the live parts of the leads on both sides of the break of circuit breaker and disconnector	200	400	1000	2000
B_1	1. Between the outer profile and the unblocked live part during equipment transportation; 2. Between unblocked live parts with intersecting power interruption maintenance at different times; 3. Between grid barrier and insulator and live part	$B_1 = A_1 + 750$ mm 750 mm is the length of the person's arm extending into the barrier by mistake and the swing length of the equipment during movement			
B_2	Between mesh barrier and live part	$B_2 = A_1 + 30$ mm $+ 70$ mm 30 mm is the horizontal construction error, and 70 mm is the length of the person's finger extending into the mesh barrier by mistake			
C	1. Between unblocked bare conductor and ground; 2. Between unblocked bare conductor and roof of building	$C = A_1 + 2300$ mm $+ 200$ mm 2300 is the height at which person raises their hands while standing, and 200 mm is the vertical construction error			
D	1. Between parallel unblocked bare conductors for power interruption maintenance at different times; 2. Between live part and edge of building	$D = A_1 + 1800$ mm $+ 200$ mm 1800 mm is the allowable movement range considering maintainers and tools, and 200 mm is the vertical construction error			
Remarks		By default, 110 kV and 220 kV are directly grounded at neutral points			

2. Bay

Bay is the smallest component of the power distribution unit, roughly corresponding to a wiring unit in the main electrical wiring. It is a complete set of electrical equipment giving priority to one primary equipment, supplemented by secondary equipment.

In a power plant or substation, a complete electrical connection part includes circuit breaker, disconnector, current transformer, voltage transformer and terminal box, called one bay. According to the different roles of the primary equipment in the bay, the function of the bay is also different, including transformer bay, bus bay, outlet bay and busbar bay.

Taking the outlet bay as an example, the outlet circuit breaker is used as the primary equipment, and this bay also contains the secondary equipment such as outlet-side disconnector, bus-side disconnector, grounding knife switch, current transformer, lightning arrester and terminal box. Another example is the transformer bay, which deems transformer as the primary equipment. This bay also contains the secondary equipment such as disconnector on each side of transformer, lightning arrester and neutral point grounding.

3. Layer

Layer refers to the layer for equipment arrangement, including single-layer arrangement, double-layer arrangement, three-layer arrangement, etc.

4. Row

Row refers to the ordering of circuit breakers in one bay, including single-row arrangement, double-row arrangement, three-row arrangement, etc.

5. Channel

To facilitate operation, maintenance and handling, maintenance channel, operation channel and explosion-prevention channel are set among different bays during the design of power distribution unit.

4.1.3 Types of Power Distribution Units

The structural form of power distribution unit is closely related to the main electrical wiring, the voltage class and the form of electrical equipment adopted, and there are many ways for classification.

1. According to the equipment location

According to the equipment location, they can be divided into indoor power distribution units and outdoor power distribution units.

(1) Indoor type: They are power distribution units arranged in building. Due to the smaller net safety distance, they can be arranged by layers to occupy less area; Maintenance, inspection and operation are all conducted indoor, thereby reducing the workload and not being affected by the weather; The outside filthy air and weather have less impact on the equipment, thus reducing the maintenance workload; The building construction needs larger investment and long construction

period, while the price of indoor equipment is generally lower than that of outdoor equipment.

(2) Outdoor type: They are power distribution units arranged out of building The workload and cost of civil construction are small, and the construction period is short; It is more convenient for expansion; The net safety distance is larger between the adjacent electrical equipment, thus facilitating live operation; It covers a larger area; As affected by the external environment, the equipment has worse operating conditions and needs higher insulation requirements; The adverse weather conditions have a certain impact on equipment maintenance, repair and operation.

In power plants or substations, 35 kV and below power distribution units are usually indoor types, and 110 kV and above power distribution units are usually outdoor types; There are also some 35 kV power distribution units arranged outdoor in rural and remote areas, and some 110 kV and 220 kV power distribution units arranged indoor as limited by environmental and geographical conditions.

2. According to the assembly method

According to the assembly method, the indoor power distribution units can be divided into assembled power distribution units and packaged power distribution units.

(1) Assembled type: power distribution units assembled for electrical equipment on site. Currently, the indoor power distribution units for heavy equipment are usually assembled type.

(2) Packaged type: power distribution units with switching devices, transformers and protective appliances having been preassembled by manufacturer according to the requirements before being transported to the site for installation. This type of power distribution unit, with compact structure, is arranged in a closed or semi-closed metal shell, thus covering a very small area; The field installation workload is small, and the construction period is short, thus facilitating relocation and expansion; The operation reliability is high, and the maintenance is easy; It consumes more materials, and the construction cost is much higher.

3–10 kV power distribution units mostly adopt indoor packaged power distribution units. If there are special requirements for 110–220 kV power distribution units, such as that outdoor air is in a serious filthy state for a long time, packaged power distribution units can also be employed.

4.1.4 Indoor Power Distribution Unit

Indoor power distribution units refer to the various types of primary and secondary electrical equipment installed indoor, thereby avoiding the effect of air pollution and adverse weather. They have the characteristics that:

(1) The allowable net safety distance is smaller, and the equipment can be arranged by layers to occupy less area.

(2) Maintenance, inspection and operation are all conducted indoor, thereby not being affected by the weather.

(3) The outside filthy air has less impact on the electrical equipment. Maintenance workload can be reduced.

(4) The building construction needs larger investment.

4.1.4.1 Types of indoor power distribution units

According to the equipment arrangement form, the indoor power distribution units can be divided into single-layer, double-layer and three-layer types.

(1) Single-layer type: They are generally power distribution units with outlet equipping no reactor. All electrical equipment is arranged in single-storey house. This arrangement form covers a large area and usually adopts a complete set of switch cabinet, mainly used for single-bus wiring, sectionalized single-bus wiring, medium and small-capacity power plants and substations.

(2) Double-layer type: They are generally power distribution units with outlet equipping reactor. All electrical equipment is arranged in two-storey building according to weight, with the heavier equipment such as circuit breaker, reactor and transformer being arranged in the first storey, and the lighter equipment such as bus and disconnector being arranged in the second storey. This arrangement form has the advantages of simpler structure, less floor area, more convenient operation and maintenance, and low comprehensive construction cost.

(3) Three-layer type: All electrical equipment is arranged in three-storey building according to weight. It is characterized by safety, reliability and small floor area, while complex structure, long construction time period, high construction cost, inconvenient operation and maintenance. Hence, the three-layer indoor power distribution units are rarely adopted in our country.

4.1.4.2 Arrangement principles of indoor power distribution units

1. General arrangement principles

(1) The power supply shall try to be arranged in the middle of each section of the bus, so that a smaller current can pass through the cross section of the bus. However, to facilitate the connection sometimes, the generator or transformer bay is arranged at the end of each section of the bus according to the arrangement of the main powerhouse or substation.

(2) Electrical equipment and conductor in a same circuit shall be arranged in one bay, thereby ensuring the convenience and safety of maintenance and limiting the scope of fault.

(3) The heavier equipment is arranged in the lower level, thereby reducing the load on the floor and facilitating installation.

(4) Under the premise of meeting the net safety distance, it needs to make full use of the bay position to reduce useless bays and decrease floor area.

(5) The equipment arrangement shall be simple and clear, exhibit symmetrical structure, and be convenient for operation and expansion.

The dimensions of equipment in the bay shall meet the minimum safe distance in 4-1. Besides, the installation and maintenance conditions of equipment should be considered to determine the exact height and width of the bay. Please refer to typical schemes in design.

2. Equipment layout

(1) Bus and disconnector.

Buses are usually arranged horizontally, vertically or in right triangles on the upper part of the power distribution unit.

Less easy to observe than vertical arrangement, the horizontal arrangement is characterized by simple building part, reduced height, and easy installation. It is widely used in substations of medium and small-capacity power plants.

The vertical arrangement can benefit from the height of building to have large phase spacing, without further depth of bay. This can save the floor space, and is easy to observe. However, the vertical arrangement features complicated structure and great height. The vertical arrangement can be used in devices with high short circuit current and up to 20 kV.

The right-angle arrangement is compact and can take full advantage of the height and depth of the indoor building. However, the three-phase arrangement is asymmetrical, to the extent that the dynamic stability and mechanical strength of each phase conductor and insulator are different in the case of external short circuit. This arrangement can be used in 6–35 kV high and medium capacity PDUs.

In addition to the minimum safe distance, the bus phase spacing must allow for the mechanical strength and installation conditions of the bus and insulator during short circuit. In the 6–10 kV devices, the phase spacing is 250–350 mm when the bus is arranged horizontally, and 700–800 mm when the bus is arranged vertically. The phase spacing is about 500 mm when the bus is arranged horizontally in the 35 kV power distribution unit.

If the main wiring is a double-bus layout, the two sets of buses should be separated by a partition wall or diaphragm; if the bus is segmented, there should also be a partition wall or diaphragm between the two sections of bus. In this way, when any bus fails, the other bus will not be affected, and it is safer when the bus is overhauled.

The bus side disconnector is usually provided under the bus. In order to prevent flashover short circuit caused by the load pulling disconnector, a fire-resistant diaphragm should be equipped between the bus and the disconnector in the indoor power distribution unit with double-bus layout. In the power distribution unit of more than two floors, the bus disconnector should be arranged separately in a compartment.

In order to ensure the safety of operators and maintainers, the indoor power distribution unit should be equipped with a complete "five-prevention" system.

(2) Circuit breaker and its operating mechanism.

Circuit breakers are usually provided in separate compartments. The oil circuit breaker compartment can be fallen into open, explosion-proof and closed type by the requirements of oil mass and fire and explosion protection. Open compartments feature non-solid diaphragms or barriers in whole or in part; the closed compartments are totally enclosed with solid walls, top cover and doors without mesh. The explosion-proof compartments are closed ones when they lead directly to the outside or are provided with explosion-proof passages.

Generally, indoor circuit breakers and oil-immersed transformers of 35 kV and below are installed in the switch cabinet or in a compartment separated by concrete and brick walls; 35–220 kV indoor circuit breakers and oil-immersed transformers shall be installed in a compartment separated by an explosion-proof diaphragm. When the total oil volume of single equipment is more than 100 kg, oil storage or oil baffle facilities shall be set up to prevent excessive combustion range after the accident oil spillage.

The circuit breaker and its operating mechanism shall be separated by a diaphragm, and the operating mechanism shall be located in and facing the channel.

(3) Transformer and lightning arrester.

Current transformers, dry type or oil immersed, can be placed in the same compartment as the circuit breaker. The wall-through current transformer should be used as a wall-through bushing as much as possible to reduce the volume and cost of the power distribution unit.

The voltage transformer is connected to the bus by the disconnector. It needs a dedicated bay, but several voltage transformers for different purposes can be installed in the same bay.

When connected with an overhead line, the bus shall be equipped with a lightning arrester. The lightning arrester can be installed at the same bay as the voltage transformer. They can share a disconnector, and be separated by a diaphragm.

(4) Reactor.

Reactors are heavy and mostly arranged on the first floor of the enclosed compartment. Reactors are arranged in three ways by their capacity: three-phase vertical, top and twin-side bottom, and three-phase horizontal arrangement.

Usually, line reactors are arranged vertically. When the rated current of the reactor exceeds 1,000 A and the reactance value exceeds 5%–6%, the vertical arrangement will be less feasible due to the heavy and bulky structure, so the top and twin-side bottom arrangement is preferred; when the rated current exceeds 1,500A, the bus sectionalizing reactor or the reactor on the LV side of the transformer should be arranged horizontally.

(5) Capacitor.

The size of the high-voltage capacitor room is mainly determined by the capacitor capacity and the requirements for the channel. The building area of the capacitor room can be about 4.5 m^2 per 100 kvar. The capacitor room must be well ventilated. If the indoor temperature cannot be kept within 40 °C, mechanical ventilators should be provided. Small-capacity capacitors may be provided in an HV power distribution unit or an unattended HV/LV distribution room.

(6) Transformer vault.

The minimum size of the transformer vault shall be determined according to the overall dimensions of the transformer and the minimum safe distance between the transformer profile and four walls of the transformer vault, and never be less than the values in Table 4-3.

Table 4-3 Minimum Safe Distance between the Transformer Profile and Four Walls of the Transformer Vault

Transformer capacity /(kV·A)	320 and below	400 to 1,000	1,250 and above
Clear distance to walls /mm	600	600	800
Clear distance to the gate /mm	600	800	1,000

The height of the transformer vault is connected with the transformer height, operation mode, ventilation conditions, etc. The floor of the transformer vault can be either raised or not raised by ventilation requirements. Whether the floor of the transformer vault is raised is determined by its ventilation conditions and ventilation area. When the areas of the inlet window and outlet window of the transformer vault cannot meet the ventilation conditions, the floor of the transformer vault must be raised. When the floor is not raised, the transformer is placed on the concrete floor, and the transformer vault is generally 3.5 to 4.8 m high; in case of raised floor, the transformer is placed on the raised floor, and below is the wind tunnel. The floor is generally raised for 0.8 m, 1.0 m, and 1.2 m, and the transformer vault is generally 4.8 to 5.7 m high.

The inlet window of the transformer vault is so low that wire gauze must be provided to prevent small animals. The outlet window is higher than the transformer, and metal shutters should be equipped to keep out the rain.

If any, two transformers in the substation are often installed separately in the transformer vault to prevent a fire in one transformer from affecting the normal operation of the other transformer.

(7) Cable structures.

Cable subways or cable trenches are used to place cables. The cable subway is a closed and narrow structure with a height of more than 1.8 m, and several layers of cable laying supports are provided on both sides to accommodate more cables. They are easily accessible for the purpose of cable laying and maintenance. The cable subways feature high costs and are generally used in the main powerhouse of a large power plant. The cable trench is a trench with a cover; the trench is about 1 m wide, and the cover must be opened for cable laying and maintenance. In case of more rain, water is easy to accumulate in the cable trench, thus bringing certain difficulties to maintenance. The cable trench features low cost and simple construction, and is mostly used in medium and small power plants and substations.

(8) Channels and exits.

The power distribution unit shall be arranged to facilitate the operation, maintenance, and handling of devices. Therefore, necessary channels must be arranged. All channels specially used for the maintenance and handling of electrical equipment are called maintenance channels; if the channel is equipped with the operating mechanism and local panel of the switching device, it is called the operation channel; the channel only connected with the explosion-proof compartment is called the explosion-proof channel. The minimum width of the channel must meet the indoor minimum safe distance requirements. Under normal circumstances, the maintenance channel should be at least 0.4 to 0.5 m wider than the maximum handling equipment, the operation channel should, at a minimum, be 1.5 to 2 m, and the explosion-proof channel 1.2 m.

The indoor power distribution unit room of different lengths must have a certain number of gateways to ensure the staff safety and easy access. The compartments with a length less than 7 m may have only one gateway, two gateways for those more than 7 m long; when the length is greater than 60 m, an additional gateway should be provided in the middle. In order to facilitate the escape of staff, the door of the power distribution unit room should be opened outwards and equipped with a spring lock; if there is a door between the adjacent power distribution unit rooms, the door should be able to open in two directions.

4.1.5 Outdoor Power Distribution Unit

At present, China's 110 kV and above voltage classes generally employ outdoor power distribution units. Outdoor power distribution units are to install electrical equipment on a foundation, support or frame in an open field, with the following characteristics:

(1) The workload and cost of civil construction are small, and the construction period is short.

(2) Convenient expansion.

(3) Large distance between adjacent equipment, convenient for hot-line work.

(4) Large footprint area.

(5) Poor operating conditions of equipment arising from the external environment, in need of reinforced insulation.

(6) A certain impact of bad weather on the maintenance and operation of equipment.

4.1.5.1 Type

According to the height of electrical equipment and bus arrangement, the outdoor power distribution unit can be divided into low, medium, semi-high and high type.

(1) Low power distribution unit: Electrical equipment is placed directly on the ground foundation, the bus is arranged low, and permanent fencing is provided around the equipment to ensure a safe distance. The low power distribution units feature low cost, large footprint, low height and poor safety.

(2) Medium power distribution unit: All electrical equipment is installed in the same level, and installed on a foundation of a certain height (2–2.5 m), so that the charged part is kept high above the ground, and the staff can move safely on the ground. The horizontal plane of the bus shall be slightly higher than that of other devices.

According to the arrangement of the disconnector, the medium power distribution unit can be divided into ordinary medium and phase split medium. The so-called phase split medium means that the disconnector is vertically retractable and directly arranged under the bus, and the rest of equipment is the same as the ordinary medium power distribution unit.

The medium power distribution unit is clearly arranged, with less misoperation, reliable operation, convenient construction and maintenance, few steel used, low cost, and large footprint area.

(3) High power distribution unit: One set of bus and its disconnector are overlapped with

another set of bus and its disconnector. Compared with the medium one, it saves about 50% of the footprint area, but consumes a lot of steel; it is expensive, difficult to operate and maintain, and very inconvenient to maintain the upper equipment.

(4) Semi-high power distribution unit: The bus is placed on a higher level, overlapping with other devices such as circuit breaker, transformer, and disconnector, and its footprint area is about 30% less than that of ordinary medium power distribution units. It is characterized by the advantages and disadvantages of high and medium power distribution units. Except for the bus side disconnector, the rest of the equipment is more convenient to maintain.

4.1.5.2 Type selection

In addition to the main electrical wiring, the selection of outdoor power distribution units has a lot to do with the site location, area, geology, terrain, and overall layout, and even subject to the supply of equipment materials, construction, operation and maintenance requirements. Therefore, the best solution should be selected through technical and economic comparison.

(1) Low power distribution unit: Despite its low construction cost, the low power distribution unit is generally not used, subject to the very large footprint area, and poor security arising from too low equipment.

(2) Medium power distribution unit: The ordinary medium power distribution unit is characterized by convenient construction, maintenance and operation, good seismic capacity, low cost, and large footprint area. This type is generally used in non-productive farmland areas, and good fields where the volume of earthwork is not large, and should be applicable to areas with high seismic intensity.

The phase split medium power distribution unit adopts rigid bus with single-column vertical retractable disconnector. The structure is simple and clear, can save a lot of frame, and occupies less area than the ordinary medium one. However, the single-column disconnector features poor seismic capacity and should not be used in areas with high seismic intensity.

(3) High power distribution unit: It covers few footprint area, saving 50% of the footprint area than the ordinary medium power distribution unit. Yet it consumes a lot of steel, and is not convenient in the maintenance and operation. It is generally used in the following circumstances: ① high-yield farmland or cities with few land and many residents; ② due to terrain constraints, narrow site or where bulk excavation, backfilling and other earth and rock works are required; ③ the existing power distribution unit to be rebuilt or expanded in the restricted space. The high power distribution unit is applicable to 220 kV and above voltage class, but not suitable for areas of high seismic intensity.

(4) Semi-high power distribution unit: The footprint is not as significant as the high type, with convenient operation, maintenance and overhaul, and the construction is less difficult and requires less steel. The semi-high type is applicable to 110 kV power distribution units.

4.1.5.3 Layout principles

1. Bus and frame

(1) The outdoor power distribution units employ two kinds of bus: flexible bus and rigid bus.

The flexible bus consists of aluminium cable steel reinforced, hose bus and bundled conductor. The three-phase horizontal arrangement is suspended on the bus frame by suspension insulators.

The rigid bus is either rectangular or tubular. The rectangular bus is used in power distribution units of 35 kV and below; the tubular bus is used in power distribution units 110 kV and above. The tubular bus is generally installed on the post insulator, the bus will not swing, the phase spacing can be reduced, and can save the footprint area by using the single-column vertical retractable disconnector. The tubular bus features a large diameter, smooth surface without burrs, and is less likely to produce corona. However, it suffers poor seismic capacity, small span, and inaccessibility for maintenance.

(2) The outdoor PDU frame is generally made of profile steel or reinforced concrete.

Steel frames are durable, mechanically strong and can be manufactured to any load and size. They are designed to fix the equipment, with strong seismic capacity and convenient transportation; the metal consumption is huge, and frequent maintenance is required for rust prevention.

Reinforced concrete can save a lot of steel, and meet a variety of strength and size requirements. It is durable, simple to maintain.

2. Power transformer

The power transformer shell is metal and not charged; thus, it is arranged on the floor and installed on the transformer foundation.

The transformer foundation is generally made of double beams with rails, and the rail spacing is equal to the center distance of the transformer roller. In order to prevent the further escalation of transformer accident as a result of transformer oil flow, the transformer with a single tank fuel of more than 1,000 kg shall be equipped with an oil reservoir or an oil protection wall under the equipment as per fire protection requirements, and its size shall be 1 m larger than the outer perimeter of equipment. The oil reservoir is generally laid with cobbles with a thickness of not less than 0.25 m.

The structures should not be less than 10 m from the contour of the outdoor oil-immersed transformer. When it is within 5 m, no doors, windows and vent holes shall be provided on external walls of the indoor power distribution unit building, the main control building and the network control building facing the oil-immersed transformer within 3 m outside the projection range of the transformer contour. In case of 5 to 10 m, class A fire doors may be equipped on the exterior walls above.

3. Other electrical equipment

(1) Circuit breaker: Circuit breakers can be arranged into single, double and triple rows by the position of circuit breakers in the power distribution unit. If arranged in three rows in the direction

of the incoming and outgoing lines, it is the three-row arrangement. When the circuit breaker is arranged on both sides of the main bus, it is the double row arrangement. If the circuit breaker is centrally arranged on one side of the main bus, it is the single row arrangement. The circuit breaker must be arranged by allowing for many factors such as main wiring, site conditions, overall layout and outlet direction.

Circuit breakers are arranged low or high by the foundation height of the circuit breaker. The low-type circuit breaker is placed on the 0.5 to 1 m concrete foundation to facilitate maintenance and seismic capacity. Allowing for the low foundation and poor safety, it must be permanently shielded from all sides, with a large footprint area, thus affecting the unobstructed channel. This method is used for low-profile power distribution units. The high-profile layout circuit breaker is placed on a high base of about 2 m, and the operating mechanism of the circuit breaker must be installed on the appropriate foundation for medium power distribution units.

(2) Disconnector and transformer: With a high-profile layout, they follow the same requirements as the circuit breaker, and the disconnector operating mechanism is installed at a certain height of its lateral foundation. For the sake of safety, each bus should be equipped with 1 to 2 sets of grounding disconnectors.

(3) Lightning arrester: either high-profile or low-profile layout. If the slender valve type arresters of 110 kV and above are installed on the 2 m foundation, the total height will be close to 6 m, thus increasing the maintenance, testing and maintenance difficulties, coupled with poor seismic capacity. Therefore, the floor type low-profile layout is used, and installed on the 0.4 m foundation, and the perimeter shielding is often provided.

(4) Cable trench: The cable trench in the outdoor power distribution unit must be arranged to mke the path of the cable shortest. The cable trench can be longitudinal or transverse by routing direction.

4.1.6 Packaged Power Distribution Units

Packaged power distribution units: Based on the main wiring requirements, the manufacturer will pre-assemble the equipment of each circuit (such as circuit breaker, disconnector, lightning arrester, transformer, and measuring instruments) in an enclosed or semi-enclosed metal housing (cabinet) to form a standard module. After delivered to the site, all modules are assembled into a complete power distribution unit, i.e., packaged power distribution unit. Its features are:

(1) Protected by a metal housing (cabinet), the electrical equipment and current carrying conductors are featured by little dust stratification, convenient maintenance, especially in dirty areas.

(2) Easy to be serialized and standardized.

(3) Good assembly quality and high operational reliability.

(4) Compact structure, reasonable layout, and few footprint area.

(5) Short project construction and small workload.

1. Type

(1) The packaged power distribution units can be open and closed type by characteristics of the cabinet.

① The open packaged power distribution unit is of exposed bus, and the units in the cabinet are not separated, with simple structure and low cost.

② Closed packaged power distribution units are also known as enclosed switch cabinets, all units are enclosed in an all-metal enclosure, but the bus, cable head, circuit breaker, and measuring meter are separated by metal compartments to escort the safe operation and prevent the escalation of accidents. It is applicable to the areas with poor working environment and high requirements.

(2) The packaged power distribution units can be stationary and mobile by fixed characteristics of the circuit breaker.

① All the electrical equipment in the stationary packaged power distribution unit is fixed in the cabinet frame. To move it, please use tools to disassemble it after power failure.

② Mobile packaged power distribution units are also known as handcart switch cabinets, since their circuit breakers, operating mechanisms and other components are mounted on a pull-out chassis for the purpose of overhauling, testing and maintenance. The disconnector uses a plug-in contact terminal instead of the traditional knife switch one.

(3) The packaged power distribution units can be single bus or double bus type by the number of buses.

(4) The packaged power distribution units can be HV or LV power distribution units by voltage class. The HV power distribution unit is also known as the HV switch cabinet, while the LV one is called the LV distribution panel.

2. HV packaged power distribution unit

The HV packaged power distribution unit is also called HV switch cabinet. With the circuit breaker as the main body, the detection instrument, protection equipment and auxiliary equipment are installed in an enclosed or semi-enclosed metal cabinet according to certain main wiring requirements. A compartment is usually formed by a cabinet. The electrical equipment in the cabinet is insulated from each other, and most of the insulating materials are made of post insulators and air. The minimum safe distance is small, and the structure can be made compact, thus saving the footprint area and materials. Based on operating experience, the HV switch cabinet is known for high reliability, easy maintenance, and simple installation, and is widely used in the 3–35 kV system.

3. Box-type substation

In the distribution system, the power supply radius of the substation is so large that the line loss increases with the power load and the line length, which will lead to low terminal voltage and greatly reduce the power supply quality. In order to reduce line loss and ensure power supply quality, it is essential to increase the power supply voltage. To this end, our country requires high voltage directly into the load center in the construction of urban and rural power supply network to increase the supply voltage from 0.4 kV to 10 kV and even 35 kV. In this way, the line loss is greatly reduced, the total copper used and total investment are also reduced, and the economic benefits are considerable. To realize high voltage directly into the load center, box-type substation is the most economical and effective packaged power distribution unit.

In the box-type substation (see Fig. 4-1), the HV switchgear, transformer and LV power distribution unit are integrated according to certain mode of connection. With packaged structure, small size, small footprint area, the box-type substation can improve power supply quality, reduce line loss, and shorten transmission cycle. Besides, it boasts easy installation, strong environmental adaptability, reliable operation, few investment, etc. It is widely used in power distribution networks.

Fig. 4-1　Box-type Substation

4. Gas insulated switchgear (GIS)

Gas insulated switchgear (GIS) is composed of circuit breaker, disconnector, grounding knife switch, transformer, lightning arrester, bus, lead-out bushing and other electrical equipment according to the wiring requirements, and fully enclosed in the grounded metal housing. The metal housing is filled with SF_6 gas, as an insulating medium and arc extinguishing medium.

模块二 高压开关柜基础知识

在本项目的学习中，学生要养成良好的安全生产意识和安全生产习惯的同时，培养团结协作的精神、一丝不苟、精益求精的工匠精神，树立勤业、乐业、爱岗敬业的职业道德。能按照标准化作业流程完成各项运检或测试工作及检修等，标准化作业流程执行力，能够准确地列出设备作用、工作原理、结构和主要运检项目。

一、高压开关柜分类

高压开关柜按断路器的固定特点可以分为固定式和手车式，按隔室特征可以分为间隔式、铠装式、箱式，按用途还可以分为环网柜或非环网柜，主要分类及其代号见表4-4。

表4-4 高压成套配电装置分类及代号

分类	户内 N		
	间隔式 J	铠装式 K	箱式 X
固定式 G	×	KGN	XGN
手车式 Y	JYN	KYN	XYN

可移动式开关柜（Y）也称为手车式开关柜，是指断路器和断路器的操动机构等部件均装设在可以推进拉出的底盘车上，便于检修、试验和维护，隔离开关采用插入式触头，代替传统的闸刀式触头。

固定式开关柜（G）是指全部电气设备均固定于柜内构架上，若要移动则必须停电后使用工具拆卸。

间隔式开关柜（J）是指各室间则采用一块或者数块绝缘板相互隔离，结构较为紧凑，但故障电弧有可能烧穿绝缘隔板，进入其它隔室，扩大事故范围。

铠装式开关柜（K）是指各室间用金属板隔离并且接地，优点可将故障电弧限制在发生电弧的隔室内，是防护功能最完善的结构。

箱式开关柜（X）间隔数目少于其他两种柜型，通常母线室、电缆室和断路器室之间没有隔板，结构设计比较简单，成本低，是一种经济型柜体的结构，但是发生故障时，事故会扩大至整个柜体，事故范围很大。

环网柜（H）在结构上，与普通开关柜没有区别，区别在于开关柜母线是否为环网的组成部分，如果其母线与其他开关柜母线组成环网，则该出线柜称为环网柜。通常环网柜的额定电流都不大，因此环网柜一般不采用复杂昂贵的断路器，而采用负荷开关-熔断器组合代替断路器。见图4-2、4-3、4-4。

图 4-2　JYN1-40.5 高压开关柜　　　　图 4-3　XGN12-12 高压开关柜

图 4-4　KYN28-12 高压开关柜

二、高压开关柜型号

高压开关柜型号有两个系列的表示方法。

第一种表示方法：

| 1 | 2 | 3 | 4 | F |

1——类型，高压开关柜统称为 G；
2——是否封闭，F 为封闭型；
3——手车式或固定式，C 为手车式，G 为固定式；
4——额定电压（kV）或设计序号；
F——防误操作型。
第二种表示方法：

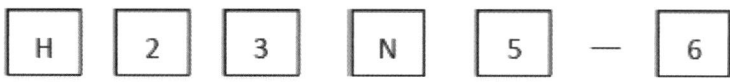

H——该开关柜用于环网，为环网柜；
2——开关柜间隔形式，间隔式为 J，铠装式为 K，箱式为 X；
3——可移动式或固定式，Y 为可移动式（手车式），G 为固定式；
N——户内式；
5——设计序号；
6——代表额定电压或者是设备最高工作电压。
例如，XGN1-10 表示该开关柜属于箱式固定式户内高压开关柜，额定电压 10 kV。

三、高压开关柜结构

虽然高压开关柜有很多种类型，但是其基本结构类似，都是包含了母线、母线侧隔离开关、断路器、出线侧隔离开关、出线电缆几个主要部件，以及避雷器、电流互感器、电压互感器、接地刀闸、测量仪表、继电保护装置等设备。

高压开关柜无论是箱式、间隔式还是铠装式，都能分割成不同隔室。

带有母线的隔室叫做母线室，母线室除了母线，可能还有母线侧隔离开关，若母线带有接地刀闸，则接地刀闸也可能设置在母线室。

带有断路器的隔室叫做断路器室，若是手车式断路器，则在断路器室还包含母线侧隔离开关和线路侧隔离开关的静触头（其动触头位于断路器手车上）。

带有出线（进线）电缆的隔室叫做电缆室，电缆室除了电缆外，还可能有避雷器、电流互感器、电压互感器、出线接地刀闸等其他一次设备。

带有控制电源的隔室通常叫做仪表室，内部装有断路器控制电源、继电保护装置、仪器仪表等低压设备。

以上四种隔室中的断路器室、母线室、电缆室都设置有专用或公用的防爆通道，也叫做压力释放通道，能够在内部出现高温高压时，释放其压力，降低开关柜爆炸可能性。

除以上四种典型隔室之外，某些高压开关柜如 XGN 柜还专门设立了操动机构室。

对于箱式高压开关柜，各个隔室之间没有明显的间隔，当一个隔室发生故障时，电弧可能会窜入其他隔室，造成事故扩大；对于间隔式高压开关柜，部分隔室之间采用了绝缘板相互隔离，对于隔室有一定的隔离作用，但是出现高温电弧时，绝缘板会融化，也会扩大事故范围；对于铠装式高压开关柜，各个隔室之间采用接地的镀锌钢板隔离，防护等级为 IP2X，防止 12 mm 的固体物质穿越隔，当一个隔室出现严重故障产生电弧时，由于隔板为接地金属，电弧接触隔板后会迅速流入大地，保证其他隔室的安全。

Module 2 Fundamentals of HV Switch Cabinet

While studying this program, students are expected to develop good awareness and habits of work safety, cultivate the spirits of unity and cooperation, meticulous workmanship and excellence, and establish the professional ethics of diligence, happy work and dedication. Students should be able to complete various operations, inspections, tests, and maintenance tasks in accordance with standardized procedures, possess the ability to execute standardized operational processes, and accurately list the equipment's functions, working principles, structures, and key inspection items.

4.2.1 Classification

The HV switch cabinet can be divided into the fixed and handcart type by fixed characteristics of circuit breaker, the compartmented, metal-clad, and cubicle type by characteristics of the compartment, and the ring main unit or non-ring main unit by use. See Table 4-4 for the main classification and code.

Table 4-4 Classification and Code of HV Packaged Power Distribution Units

Classification	Indoor N		
	Compartmented J	Metal-clad K	Cubicle X
Fixed G	×	KGN	XGN
Handcart Y	JYN	KYN	XYN

Mobile switch cabinet (Y) is also known as handcart switch cabinet, since the circuit breaker, operating mechanism and other components are mounted on a pull-out chassis for the purpose of overhauling, testing and maintenance. The disconnector uses a plug-in contact terminal instead of the traditional knife switch one.

In the fixed switch cabinet (G), all the electrical equipment is fixed in the cabinet frame. To move it, please use tools to disassemble it after power failure.

Compartmented switch cabinet (J) refers to the isolation of compartments using one or more insulating boards to make the structure compact; the fault arc may burn through the insulation partition and enter other compartments, widening the scope of accident.

Metal-clad switch cabinet (K) means that compartments are isolated with a metal sheet and grounded. The fault arc can be confined to the compartment where the arc occurs, making it the most perfect structure of protection.

Cubicle switch cabinet (X) has fewer compartments than the other two types. Usually there is no diaphragm among the bus compartment, cable compartment and circuit breaker compartment. With simple structure and low cost, it is an economical cabinet. However, a failure will expand to the entire cabinet, resulting in a large scope of accident.

Ring main unit (H) is not structurally different from ordinary switch cabinet, but whether the switchgear bus is part of the ring network. If its bus forms a ring net with other switch cabinet buses, the outgoing cabinet is called a ring main unit. Not carrying large rated current, the ring main unit generally does not use complex and expensive circuit breakers, but the load switch-fuse combination to replace the circuit breaker. see Fig. 4-2, 4-3, 4-4.

Fig. 4-2　JYN1-40.5 HV Switch Cabinet　　　Fig. 4-3　XGN12-12 HV Switch Cabinet

Fig. 4-4　KYN28-12 HV Switch Cabinet

4.2.2 Model

There are two ways to express the model of HV switch cabinets.

The first representation:

| 1 | 2 | 3 | 4 | F |

1-type, HV switch cabinets are collectively referred to as G;

2-enclosed or not, F = enclosed;

3-handcart or fixed, C = handcart, G = fixed;

4-rated voltage (kV) or design sequence;

F-stands for anti-misoperation.

The second representation:

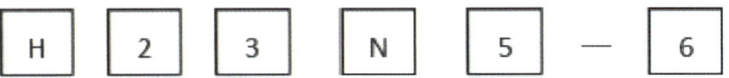

H-represents the switch cabinet for ring net, being ring main unit;

2-the compartmented form of switch cabinets, compartmented (J), metal-clad (K), and cubicle (X);

3-mobile or fixed, Y = mobile (handcart), G = fixed;

N-indoor;

5-design sequence;

6-rated voltage or the maximum operating voltage of the device

For example, XGN1-10 indicates that the switch cabinet is a cubicle fixed indoor HV switch cabinet with a rated voltage of 10 kV.

4.2.3 Structure

Despite many types, HV switch cabinets feature similar basic structure. They contain main components, such as bus, bus side disconnector, circuit breaker, outgoing side disconnector, and outlet cable, as well as lightning arrester, current transformer, voltage transformer, grounding knife switch, measuring meter, relay protection device, etc.

The HV switch cabinets, cubicle, compartmented or metal-clad, can be divided into different compartments.

A compartment with buses is called bus compartment. In addition to the bus, the bus compartment may contain a bus side disconnector. If any, the grounding knife switch may be located in the bus compartment.

A compartment with a circuit breaker is called the circuit breaker compartment. In case of a handcart type, the circuit breaker compartment further contains the bus side disconnector and the static contact terminal of the line side disconnector (the moving contact terminal is located on the handcart of the circuit breaker).

A compartment with outgoing (incoming) cables is called the cable compartment. Besides, the cable compartment may contain other primary equipment such as lightning arrester, current transformer, voltage transformer, and outlet grounding knife switch.

A compartment with the control power supply is usually called the instrument compartment, which is internally equipped with the circuit breaker control power supply, relay protection device, instruments and other LV devices.

The circuit breaker compartment, bus compartment and cable compartment are provided with explosion-proof channels for private or public use, also known as pressure relief channels. They can release the pressure in case of high temperature and high pressure inside, thus reducing the possibility of switch cabinet explosion.

In addition to the above four typical compartments, some HV switch cabinets such as XGN have a special operating mechanism compartment.

The cubicle HV switch cabinet has no obvious bay between compartments. When any compartment fails, the arc may escape into the other compartments, causing an extended accident. The compartmented HV switch cabinets are isolated by insulating boards, having a certain isolation effect on the compartment. However, the high-temperature arc will melt the insulating board, thus widening the scope of accident. For metal-clad HV switch cabinets, compartments are isolated by grounded galvanized steel sheets with IP2X protection to prevent 12 mm solids from passing through the compartments. When a compartment malfunctions badly and gives rise to arc, the arc will flow rapidly into the ground after touching the diaphragm (grounded metal), thus ensuring the safety of other compartments.

模块三　KYN28-12 高压开关柜

KYN28-12 铠装式金属封闭高压开关设备（以下简称 KYN28 高压开关柜）适用于三相交流 50 Hz、3.6～12 kV 单母线及单母线分段电力系统，具有"五防"功能。开关柜的可移开部分可配置真空断路器或熔断器等元器件。开关柜外壳的防护等级为 IP4X，当断路器室门打开、手车移开时，防护等级为 IP2X。

IP4X 指能阻挡直径大于 1 mm 的固体物质进入，IP2X 指能组织直径大于 12 mm 的固体进入。根据开关柜额定电流/额定短路开断电流确定了六种参数序列：630 A/20 kA、1250 A/25 kA、1250 A/31.5 kA、2500 A/31.5 kA、3150 A/40 kA、4000 A/40 kA。

一、KYN28 高压开关柜结构（出线柜）

KYN28 高压开关柜按功能分，可以分为出线柜、进线柜、分段柜、电容器柜、站用变柜、计量柜、互感器柜等，都是由柜体和可移动的手车组成，但是由于功能不一样，结构上也有有一定差别。其中出线柜和进线柜是基本柜方案，其他功能的柜体均属于派生方案。本章以出线柜为例，介绍 KYN28 高压开关柜结构以及功能。

KYN28 高压开关柜出线柜柜体包含断路器室、母线室、电缆室、仪表室四大部分，以及镀锌钢板外壳、控制面板等附件（见图 4-5）。

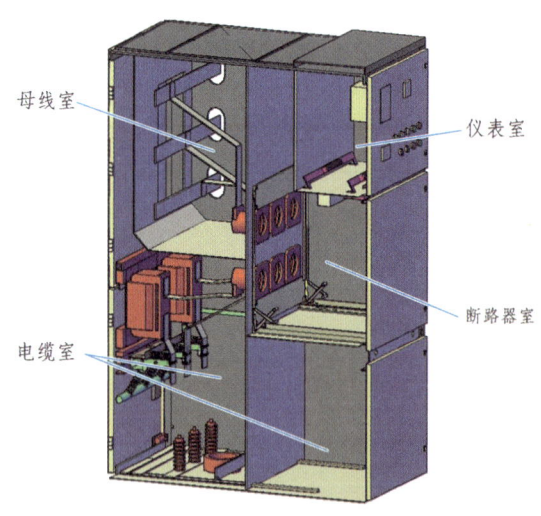

图 4-5　KYN28-12 高压开关柜进线柜主体结构

1. 柜体、外壳和隔板

开关柜的高度统一为 2240 mm；开关柜宽度根据分支母线电流和开关柜作用不同而不同，有 800 mm、1000 mm、1200 mm 三种，电缆进线开关柜深度通常为 1450 mm，架空进线开关柜深度统一为 1750 mm。

开关柜的柜门由冷轧钢板制成，表面做喷塑处理，柜体框架采用镀锌钢板，具有很强的抗氧化、耐腐蚀功能，且刚度和机械强度比普通低碳钢板高。由于柜体外壳为全金属，并且采用明显两点接地设计，内部并未充装其他绝缘气体，柜内完全以空气作为绝缘介质，各相导体之间以及对地距离应当满足大于最小安全净距 125 mm 的基本要求。

母线室、断路器室、电缆室三个隔室的顶部装有共用的压力释放通道。压力释放板通常安装在柜体顶部，一侧安装铁螺钉（铰链），一侧安装尼龙螺钉，出现内部故障时，高压室内气压升高，尼龙螺钉受热融化，此时压力室放板仅有一侧铁螺钉连接（铰链），而由于正面、背面柜门可靠密封，高压气体将冲开压力释放板，并从柜体顶部释放出来。

相邻的开关柜由各自的侧板隔开，并柜后仍有空气缓冲层，可以防止开关柜被故障电弧贯穿熔化。仪表室装配成独立隔室，与高压区域（断路器室、母线室、电缆室）分隔开。水平隔板将断路器室和母线室、电缆室隔开，隔板采用接地的镀锌钢板，即使断路器手车移开（此时活门会自动关闭），也能防止操作者触及母线室和电缆室内的带电部分。卸下紧固螺栓就可移开水平隔板，便于电缆密封终端的安装。

2. 开关柜面板

开关柜各个隔室均带有铰链结构的面板，采用冷轧钢板制成，均采用大于 4 mm^2 的具有透明绝缘护套的多股软铜线连接柜体接地。正面从上至下有仪表室面板、断路器室面板、电缆室面板，背面从上至下有母线室面板、电缆室面板。

仪表室面板上主要有电压、电流检测仪表，开关柜就地电动控制按钮，开关柜信号灯等，起到信号监视和就地电动操作的作用。

断路器室面板上主要有当前开关柜的接线图、双重名称、铭牌、手车手柄插口和断路器观察窗。部分公司的该产品也会将断路器机械分合闸按钮设计在断路器室面板上，使得断路器只能在关门时操作，更加安全。

前下柜门面板主要有一个电缆室观察窗和柜门紧急解锁孔，电缆室后柜门面板有观察窗和柜内照明开关。母线室面板仅有当前开关柜的接线图和双重名称。若该柜体无接地刀闸，则为了防止误入带电间隔，也可在电缆室柜门装设电磁锁。

3. 仪表室

仪表室也可以叫做或者低压室，位于开关柜正面的最上方。仪表室包含开关柜的电流、电压测量仪表，继电保护装置，电动控制按钮和信号指示器，这些低压设备镶嵌在面板上。柜内还装设有断路器的储能、合闸电源，开关柜的加热、照明电源，以及接线端子和线槽。仪表室内部空间非常充裕，与断路器室之间有控制线穿越孔，以便控制电源的联接。

4. 断路器手车与断路器室

与 KYN28-12 配套的断路器通常为 ZN63（VS1）断路器，断路器装设在钢制的底盘车上合为一体，也称为手车。底盘车上除了装配断路器之外，还可以装配其他设备如熔断器、隔离开关等，使用灵活，装配后，均称为手车。手车上带有滑轮，能够在断路器室导轨上运动，在"工作""试验"两个不同位置之间移动，如图 4-6 所示。

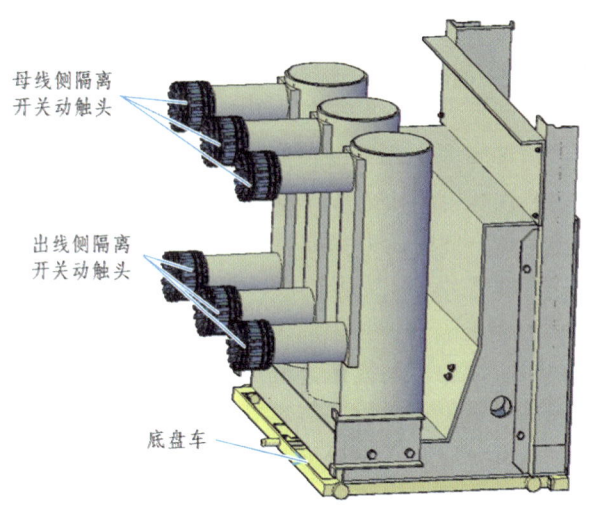

图 4-6 ZN63 断路器手车

ZN63 断路器后方有 6 个触头，上面一排触头为母线侧隔离开关动触头，下面一排触头为出线侧隔离开关动触头。当手车在断路器室内移动至"工作"位置时，母线侧、出线侧隔离开关动触头分别插入两组隔离开关静触头，此时断路器两侧隔离开关为"合闸"状态；当手车在断路器室内移动至"试验"位置时，母线侧、出线侧隔离开关动触头与两组隔离开关静触头分离，此时断路器两侧隔离开关为"分闸"状态。手车移动至"试验"位置后，还可以使用转运小车，将断路器完全拉出开关柜，称为"检修"位置，该状态下，断路器可以在转运小车上移动，检修断路器时可以完全不影响其他设备运行。手车与开关柜之间的信号、保护和控制线，用一个控制线插头（航空插头）连接，设置在断路器顶部。

断路器室四面被敷铝锌板包裹，与其他隔室隔离，仅有数毫米宽的通风通道。底部带有断路器手车外壳的接地体，与手车底盘相接，保证断路器外壳电位为地电位。隔离开关静触头外部带有活门挡板，当断路器离开"工作"位置时，金属活门挡板会自动降下，挡住带电的隔离开关静触头，以防止手车处于"检修"位置时，人员触碰带电的隔离开关静触头。金属活门挡板以及它的金属部件应按规范与接地体相连。在断路器室顶部，有控制线插座（航空插座），作为手车与开关柜的控制信号、保护控制连接。

5. 母线室

各个开关柜通过母线室做到电气连接，母线从一个开关柜的母线室通过穿墙套管引至另一个开关柜的母线室，通过柜内支柱绝缘子进行固定。矩形的分支母线直接用螺栓联接到主母线上，不需任何连接夹。所有母线和分支母线都选用全圆边形铜排。分支母线与主母线接触面有镀银处理，降低接触电阻，并且配有永久性相色标识。套管板和套管将柜与柜之间的母线隔离起来，并有支撑作用。对电动应力大的开关柜，一般需要这种支撑。见图 4-7。

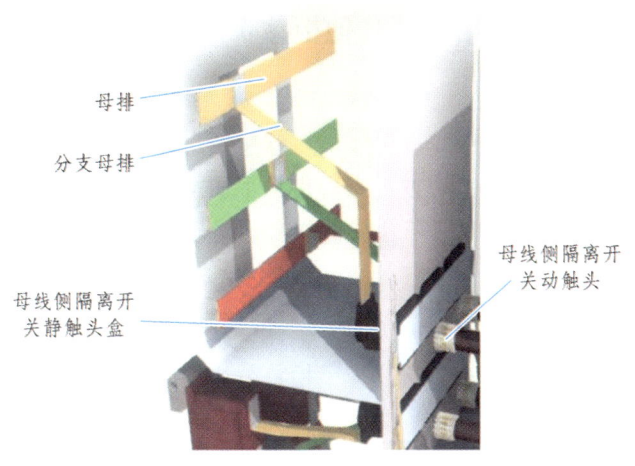

图 4-7 KYN28-12 高压开关柜母线室

母线室母线有垂直"一字"形排列，和"品字"形排列两种排列方式。

6. 电缆室

对于出线柜而言，电缆室除了包含出线电缆外，还包括出线电流互感器、出线避雷器以及出线接地刀闸。当电缆室门打开后，要求有足够的空间供施工人员进入柜内安装电缆。盖在电缆入口处的底板可采用非导磁的不锈钢板，是开缝的，可拆卸的，便于现场施工。底板中穿越一、二次电缆的变径密封圈开孔应与所装电缆相适应，以防小动物进入。对于湿度较大的电缆沟，为防止电缆沟湿气进入开关柜，一、二次电缆引出孔洞应封堵良好。KYN28-12 高压开关柜电缆室如图 4-8 所示。

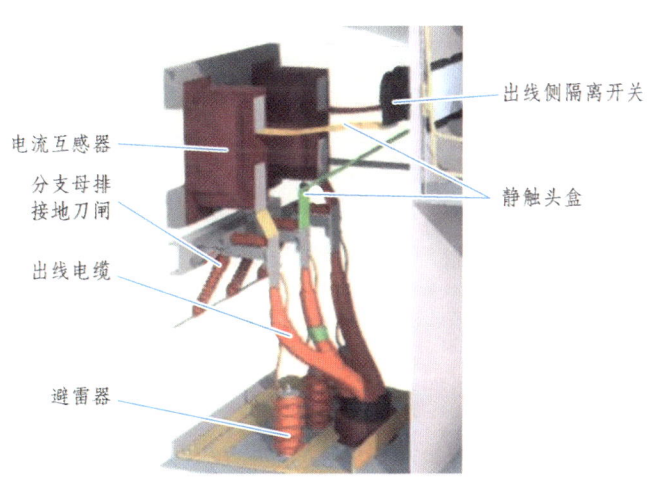

图 4-8 KYN28-12 高压开关柜电缆室

7. 泄压通道

国家规定高压开关柜除仪表室外，断路器室、母线室、电缆室均应设有符合要求的排气通道和泄压装置。KYN28-12 中采用顶部泄压通道，当产生内部电弧时，泄压通道将被自动打开，释放内部压力。

二、KYN28 高压开关柜结构（其他柜体）

1. 进线柜

与出线柜相比，进线柜电缆室应有母线电压互感器，无接地刀闸；若进线侧为架空进线时，电缆室无电缆，进线从母线室后方进入，再从上而下进入电缆室，与母线隔离开关静触头相连，此柜体由于母线室与进线同在开关柜背面上部，所以深度比电缆进线柜更深。

2. 分段柜

分段柜是用于单母线分段接线，分为两个柜体，其中一个断路器室内设置分段断路器手车，另外一个仅设置隔离开关手车，这是为了检修分段柜时，母线断停，缩小停电范围。除此之外，分接，下母线室中母线经过穿墙套管连接，如图4-9。

段柜连接两根分段母线，无进出线，所以没有电缆室，背面上、下均为母线室，两柜体上母线室中母线无连。

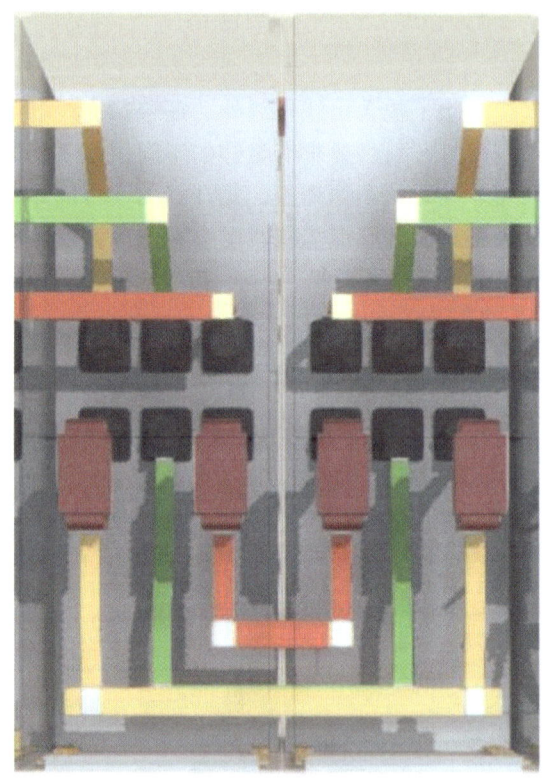

图 4-9　分段柜母线走向

Module 3　KYN28-12 HV Switch Cabinet

KYN28-12 armored metal-enclosed HV switchgear (hereinafter referred to as KYN28 HV switch cabinet) is suitable for three-phase AC 50 Hz, 3.6 ~ 12 kV single bus and single bus segment power system, with "five-prevention" function. The removable part of the switch cabinet can be configured with components such as vacuum circuit breaker or fuse. The IP grade of the switch cabinet housing is IP4X, and IP2X when the circuit breaker compartment is door-opened and the handcart is removed.

IP4X can block the entry of solid substances with a diameter greater than 1 mm, and IP2X can prevent the entry of solid substances with a diameter greater than 12 mm. According to the rated current/rated short-circuit breaking current of switch cabinets, six parameter sequences are determined: 630 A/20 kA, 1,250 A/25 kA, 1,250 A/31.5 kA, 2,500 A/31.5 kA, 3,150 A/40 kA, 4,000 A/40 kA.

4.3.1　Structure (Outgoing Cabinet)

By function, KYN28 HV switch cabinet can be divided into an outgoing cabinet, an incoming cabinet, a sectional cabinet, a capacitor cabinet, a station transformer cabinet, a measuring cabinet, a transformer cabinet, etc. Composed of a cabinet and a movable handcart, they are somewhat different in structure for specific functions. The outgoing cabinet and incoming cabinet are basic cabinets, and the other functional cabinets are derived series. This chapter takes the outgoing cabinet as an example to describe the structure and functions of the KYN28 HV switch cabinet.

The outgoing cabinet body includes a circuit breaker compartment, a bus compartment, a cable compartment, and an instrument compartment, as well as accessories such as galvanized steel shell and control panel. see Fig. 4-5.

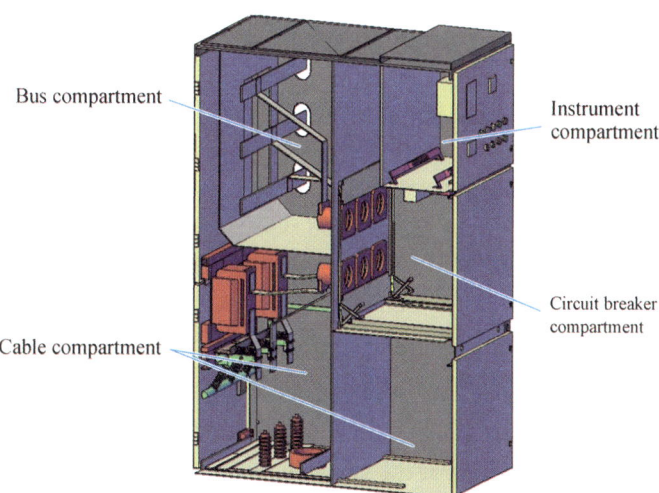

Fig. 4-5　Major Structure of Incoming Cabinet for KYN28-12 HV Switch Cabinet

1. Cabinet, enclosure and diaphragm

The switch cabinet is designed to be 2,240 mm high; its width varies according to the branch bus current and switch cabinet function: 800 mm, 1,000 mm and 1,200 mm. The cable incoming switch cabinet is usually 1,450 mm deep, and the overhead incoming switch cabinet is designed to be 1,750 mm.

The door of the switch cabinet is made of cold-rolled steel sheet, and the surface is sprayed with plastic. Made of galvanized steel sheet, the cabinet frame features strong oxidation and corrosion resistance, and higher rigidity and mechanical strength than ordinary low carbon steel sheets. Since the cabinet enclosure is all-metal and adopts the clearly two-point grounding design, the interior is not filled with other insulating gases, and the cabinet is completely insulated by air. The distance between the conductors of each phase and the ground clearance should meet the basic requirements of greater than the minimum safe distance (125 mm).

The tops of the bus compartment, circuit breaker compartment and cable compartment are provided with a shared pressure release channel. The pressure relief plate is usually mounted on the top of the cabinet with iron screws (hinges) on one side and nylon screws on the other. In case of internal failure, the pressure rises in the high pressure chamber increases, and the nylon screw is melted by heat. At this time, the pressure relief plate has only one side of the iron screw connected (hinge). Since the front and back cabinet doors are reliably sealed, the high-pressure gas will burst the pressure relief plate and release from the top of the cabinet.

Adjacent switch cabinets are separated by their respective side plates, and there is still an air buffer layer after the cabinets are assembled, which can prevent the switch cabinets from being melted by the fault arc. The instrument compartment is designed to be an independent compartment, separated from the HV zone (circuit breaker compartment, bus compartment, and cable compartment). A horizontal diaphragm separates the circuit breaker compartment from the bus compartment and cable compartment. The diaphragm is the grounded galvanized steel sheet. Even if the circuit breaker handcart is moved away (when the valve automatically closes), the operator is prevented from reaching the live parts of the bus compartment and cable compartment. The horizontal diaphragm can be removed by removing the fastening bolts so as to facilitate the installation of the cable sealing terminal.

2. Switch cabinet panel

Each compartment of the switch cabinet is equipped with hinged panels made of cold-rolled steel sheets and grounded by multiple strands of annealed copper wire with transparent insulating sheath larger than 4 mm^2. There are instrument compartment panel, circuit breaker compartment panel, and cable compartment panel from top to bottom on the front, and bus compartment panel and cable compartment panel from top to bottom on the back.

On the panel of the instrument compartment are mainly equipped the voltage and current detectors, local electric control buttons of the switch cabinet, and signal lights of the switch cabinet, which play the role of signal monitoring and local electric operation.

The panel of circuit breaker compartment is mainly equipped with the wiring diagram of the current switch cabinet, double name, nameplate, handcart handle jack, and circuit breaker sight glass. The product of some companies will design the mechanical opening and closing button of the circuit breaker on the panel of circuit breaker compartment, so that the circuit breaker can only be operated when the door is closed, which is more secure.

The front lower cabinet door panel is mainly provided with a sight glass of the cable compartment and an emergency release hole of the cabinet door, while the rear cabinet door panel is provided with an sight glass and a lighting switch in the cabinet. The bus compartment panel has the wiring diagram and dual designations of the current switch cabinet only. If the cabinet is not provided with a grounding knife switch, a magnetic lock may be installed on the door of the cable compartment to prevent stray into the energized compartments.

3. Instrument compartment

The instrument compartment can also be called the LV compartment, located at the top of the front of the switch cabinet. The instrument compartment contains current and voltage measuring meters for the switch cabinet, relay protection devices, electric control buttons and signal indicators. These LV devices are embedded in the panel. The cabinet is further equipped with circuit breaker energy storage, closing power supply, switch cabinet heating, lighting source, wiring terminal and trunking. The instrument compartment has plenty of room, and has a control line crossing hole with the circuit breaker compartment to control the connection of the power supply.

4. Circuit breaker handcart and circuit breaker compartment

The circuit breaker for KYN28-12 is usually ZN63 (VS1) circuit breaker, which is integrated on a steel chassis, also known as a handcart. In addition to the circuit breaker, the chassis can be provided with other equipment such as fuse and disconnector. Featured by flexible use. After assembly, they are called the handcart. The handcart is equipped with a pulley, which can move on the slideway of the circuit breaker compartment and between the two positions of "Work" and "Test". As shown in Fig. 4-6.

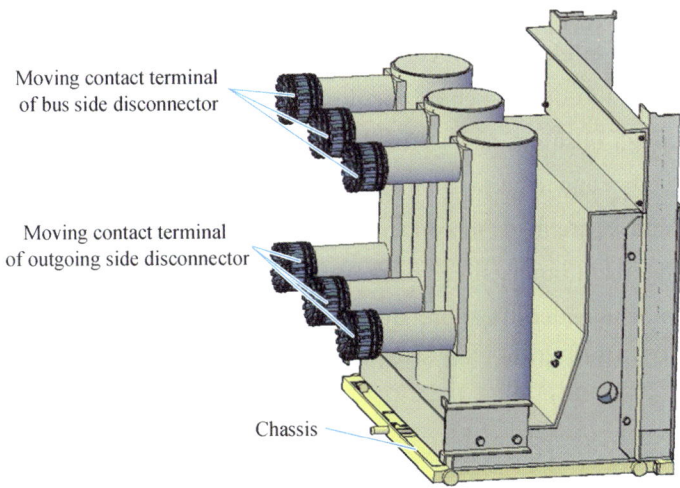

Fig. 4-6 Handcart of ZN63 Circuit Breaker

There are six contact terminals at the rear of the ZN63 circuit breaker. The upper row is the moving contact terminals of bus side disconnector, and the lower row the moving contact terminals of outgoing side disconnector. When the handcart moves to the "Work" position in the circuit breaker compartment, the moving contact terminals of the bus side disconnector and the outgoing side disconnector are inserted into the static contact terminals of the disconnectors, respectively; the disconnectors on both sides of the circuit breaker are in the "closing" state. When the handcart moves to the "Test" position in the circuit breaker compartment, the moving contact terminals of the bus side disconnector and the outgoing side disconnector are separated from the static contact terminals; the disconnectors on both sides of the circuit breaker are in the "opening" state. After the handcart moves to the "Test" position, the transfer trolley can be used to completely pull the circuit breaker out of the switch cabinet, which is called the "Maintenance" position. In this state, the circuit breaker can be moved on the transfer trolley, and the operation of other equipment is completely unaffected when the circuit breaker is maintained. The signal, protection and control lines between the handcart and the switch cabinet are connected by a control wire plug (aviation plug), which is set at the top of the circuit breaker.

The circuit breaker compartment is wrapped with aluminum and zinc panels on all sides and isolated from other compartments, leaving a ventilating duct of only a few millimeters wide. A grounding body with a circuit breaker handcart housing at the bottom is connected to the chassis so that the circuit breaker enclosure is of ground potential. The static contact terminal of disconnector is provided with a valve damper outside. When the circuit breaker leaves the "Work" position, the metal valve damper will lower automatically to block the live static contact terminal of disconnector to prevent the human touch while the handcart is in the "Maintenance" position. The metal valve damper and its metal parts shall be connected to the grounding body as required. At the top of the circuit breaker compartment, there is a control wire plug (aviation plug), which is used as the handcart to be connected with the control signal and protection control of the switch cabinet.

5. Bus compartment

Each switch cabinet is electrically connected through the bus compartment. The bus is led from the bus compartment of one switch cabinet to the bus compartment of the other switch cabinet through the wall-through bushing, and is fixed by the post insulator in the cabinet. Rectangular branch buses are directly connected to the main bus with bolts, in no need of any connecting clips. All buses and branch buses are made of copper busbar with full round edges. The contact surface of the branch bus and the main bus is silver plated to reduce the contact resistance, and is equipped with permanent phase color identification. Bushing plates and bushing isolate the buses between cabinets and serve as support. This kind of support is generally required for switch cabinets with high electric stress. see Fig. 4-7.

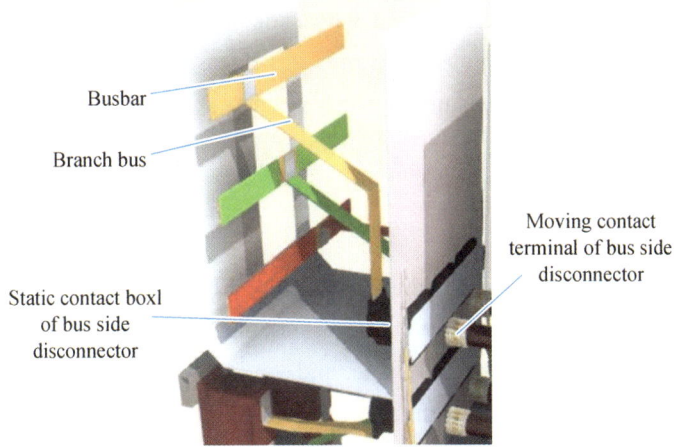

Fig. 4-7 Bus Compartment of KYN28-12 HV Switch Cabinet

The bus compartment has its bus in the following structure: vertical "in-line" and "top and twin-side bottom" arrangement.

6. Cable compartment

For the outgoing cabinet, the cable compartment contains the outlet cable, the outgoing current transformer, the outgoing lightning arrester, and the outgoing grounding knife switch. When opened, the cable compartment must have enough space for construction workers to enter the cabinet for cable laying. The bottom plate covering at the cable entrance can be made of non-magnetic stainless steel sheet, which is slotted and removable for the sake of site construction. The opening of reducer seal ring crossing the primary and secondary cables in the base slab shall be compatible with the installed cables to keep small animals out. For the cable trench with high humidity, the primary and secondary cable outlet holes should be well sealed to prevent the moisture from entering the switch cabinet. As shown in Fig. 4-8.

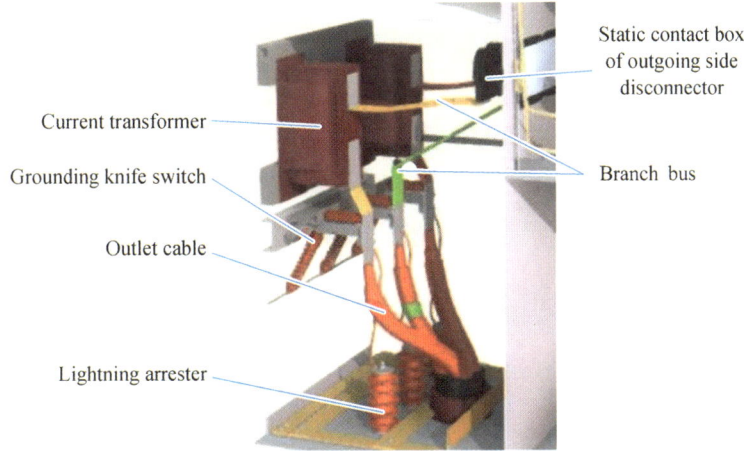

Fig. 4-8 Cable Compartment of KYN28-12 HV Switch Cabinet

7. Pressure releasing pathway

Except for the instrument compartment, the circuit breaker compartment, the bus compartment and the cable compartment must be equipped with satisfactory exhaust ducts and pressure relief devices as required. KYN28-12 employs a top pressure releasing pathway. In case of internal arc, the pressure releasing pathway will be automatically opened to release the internal pressure.

4.3.2 Structure (Other Cabinets)

1. Incoming cabinet

Compared with outgoing cabinet, the cable compartment of incoming cabinet shall be equipped with the bus voltage transformer but no grounding knife switch. If the incoming line side is overhead, the cable compartment has no cable, and the incoming cable enters the rear of the bus compartment, and then into the cable compartment from the top down, and connects to the static contact terminal of the bus disconnector. The cabinet is deeper than the incoming cabinet because both the bus compartment and the incoming line are on the upper back of the switch cabinet.

2. Sectional cabinet

It is used for sectionalized single-bus configuration, and divided into two cabinets. One circuit breaker compartment is equipped with a section circuit breaker handcart, and the other is equipped with a disconnector handcart only. This is to cut off the bus in the maintenance of the sectional cabinet, thus narrowing the power outage. In addition, tap and connect the bus in the lower bus compartment through a wall-through bushing, as shown in Fig. 4-9.

The sectional cabinet is connected with two sectionalized buses, no incoming and outgoing lines, so there is no cable compartment. The upper and lower parts of the back are bus compartments, and the buses in the bus compartments of the two cabinets are not connected

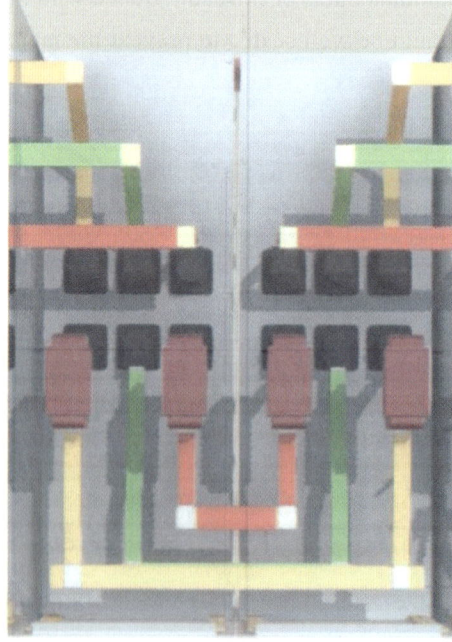

Fig. 4-9 Strike of the Bus in the Sectional Cabinet

模块四　开关柜的"五防"联锁

一、"五防"联锁概念

电力系统运行中往往会产生各种事故或故障,而事故产生的最大原因为电气误操作。而电气误操作的发生,直接威胁了人身安全、电网安全和设备安全。

例如断路器合闸状态,线路处于送电状态,直接拉开线路隔离开关(带负荷拉隔离开关),则会导致隔离开关出现五防熄灭的电弧,若设备为封闭式开关柜时,甚至会导致开关柜爆炸,如图 4-10 所示。

产生电弧(左);开关柜发生爆炸(中、右)

图 4-10　带负荷拉隔离开关

根据《国家电网公司防止电气误操作安全管理规定》《电气装置安装工程高压施工及验收规范》,高压开关柜应具备防止电气误操作的"五防"功能。

该规定从人、物、环境等多方面采取措施,降低了电气误操作事故的发生,其中在"物"方面采取的措施便包括防误装置措施。电气设备中的防误装置也简称"五防","五防"功能包括以下五点:

(1) 防止误分、误合断路器。
(2) 防止带负荷拉、合隔离开关或手车触头。
(3) 防止带电挂(合)接地线(接地刀闸)。
(4) 防止带接地线(接地刀闸)合断路器(隔离开关)。
(5) 防止误入带电间隔。

"五防"功能除"防止误分、误合断路器"仅采取提示性措施外,其余"四防"功能均采用强制性防止电气误操作措施。所谓强制措施,是指设备的电动操作、控制回路中串联以闭锁回路控制的接电、锁具,或者是在设备机械部件上加装符合"五防"原理控制的机械锁具。目前电气设备的"五防"联锁功能通常有机械联锁、程序锁、电磁锁、电气闭锁和微机闭锁等实现方式,根据具体设备的尺寸、位置、功能等不同而采用不同的方式实现"五防"功能。

二、KYN28-12 高压开关柜的主要"五防"联锁功能

为了更好理解和掌握开关柜的"五防"功能，需要了解断路器手车底盘车的构造及工作原理，如图 4-11 所示。

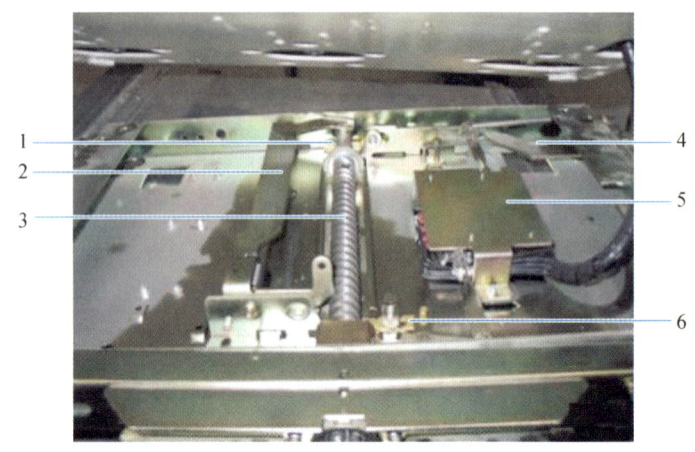

1—底盘车位置限位器；2—合闸位联锁板；3—丝杠；4—底盘车与接地刀闸联锁板；
5—电气联锁位置开关；6—底盘车"工作"位置限位器。

图 4-11 底盘车构造

1. 防止误分合断路器

根据《国家电网公司防止电气误操作安全管理规定》，防止误分合断路器可以提示性措施，而非强制性措施，所以针对 KYN28-12 高压开关柜所配备的断路器，仅有部分厂家采用了"闭锁电磁铁"来防止在控制回路不带电的时候，合闸断路器。

控制回路带电时，当断路器合闸于故障回路，继电保护会立刻动作，让该断路器跳闸，迅速切除故障。而控制回路停电时，当手动合闸断路器与故障回路，继电保护装置由于停电而无法动作，导致无法迅速切除故障回路，则会导致上级开关跳闸，扩大事故范围。

"闭锁电磁铁"装设在断路器操动机构中，位于机械"合闸"按钮上方。当控制回路带电时，手动"合闸"按钮可以自由按动，而当控制回路停电时，手动"合闸"按钮被电磁铁闭锁，无法按下，也无法合闸，如图 4-12 所示。

2. 防止带负荷拉、合隔离开关或手车触头

KYN28-12 高压开关柜的母线侧、出线侧隔离开关为插入式隔离开关，其动触头位于断路器尾部。当断路器手车处于"试验"位置时，隔离开关处于分闸，当断路器手车处于"工作"位置时，隔离开关处于合闸。

如果断路器处于"合闸"位置，将手车从"工作"位置拉至"试验"位置时，就造成带负荷分隔离开关，反之为带负荷合隔离开关。在断路器手车的底盘车中，有强制机械联锁装置，防止操作人员出现此类误操作。

（a）控制回路不带电　　　　　　（b）控制回路带电

图 4-12　闭锁电磁铁

断路器手车在"工作"位置，并且断路器在合位时，此时圈注区域在断路器机构联锁作用下，联锁板压下，联锁板卡住丝杠末端限位块，丝杠无法转动，从而底盘车无法从"工作"位置拉出，如图 4-13 所示。

断路器手车处于"试验"位置，并且断路器在合闸状态，手车底盘中的联锁板被断路器合闸闭锁装置的连板压下，联锁板前端的限位孔卡住丝杠末端限位块，将无法推进断路器手车，如图 4-13 所示。

（a）手车"工作"位置闭锁

（b）手车"试验"位置闭锁　　　　　　（c）手车"试验"位置解锁

图 4-13　底盘车丝杆闭锁装置

除以上两种情况，还有两种特殊情况会出现带负荷拉合隔离开关。当断路器手车既未处于"试验"位置，也未处于"工作"位置，而处于两者之间的"中间"位置时，合闸断路器，再将断路器推入"工作"位置，同样属于带负荷合隔离开关。为此，当断路器手车离开"工作""试验"位置时，底盘车会带动断路器内闭锁装置，闭锁"合闸"按钮，如图 4-14 所示。

（a）"试验"或"工作"解锁状态　　　　　　（b）"中间"位置闭锁状态

图 4-14　"中间"位置断路器合闸闭锁

还有一种情况就是当断路器手车处于"工作"位置并且断路器位于合闸状态时，工作人员拉动底盘车把手，将手车整体拉出柜体，会导致带负荷拉隔离开关。所以当手车一旦离开"试验"位置，底盘车把手将闭锁，工作人员无法再拉动底盘车，如图 4-15 所示。

图 4-15　底盘车闭锁

3. 防止带电合接地刀闸和防止带接地刀闸送电

当隔离开关在合闸位置，即断路器手车处于"工作"位置时，断路器也处于合闸状态，此时合上出线侧接地刀闸，会导致 10 kV 母线直接接地，轻则跳闸导致断路器跳闸母线停电，重则导致出线柜电缆室发生爆炸造成严重事故。同理当出线侧接地刀闸在"合闸"位置时，将断路器手车推至"工作"位置，再合上断路器，也会造成母线直接接地。

KYN28-12 高压开关柜和断路器手车采用了一系列机械联锁装置，防止带电合接地刀闸和防止带接地刀闸送电。

当断路器手车在导轨上从"试验"位置离开，向"工作"位置移动时，导轨上接的闭锁挡片会被底盘车侧板挡住，使接地刀闸挡板无法打开，也无法操作接地刀闸，如图 4-16 所示。这样就防止了断路器手车离开"试验"位置向"工作"位置移动时，或者处于"工作"位置时，合接地刀闸导致母线接地。

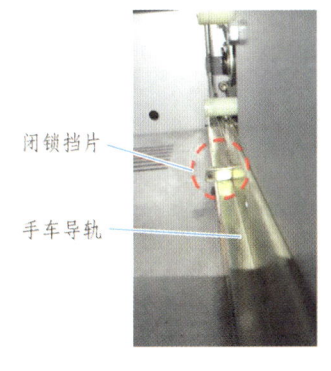
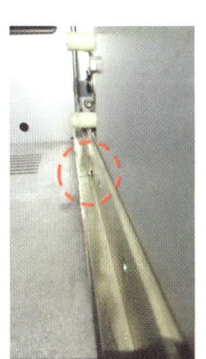
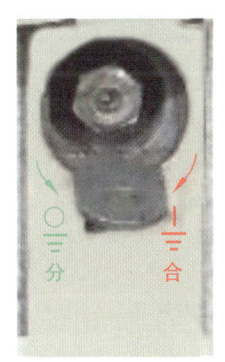

（a）手车闭锁　　　（b）手车解锁　　（c）接地刀闸挡板关闭　（d）接地刀闸挡板开启

图 4-16　手车与接地刀闸的闭锁

当接地刀闸处于"合闸"位置时，接地刀闸挡板无法关闭，断路器手车导轨上的联锁挡板伸出，挡住底盘车，手车无法在手车导轨上从"试验"位置移动至"工作"位置。

接地刀闸对手车的闭锁和手车位置对接地刀闸的闭锁采用同一套机械系统，达到双向闭锁功能。

4. 防止误入带电间隔

户内成套式高压开关设备由于结构紧凑，安全净距较小，误入带电间隔往往会造成人身伤亡事故。

如图 4-17 所示，接地刀闸处于分闸位置时，接地刀闸机构主轴上的联锁钩应能准确联锁，即锁套上的偏心锁钩旋转 90°（拉杆式接地刀闸）或 180°（伞齿轮式接地刀闸）卡住电缆室门，此时电缆室门应不能打开。

（a）接地刀闸合位，解锁　　　　　　　（b）接地刀闸分位，闭锁

图 4-17　接地刀闸对后电缆室柜门闭锁

为防止电缆室柜门解锁后，再次分开接地刀闸，让出线失去接地保护，所以电缆室门未关闭好时如图 4-18（a）所示，接地刀闸连杆上的凸台被开关柜体卡住，如图 4-18（b）所示，接地刀闸机构主轴不能转动，不能进行分闸；只有当电缆室门关闭好，柜门上的按钮被柜门按下后，连杆才可以转动如图 4-18（c）所示，接地刀闸才能分闸，可有效防止工作人员误入带电的电缆室。

（b）接地刀闸闭锁

（a）电缆室后柜门打开　　　　　　　（c）接地刀闸解锁

图 4-18　电缆室后柜门对接地刀闸闭锁

三、KYN28-12 其他特殊闭锁功能

除上述基本的"五防"联锁外，标准化设计的开关柜还应将以下闭锁功能作为常规配置。

（1）除仪表室外所有隔室柜门以及接地刀闸操作孔增挂"五防"锁。

（2）断路器手车离开"工作"位置时，活门挡板自动降下。

（3）断路器室面板上增设"紧急分闸"机构，并且不能带挂锁功能。

（4）断路器室柜门强制关闭闭锁装置：手车必须处于"试验"位置时，断路器室柜门才能打开；断路器室柜门打开后，手车不能移动。

（5）需要打开后门时，下柜门未开启时，上柜门无法开启。

（6）前下柜门未关闭，接地刀闸无法分闸，手车无法进入"工作"位置。

（7）接地刀闸未合闸，前下柜门无法开启。

（8）接地刀闸分、合闸到位后，才能正常取出手柄。

Module 4 "Five-prevention" Interlocking of the Switch Cabinet

4.4.1 Concept of "Five-prevention" Interlocking

Various accidents or failures often occur in the operation of power system, and electrical misoperation is the chief culprit. The electrical misoperation directly threatens personal, power grid and equipment safety.

For example, when the circuit breaker is in closed status, the line is supplying power. Switching the line disconnector directly (switching the disconnector with load) will cause the disconnector to generate the five-prevention quenching arc. If the device is an enclosed switch cabinet, it may explode, as shown in Fig. 4-10.

  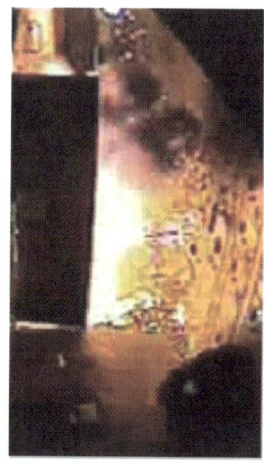

Generating arc (left); Explosion of switch cabinet (middle, right)

Fig. 4-10　Switching Disconnector with Load

In accordance with the *Regulations of State Grid Corporation of China on the Safety Management for Prevention of Electrical Misoperation*, the *Code for High Voltage Construction and Acceptance of Electrical Equipment Installation Engineering*, the HV switch cabinet should have the "five-prevention" function against electrical misoperation.

These Regulations take measures from such aspects as people, things and environment to minimize electrical misoperation accidents. The measures in terms of "things" include anti-misoperation device. The anti-misoperation device in electrical equipment is also referred to as "five preventions", specifically including the following:

(1) Prevent the circuit breaker from opening and closing accidentally.

(2) Prevent the disconnector or handcart contact terminal from opening and closing with load.

(3) Prevent switching on grounding wire (grounding knife switch) with load.

(4) Prevent switching on the circuit breaker (disconnector) with grounding wire (grounding

knife switch).

(5) Prevent the straying into energeized compartment.

Except the suggestive measures to "prevent the circuit breaker from opening and closing accidentally", the other "four preventions" take compulsory measures to prevent electrical misoperation. The so-called compulsory measures refer to the series connection of power feeding and lock in the electric operation and control circuit of equipment to lock the circuit, or the additional provision of mechanical lock in line with the "five-prevention" principle on the mechanical parts of equipment. At present, the "five-prevention" interlocking function of electrical equipment is usually achieved by mechanical interlocking, program locking, electromagnetic locking, electrical locking and microcomputer locking. Different ways are adopted to achieve the "five-prevention" function by size, location and function of specific equipment.

4.4.2 Main "Five-prevention" Interlocking Function of KYN28-12 HV Switch Cabinet

To better understand and master the "five-prevention" function of the switch cabinet, it is necessary to know the structure and working principle of the circuit breaker handcart chassis, as shown in Fig. 4-11.

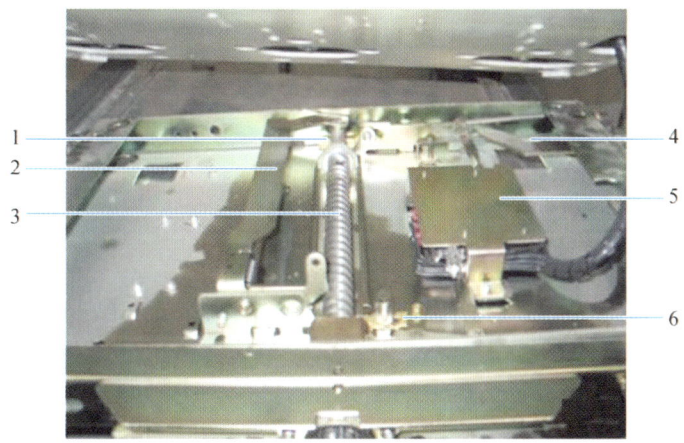

1-Chassis position limiter; 2-Closing position lockpatch; 3-Leadscrew; 4-Chassis and grounding knife switch lockpatch; 5-Electrical interlock position switch; 6-Chassis "Work" position limiter.

Fig. 4-11　Chassis Construction

1. Prevent the circuit breaker from opening and closing accidentally

In accordance with the *Regulations of State Grid Corporation of China on the Safety Management for Prevention of Electrical Misoperation*, suggestive measures (not compulsory measures) are taken to prevent the circuit breaker from opening and closing accidentally. Therefore, for the circuit breaker in KYN28-12 HV switch cabinet, only some manufacturers employ a "latching electromagnet" to prevent the circuit breaker from closing when the control circuit is not charged.

When the control circuit is energized, and the circuit breaker is closed on the fault circuit, the relay protection will immediately act to make the circuit breaker trip and quickly remove the fault.

In case of outage to the control circuit, when the circuit breaker and the fault circuit are manually closed, the relay protection device fails to function due to power failure, unable to remove the fault circuit quickly. This will cause the upstream switch to trip and widen the scope of accident.

The "latching electromagnet" is installed in the circuit breaker operating mechanism, and located above the mechanical "closing" button. When the control circuit is energized, the manual "closing" button can be freely pressed. When the control circuit is powered off, the manual "closing" button is locked by the electromagnet and cannot be pressed or closed, as shown in Fig. 4-12.

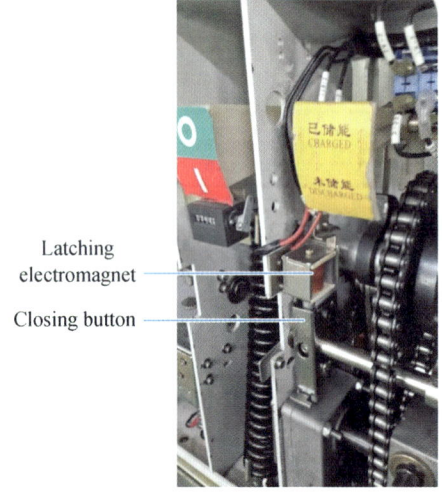

(a) Control circuit not energized   (b) control circuit being energized

Fig. 4-12   Latching Electromagnet

2. Prevent the disconnector or handcart contact terminal from opening and closing with load

The bus side, outgoing side disconnectors of KYN28-12 HV switch cabinet are plug-in ones, and their moving contact terminals are located at the tail of the circuit breaker. When the circuit breaker handcart is in the "Test" position, the disconnector is in the opening status; when the circuit breaker is in the "Work" position, the disconnector is in the closing status. If the circuit breaker is in the "closing" position, when the handcart is switched from the "Work" position to the "Test" position, it will cause the opening of disconnector with load, and the closing of disconnector with load by switching from "Test" to "Work". The chassis of the circuit breaker handcart is equipped with a forced mechanical interlocking device to prevent such misoperation by the operator.

When the circuit breaker handcart is in the "Work" position, and the circuit breaker is in the closing position, the circled area is under the interlocking action of the circuit breaker mechanism, the interlocking plate presses down to clamp the leadscrew end stop block, and the leadscrew cannot turn, so the chassis cannot be pulled out of the "Work" position, as shown in Fig. 4-13.

When the circuit breaker handcart is in the "Test" position and the circuit breaker is in the closing status, the interlocking plate in the chassis of the circuit breaker is pressed down by the connecting plate of the circuit breaker closing and locking device, and the limit hole at the front end of the interlock plate jams the leadscrew end stop block, making it unable to push the circuit breaker handcart, as shown in Fig. 4-13.

(a) Handcart "Work" position latched

(b) Handcart "Test" position latched

(c) handcart "Test" position unlocked

Fig. 4-13 Chassis Leadscrew Locking Device

In addition to the above two cases, there are two special cases that will cause the switching of disconnector with load. When the circuit breaker handcart is neither in the "Test" nor the "Work" position, but in the "middle" position, close the circuit breaker, and then switch the circuit breaker to the "Work" position, which is also switching the disconnector with load. For this reason, when the circuit breaker handcart leaves the "Work" and "Test" positions, the chassis will drive the locking device in the circuit breaker to lock the "closing" button, as shown in Fig. 4-14.

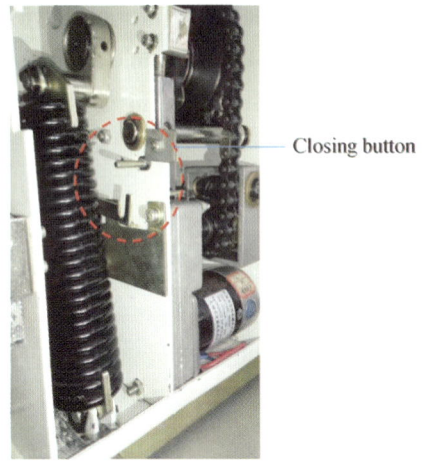 

(a) "Test" or "Work" unlocked state   (b) "Middle" position latched

Fig. 4-14   Closing and Locking of the Circuit Breaker in the "Middle" Position

There is another case that when the circuit breaker handcart is in the "Work" position and the circuit breaker is in the closed status, the staff holds the chassis handlebar and pulls the handcart out of the cabinet. This will switch the disconnector with load. Therefore, once the handcart leaves the "Test" position, the chassis handlebar will be latched to keep the chassis fixed, as shown in Fig. 4-15.

Fig. 4-15   Chassis Locking

3. Prevent switching on grounding knife switch with load and power transmission with grounding knife switch

When the disconnector is in the closed position, that is, the circuit breaker handcart is in the "Work" position, and the circuit breaker is also in the closed status. At this time, closing the grounding knife switch on the outgoing side will lead to the direct grounding of the 10 kV bus. The consequence will range from the tripping operation (to trigger the circuit breaker tripping and bus outage) to the explosion in the cable compartment of the outgoing cabinet. Likewise, when the grounding knife switch on the outgoing side is in the "closed" position, push the circuit breaker handcart to the "Work" position and then close the circuit breaker. This will also cause the bus to be directly grounded.

KYN28-12 HV switch cabinet and circuit breaker handcart use a series of mechanical interlocking devices to prevent switching on grounding knife switch with load and power transmission with grounding knife switch.

When the circuit breaker handcart is moved from the "Test" position to the "Work" position on the slideway, the latching catch connected to the slideway will be blocked by the side panels of the chassis, so that the grounding knife switch cannot be opened or operated, as shown in Fig. 4-16. This prevents the circuit breaker handcart from moving from the "Test" position to the "Work" position, or when in the "Work" position, switching on the grounding knife switch keeps the bus grounded.

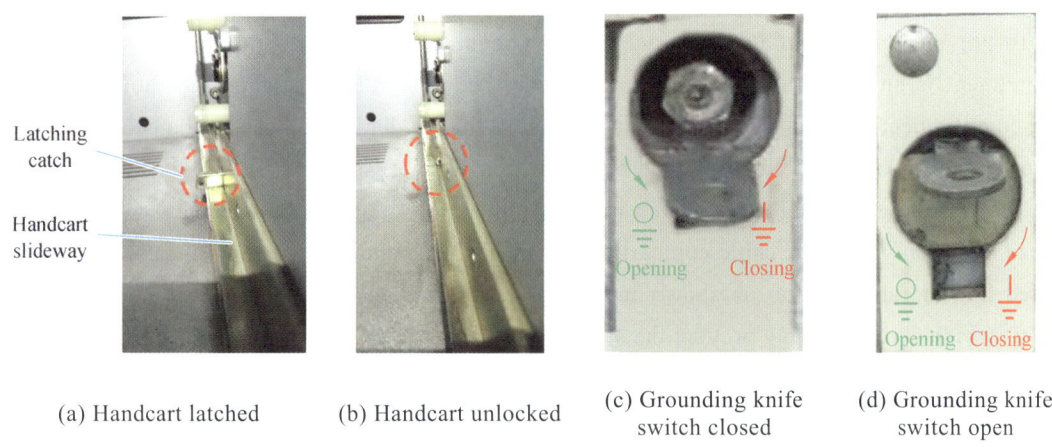

(a) Handcart latched   (b) Handcart unlocked   (c) Grounding knife switch closed   (d) Grounding knife switch open

Fig. 4-16 Locking of Handcart and Grounding Knife Switch

When the grounding knife switch is in the "closed" position, the grounding knife switch damper cannot be closed, the interlocking baffle on the circuit breaker handcart slideway extends to block the chassis, so that the handcart cannot move from the "Test" position to the "Work" position on the handcart slideway.

The same mechanical system is used for the latching of the grounding knife switch to handcart and the latching of the handcart position to the grounding knife switch, thus achieving the bi-directional locking function.

4. Prevent the straying into energized compartment

The indoor packaged HV switchgear features compact structure and small safe distance, and straying into energized compartments often results in personal injury and even death.

As shown in Fig. 4-17, when the grounding knife switch is in the open position, the interlocking hook on the main shaft of the grounding knife switch mechanism should be able to interlock accurately, that is, the eccentric hook on the lock sleeve is rotated 90° (tie rod grounding knife switch) or 180° (bevel gear grounding knife switch) to get the door of cable compartment stuck.

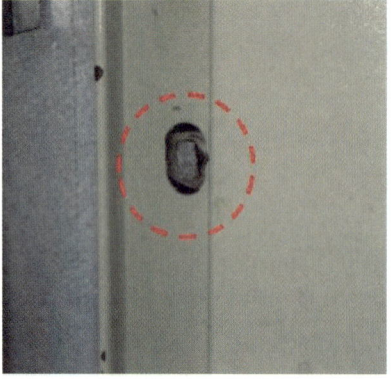 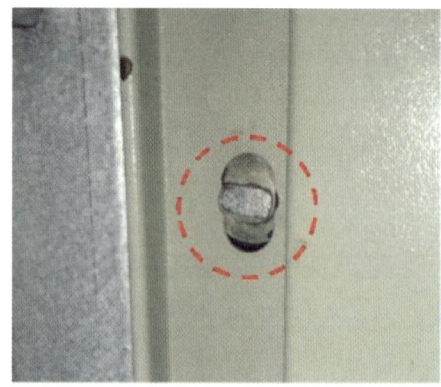

(a) The grounding knife switch closed, unlocked (b) The grounding knife switch opening, latched

Fig. 4-17 Latching of Grounding Knife Switch to the Rear Cabinet Door of Cable Compartment

To prevent the grounding knife switch from opening again after the cabinet door of cable compartment is unlocked, so that the outgoing lines are not grounded, when the cabinet door of cable compartment is not closed properly, as shown in Fig. 4-18 (a), the boss on the connecting rod of the grounding knife switch is jammed by the switch cabinet, as shown in Fig. 4-18 (b), and the main shaft of the grounding knife switch mechanism cannot rotate nor be opened. The connecting rod can rotate only after the cable compartment door is closed and the button on the cabinet door is pressed by the cabinet door, as shown in Fig. 4-18 (c), and the grounding knife switch can be opened, which can effectively prevent the staff from straying into the live cable compartment.

(b) The grounding knife switch latched

(a) The rear cabinet door of cable compartment open

(c) The grounding knife switch unlocked

Fig. 4-18 Latching of the Rear Cabinet Door of Cable Compartment to the Grounding Knife Switch

4.4.3　Other Special Blocking Functions of KYN28-12

In addition to the above basic "five-prevention" interlocking, the switch cabinet of standardized design should have the following locking functions.

(1) The "five-prevention" locks are installed on all compartment doors and grounding knife switch operating holes except the instrument compartment.

(2) When the circuit breaker handcart leaves the "Work" position, the valve damper lowers automatically.

(3) The "emergency opening" mechanism is added on the panel of the circuit breaker circuit breaker compartment, but no padlock function.

(4) The locking device to force the cabinet door of circuit breaker compartment to be closed: When the handcart must be in the "Test" position, the cabinet door of circuit breaker compartment can be opened; the handcart cannot move after the cabinet door of circuit breaker compartment is opened.

(5) When the rear door is to be opened, the upper cabinet door cannot be opened if the lower cabinet door is not opened.

(6) The front lower cabinet door is not closed, the grounding knife switch cannot be opened, and the handcart cannot enter the "Work" position.

(7) The grounding knife switch is not closed, and the front lower cabinet door cannot be opened.

(8) The handle can be taken out properly only after the grounding knife switch is opened and closed in place.

任务一 KYN28-12 高压开关柜整体检查与维护

一、工作任务

高压开关柜（出线柜）整体检查与维护，包括对开关柜的各个隔室、每个隔室内部的元件以及之间的连锁装置进行检查和维护。检查与维护项目包括：开关柜箱体、母线室、断路器室、断路器手车、电缆室、仪表室以及隔室内部元件和元件之间的联锁装置等。

二、引用标准

（1）《电气装置安装工程高压电器施工及验收规范》（GB 50147—2010）。
（2）《国家电网公司五项通用制度变电检修管理规定》（第 5 分册开关柜检修细则）。
（3）《国家电网公司变电运维管理规定》。
（4）《国家电网公司变电运修管理规定》。
（5）《国家电网公司电力安全工作规程》（变电部分）。
（6）《四川省电力公司高压开关柜现场维护导则》。

三、工作要求

（1）作业为室内停电作业，无天气要求。
（2）被检修设备与其他带电设备均应使用围栏隔离，面向通道处设置唯一出入口。
（3）被检修间隔要求母线与出线电缆均停电并且接地。
（4）作业人员精神状态良好，熟悉工作中安全措施、技术措施以及现场工作危险点。
（5）实训现场要求按生产现场规范布置安全措施，并严格执行标准化作业。
（6）作业人员应规范穿戴劳动保护用品，做好安全防护。

四、工作准备

被检修设备一次接线图见图 4-19。

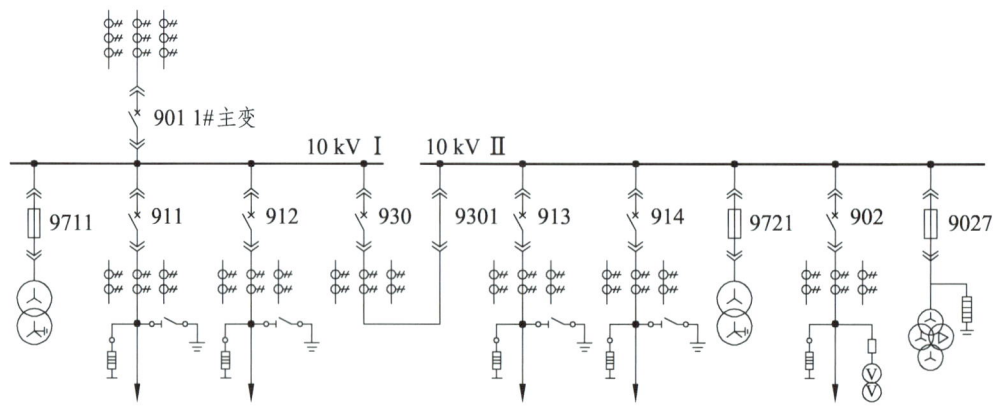

图 4-19 被检修设备一次接线图

(一)危险点及预控措施

1. 高压触电

危险点:10 kV Ⅱ 母停电,10 kV Ⅰ 母整体视为带电,902 进线柜电缆室视为带电,930 分段柜视为带电,误入带电间隔可能导致高压触电。

预控措施:用围栏将被检修间隔与相邻带电设备(间隔)隔离,并且向作业现场内悬挂适量"止步,高压危险"标识牌,在通道处设置唯一出入口,悬挂"从此进出"标识牌;被检修间隔母线侧与出线电缆需各装设一组三相短路接地线;902 开关柜手车、930 开关柜手车需拉出至"检修"位置;902、930 开关柜门上悬挂"止步,高压危险"标识牌;工作时至少需要两人,一人监护一人操作,听工作负责指挥。

2. 低压触电

危险点:控制回路、合闸回路、照明回路低压断路器进线侧(上端)均低压电;合上空开后,断路器和开关柜二次回路均视为带电状态,触摸可能造成低压触电。

预控措施:若控制、储能低压断路器上悬挂了"禁止合闸,有人工作"标识牌时,工作负责人需要向运行人员申请更变安全措施得到许可后,方能合闸;勿将手指伸入低压断路器上端头;拔出航空插头前,需先将低压电源断开,拔出航空插头后不允许合上低压断路器;未得到负责人的许可,不允许打开断路器面板;工作时至少需要两人,一人监护一人操作,听工作负责指挥。

3. 机械伤害

危险点:断路器分合、接地刀闸分合时未通知其他工作班成员,可能会对工作班成员造成机械伤害;将断路器从"试验"位置转移至"检修"位置或将断路器从"检修"位置转移至"试验"位置时,手车可能发生倾倒,对操作者造成机械伤害。

预控措施:断路器分合、接地刀闸分合时,需要大声呼唱,得到工作班所有人员大声回应之后,方可操作;操作接地刀闸前,需要特别注意开关柜下柜门内是否有人员正在操作;未得到负责人的许可,不允许打开断路器面板;将断路器从"试验"位置转移至"检修"位置或将断路器从"检修"位置转移至"试验"位置时,需要用锁扣将转运小车锁在开关柜上,防止小车倾倒;工作时至少需要两人,一人监护一人操作,听工作负责指挥。

4. 设备损坏

危险点:未按照要求分合断路器、分合接地刀闸、野蛮操作,或者操作时遗留工具在柜体内,均会对设备造成不同程度的损坏。

预控措施:禁止野蛮操作,操作过程中,禁止遗留任何工具于柜内,若出现不能操作的现象,应当立即停止,不得使用蛮力,立即汇报负责人;工作时至少需要两人,一人监护一人操作,听工作负责指挥。

(二)工器具及材料选择

KYN28-12 高压开关柜整体检查及维护工器具及材料见表 4-5。

表 4-5　KYN28-12 高压开关柜整体检查及维护工器具及材料

类别	名称	规格型号	数量	备注
专用工具	地刀操作把手		1 把	
	储能把手		1 把	
通用工具	手车把手		1 把	
	内六角扳手		1 套	
	一字螺丝刀	150 mm	1 把	
	十字螺丝刀	150 mm	1 把	
	万用表		1 个	
	锉刀		1 把	
	活动扳手	200 mm	1 把	
	铜刷		1 把	
二硫化钼	润滑脂		0.5 kg	
	材料防锈漆		0.5 kg	
	凡士林		0.5 kg	

（三）作业人员分工

KYN28-12 高压开关柜整体检查及维护人员分工见表 4-6。

表 4-6　KYN28-12 高压开关柜整体检查及维护人员分工

序号	工作岗位	数量	职责
1	工作负责人	1	负责本次工作的人员分工、现场查勘、作业方案制定、召开班前会、作业过程中安全监督、工作中突发状况的处理、工作质量的监督、班后会总结以及现场的指挥
2	操作人员	1	负责开关柜的操作、检查、维护、记录
3	辅助操作人员（可无）	1	辅助操作人员进行检查、记录

五、工作程序

KYN28-12 高压开关柜整体检查及维护操作流程见表 4-7。

表 4-7 KYN28-12 高压开关柜整体检查及维护操作流程

序号	步骤	检查项目	维护内容
1	工作前准备工作	1. 检查工器具是否齐全，检查工器具外观和试验合格； 2. 工作负责人同工作许可人巡视待检修设备，确认工作票所列安全措施已经正确执行，安全措施是否完备，现场是否具备开工条件，必要时进行补充； 3. 执行工作许可手续； 4. 对工作班成员召开班前会； 5. 抄写设备铭牌参数	1. 工器具无损伤、变形、失灵现象，需要试验的工器具合格证在有效期内； 2. 巡视现场时禁止无关人员进入现场； 3. 班前会应包含工作地点双重名称；工作时间与内容；工作分工；工作危险点及预控措施；停电范围及工作现场安全措施； 4. 全体工作成员应当正确穿戴安全帽、工作服、工作鞋、劳保手套等劳动保护用品
2	确认设备状态	负责人带领工作班成员确认设备处于分闸未储能，接地刀闸处于合位	
3	箱体的检查与维护	1. 检查箱箱体漆面是否变色、鼓包、起皮； 2. 检查箱箱体表面油漆是否完好，是否锈蚀； 3. 检查箱箱体外部螺栓、销钉是否缺失、松动、脱落； 4. 检查箱门锁、把手、操作孔是否完好； 5. 检查箱体表面安装的观察窗玻璃有无裂纹、爆裂； 6. 检查紧急解锁挡板是否完好； 7. 检查泄压通道结构是否正常； 8. 检查设备双重名称和铭牌是否完整、齐全； 9. 检查柜门接地、密封条是否完好； 10. 检查各隔室照明是否正常	1. 保持开关室内的通风良好，便于湿气排除，营造良好的设备运行环境； 2. 用铜刷清除锈蚀柜面和其它钢质部件上的铁锈，并除去油污，然后立即涂上防锈底漆； 3. 对松动的螺栓、销钉进行紧固，缺失的螺栓销钉进行补缺； 4. 对有损坏的窗口玻璃、信号灯、铭牌等附件进行更换； 5. 更换损坏部件，对无法立即更换、维修的非关键损坏部件进行记录
4	母线室的检查与维护	1. 检查母线的绝缘支撑有无闪络、放电痕迹，是否松动； 2. 检查母线连接螺栓是否缺失、松动、脱落；	1. 绝缘材料表面有污物时，用软布擦去灰尘，用轻度碱性的清洁剂，擦去黏性或油脂性脏物，然后用清水擦干净，再进行干燥处理；

续表

序号	步骤	检查项目	维护内容
4	母线室的检查与维护	3. 检查母线热缩套是否破裂,相序标识是否完整; 4. 检查母线是否有松动,是否有发热、放电痕迹; 5. 检查室内有无异物; 6. 检查绝缘件表面有无脏污	2. 若发现绝缘件上有麻孔、裂纹、炸裂等迹象应立即更换; 3. 清除内部异物,使用对有松动、脱落的母线连接螺栓进行紧固; 4. 热缩套有皲裂、变色、脱落现象及时更换; 5. 若发现有放电痕迹,应查明原因,然后清洁或更换相关部件; 6. 更换损坏部件,对无法立即更换、维修的非关键损坏部件进行记录
5	断路器室的检查与维护	1. 检查手车内有无杂物,手车室内壁有无烧蚀痕迹,有无异味; 2. 检查触头盒及静触头,看触头盒有无裂纹、放电烧蚀迹象,静触头固定是否牢固,有无氧化、烧蚀现象; 3. 检查活门联锁机构是否可靠,尼绒滚轮有无变形损坏,各个连接部位的轴销是否齐全有无损坏; 4. 检查手车导轨的平直度,看有无变形现象; 5. 检查手车接地触头是否接地良好; 6. 检查手车位置与航空插头联锁是否正常	1. 使用软布沾取轻度碱性的清洁剂清洁触头盒及静触头表面,然后用清水擦干净,再进行干燥处理,并在触头表面涂凡士林,触头表面若有镀银层应使用百洁布蘸酒精进行轻清擦,对触头烧蚀表面烧蚀深度超过 1 mm 的需要进行更换; 2. 使用清洁的软布清擦绝缘件,保证表面的整洁度; 3. 对要求灵活转动部位使用二硫化钼进行润滑; 4. 清除内部异物,若导轨上有毛刺阻碍手车的正常运动可使用锉刀将毛刺剔除; 5. 更换损坏部件,对无法立即更换、维修的非关键损坏部件进行记录
6	电缆室的检查与维护	1. 检查内部有无异物; 2. 检查各连接螺栓及设备固定螺栓有无缺失、松动、脱落现象; 3. 检查接地刀闸的拉合是否灵活,是否到位,触头、触指有无烧蚀,弹簧弹性是否良好; 4. 检查各个转动关节的连接销钉是否齐全,固定的销钉的开口销、卡簧销有无脱落、丢失情况;	1. 对松动的螺栓进行紧固,缺失的螺栓销钉进行补缺; 2. 使用软布沾取轻度碱性的清洁剂清洁接地刀闸触头、触指表面,然后用清水擦干净,再进行干燥处理,并在表面涂凡士林; 3. 使用清洁的软布清擦绝缘件,保证表面的整洁度;

续表

序号	步骤	检查项目	维护内容
6	电缆室的检查与维护	5. 检查支持绝缘子、绝缘护套、热缩套等有无凝露，有无放电烧蚀、开裂、脱落情况； 6. 检查加热器工作是否正常； 7. 检查电流互感器、避雷器等电器设备外观有无变色、开裂、烧蚀迹象，接线是否牢固； 8. 检查电缆封堵泥是否完好； 9. 检查相序标识完整	4. 将柜内杂物清出，对在其中发现的遗留物品如螺栓等要检查其来源，无用的取出，有用的复位； 5. 发现放电、闪络痕迹，应当找出原因； 6. 更换损坏部件，对无法立即更换、维修的非关键损坏部件进行记录
7	仪表室的检查与维护	1. 检查储能、控制、加热照明电源低压断路器是否正常工作； 2. 检查端子排有无松动，柜内有无异物； 3. 改变手车位置、断路器状态、接地刀闸状态，检查信号、位置指示灯指示是否正确，有无损坏； 4. 检查二次线线头是否松动、脱出； 5. 检查温湿度控制器能否正常工作	1. 清扫低压室内部，保持内部的清洁； 2. 对端子排、二次线进行紧固； 3. 更换损坏部件，对无法立即更换、维修的非关键损坏部件进行记录。 注意：工作票中若有"断开待检修设备控制回路、储能回路电源并悬挂'禁止合闸，有人工作'"这一项，则必须经许可人同一后，方可合控制回路、储能回路电源
8	手车的检查与维护	1. 手车外观是否正常； 2. 检查导电回路的连接是否紧固；触头与触臂连接螺栓是否紧固； 3. 检查底盘车滚轮有无形变； 4. 检查绝缘件是否清洁、状态是否良好有无放电、闪络痕迹； 5. 检查梅花触头有无氧化、烧蚀痕迹； 6. 检查梅花触头夹紧弹簧弹性是否良好； 7. 检查断路器铭牌是否完好，分合闸及储能标识是否清楚，面板无变形破损	1. 使用软布沾取轻度碱性的清洁剂清洁触头表面，然后用清水擦干净，再进行干燥处理，并在触头表面涂医用凡士林； 2. 使用清洁的软布清擦绝缘件，保证表面的整洁度； 3. 对产生滚动与滑动摩擦的接触面及设备进行润滑
9	现场恢复至初始状态	1. 关闭所有柜门； 2. 断开控制回路、合闸回路电源； 3. 将"禁止合闸，有人工作"标识牌挂回控制回路、合闸回路处	
10	工作终结	1. 清理作业现场，做到"工完料尽场地清"； 2. 召开班后会； 3. 终结工作票	严禁负责人前去终结工作票的同时清理场地

Task 1　Integral Inspection and Maintenance of KYN28-12 HV Switch Cabinet

1.1　Work Tasks

The integral inspection and maintenance of the HV switch cabinet (outgoing cabinet) include the inspection and maintenance to each compartment of the switch cabinet, the components of each compartment and the interlocking devices between these components. Inspection and maintenance items include: switch cabinet box, bus compartment, circuit breaker compartment, circuit breaker handcart, cable compartment, instrument compartment, and interlocking device between components in the compartment.

1.2　References

(1) *Code for Construction and Acceptance of High-voltage Electrical Apparatus of Electric Equipment Installation Engineering* (GB 50147-2010).

(2) *Five General Regulations of State Grid Corporation of China on Substation Maintenance Management* (Volume 5: Rules for Maintenance of Switch Cabinets).

(3) *Regulations of State Grid Corporation of China on Management of Substation Operation and Maintenance.*

(4) *Regulations of State Grid Corporation of China on Management of Substation Operation and Maintenance.*

(5) *Electric Power Safety Working Regulations (Power Transformation) of State Grid Corporation.*

(6) *Field Maintenance Guidelines for High-voltage Switch Cabinets of Sichuan Electric Power Company.*

1.3　Work Requirements

(1) The operation belongs to indoor power interruption operation, without weather requirements.

(2) Equipments under maintenance are to be isolated from other electrified equipments by a fence. The only exit is to be provided at the place oriented towards the passage.

(3) For the maintenance bay, the bus and outlet cable are required to be powered off and grounded.

(4) Operators are to be in good mental state, and are aware of safety measures, technical measures and hazards to site operation during operation.

(5) Take safety measures on practical training site as per regulations on production site, and strictly implement standard operation.

(6) Operators are requested to wear labor protection appliances to ensure safety protection.

1.4 Preparation for Work

Primary wiring diagram of the equipment under maintenance is shown in Fig. 4-19.

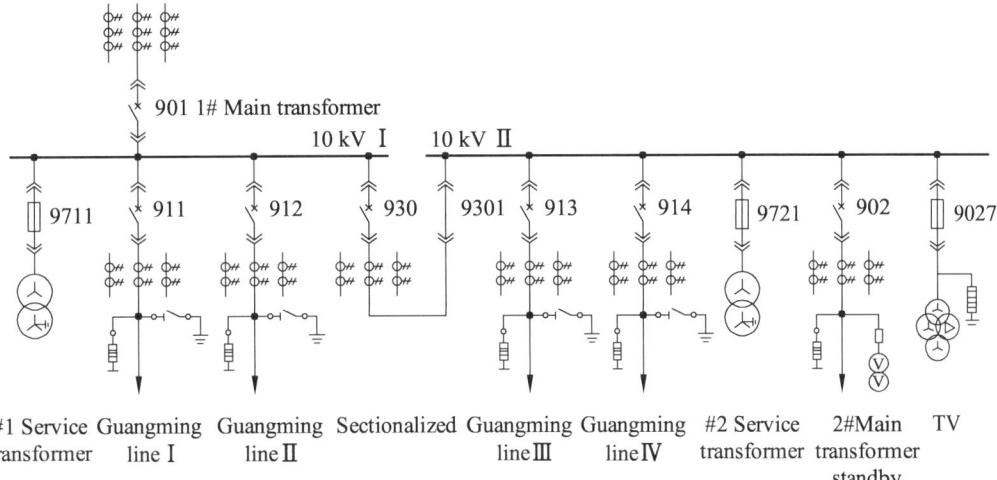

Fig. 4-19　Primary Wiring Diagram of the Equipment under Maintenance

1.4.1 Hazards and Preventive and Control Measures

1. High-voltage electric shock

Hazards: 10 kV bus II outage, 10 kV bus I electrically charged as a whole, the cable compartment of 902 incoming cabinet as charged, and 930 sectional cabinet as charged; straying into the energized compartments may give rise to high-voltage electric shock.

Preventive and control measures: Use fence to isolate maintenance interval from adjacent live equipments, and provide such sign boards as "Stop! High Voltage, Danger!" on operation site. The only exit is to be set at the passage, which is to be provided with the sign board indicating "Entrance/Exit"; The bus side of the maintenance bay and the outgoing cable shall be equipped with a set of three-phase short-circuit grounding wires. Switch cabinet handcarts 902 and 930 should be pulled out to the "Maintenance" position; the sign boards "Stop! High Voltage, Danger!" are hung on the doors of 902 and 930. At least two persons are required during work, one for supervision and the other for operation under instructions of person in charge of work.

2. Low-voltage electric shock

Hazard: there is low-voltage current in the incoming side (upper end) of low-voltage circuit breaker of control circuit, closing circuit and lighting circuit. After closing the air switch, the secondary circuits of circuit breaker and switch cabinet are all considered as live circuits, and touching the circuits may cause low-voltage electric shock.

Preventive and control measures: If a "No Closing, Work in Progress" sign board is suspended on the LV circuit breaker for control and energy storage, the person in charge of work must request for safety precautions change from the operator before closing; do not put your fingers into the upper end of the LV circuit breaker; before pulling out the aviation plug, disconnect the LV power supply, and do not close the LV circuit breaker after pulling out the aviation plug; do not open the circuit breaker panel without the permission of the person in charge. At least two persons are required during work, one for supervision and the other for operation under instructions of person in charge of work.

3. Mechanical injury

Hazard: during opening and closing of circuit breaker and grounding knife switch, if notification is not given to other members of the work team, the members may suffer mechanical injury; When the circuit breaker is moved from the "Test" position to the "Maintenance" position or the circuit breaker is moved from the "Maintenance" position to the "Test" position, the handcart may be tripped, causing mechanical injury to the operator.

Preventive and control measures: During opening and closing of circuit breaker and grounding knife switch, it is necessary to shout and sing loudly, and to give operation only after receiving the loud response from all members of the work team. Before operating the grounding knife switch, please note that someone may be operating in the lower cabinet door of the switch cabinet. When the circuit breaker is moved from the "Test" position to the "Maintenance" position or the circuit breaker is moved from the "Maintenance" position to the "Test" position, a lock is used to lock the transfer trolley on the switch cabinet to prevent it from toppling. At least two persons are required during work, one for supervision and the other for operation under instructions of person in charge of work.

4. Equipment damage

Hazard: the equipment may be damaged to varying degrees if opening and closing of circuit breaker and grounding knife switch are not in accordance with requirements, or rough handling is adopted, or tools are left in the cabinet during operation.

Preventive and control measures: rough handling is prohibited. It is forbidden to leave any tool in the cabinet during operation. If there is a phenomenon that operation cannot be given during the operation, the operation shall be stopped immediately, no brute force shall be used, and report shall be sent to the person in charge immediately; At least two persons are required during work, one for supervision and the other for operation under instructions of person in charge of work.

1.4.2　Work Tools and Material Selection

See Table 4-5 for the tools and materials for integral inspection and maintenance of KYN28-12 HV switch cabinet.

Table 4-5　Tools and Materials for Integral Inspection and Maintenance of KYN28-12 HV Switch Cabinet

Category	Name	Specification and model	Quantity	Remarks
Specialized tools	Grounding knife-switch operating handle		1	
	Energy storage handle		1	
General tools	Handcart handlebar		1	

Continued

Category	Name	Specification and model	Quantity	Remarks
General tools	Allen wrench		1 set	
	Slotted screwdriver	150 mm	1	
	Cross screwdriver	150 mm	1	
	Multimeter		1 drum	
	File		1	
	Monkey wrench	200 mm	1	
	Copper brush		1	
Material	Molybdenum disulphide grease		0.5 kg	
	Antirust paint		0.5 kg	
	Vaseline		0.5 kg	

1.4.3 Division of Labor among Operators

See Table 4-6 for the personnel allocation for integral inspection and maintenance of KYN28-12 HV switch cabinet.

Table 4-6 Personnel Allocation for Integral Inspection and Maintenance of KYN28-12 HV Switch Cabinet

S/N	Job	Quantity	Responsibilities
1	Person in charge of work	1	Responsible for the personnel allocation, site survey, stipulation of operation scheme, pre-shift meeting, safety supervision during operation, handling of emergencies during work, supervision of work quality, summary of post-shift meeting and field command
2	Operator	1	Responsible for the operation, inspection, maintenance and recording of the switch cabinet
3	Auxiliary operators (optional)	1	Responsible for inspection and recording

1.5 Working Procedures

See Table 4-7 for the operation procedures for integral inspection and maintenance of KYN28-12 HV switch cabinet.

Table 4-7 Operation Procedures for Integral Inspection and Maintenance of KYN28-12 HV Switch Cabinet

S/N	Procedure	Inspection items	Maintenance content
1	Preparations before work	1. Check if all instruments and tools are complete, and if their appearance and test are acceptable; 2. The person in charge of work and work permitter shall make an tour inspection for equipments under maintenance, and confirm all safety measures as listed by work ticket have been properly implemented. It is also necessary to check if safety measures are complete, and if the site is provided with conditions for commencement of work, and make supplements as required;	1. Instruments and tools are free of damage, deformation and malfunction. Qualification certificates for instruments and tools to be tested are within the term of validity; 2. Prevent other persons from entering the site during tour inspection; 3. Toolbox meeting shall cover dual designations of work place; Working hours and contents; Work division; Working hazards as well as preventive and control measures; Power-cut scope and safety measures on work site;

Continued

S/N	Procedure	Inspection items	Maintenance content
1	Preparations before work	3. Implement work permit procedures; 4. Call in toolbox meeting participated by members of work team; 5. Record parameters on the equipment nameplate	4. All work members shall properly wear such labor protection appliances as safety helmet, working clothes, working shoes and protective gloves
2	Confirm equipment status	The person in charge shall lead members of work team to confirm that the equipment is in the opening status without energy storage, and the grounding knife switch in the closed status.	
3	Box inspection and maintenance	1. Check the paint surface of the hatch box for any discoloring, bulging, and peeling; 2. Check the paint on the surface of the hatch box for any damage and rust; 3. Check the external bolts and pins of the hatch box for any missing, loosening or falling off; 4. Check the door lock, handle and operating hole of the hatch box for any damage; 5. Check the sight glass mounted on the surface of the hatch box for any crack and bursting; 6. Check the emergency release baffle for any damage; 7. Check the structure of pressure relief channel for anomalies; 8. Check the dual designations and nameplate of equipment for any missing; 9. Check the grounding and the sealing strip of the cabinet door for any damage; 10. Check each compartment for normal lighting	1. Keep the switchgear room well-ventilated, facilitate the removal of moisture, and build a good operating environment for equipment; 2. Remove rust from cabinet surfaces and other steel parts with a copper brush, and remove the oil stains, and then immediately apply the anti-rust primer; 3. Tighten loose bolts and pins, and make up missing bolts and pins; 4. Replace damaged sight glass, signal light, nameplate and other accessories; 5. Replace damaged parts and record non- critical damaged parts that cannot be immediately replaced or repaired
4	Inspection and maintenance of bus compartment	1. Check the insulation support of the bus for any flashover, discharge and looseness; 2. Check whether the bus connecting bolts are missing, loose or falling off; 3. Check whether the bus heat-shrinkable sleeve is broken and whether the phase sequence mark is complete; 4. Check the bus for loosening, any traces of heat and discharge; 5. Check the compartment for foreign bodies; 6. Check insulating part surfaces for any dirt	1. In case of any dirt on the surface of the insulating material, wipe off the dust with a soft cloth, wipe away sticky or greasy dirt with a mildly alkaline detergent, then wipe it with water, and dry it; 2. Replace the insulating parts immediately in case of air checks, cracks, rupture and other traces found; 3. Remove foreign bodies from the interior and tighten the bus connecting bolts that are loose or fallen off; 4. Replace the heat-shrinkable sleeves timely in case of any crack, discoloration and/or shedding; 5. Identify the cause for any discharge and then clean or replace the components; 6. Replace damaged components and keep a record of non-critical damaged components that cannot be immediately replaced for maintenance

Continued

S/N	Procedure	Inspection items	Maintenance content
5	Inspection and maintenance of circuit breaker compartment	1. Check the handcart for any debris, and the interior wall of the handcart for any ablation and peculiar smell; 2. Check the contact box for any cracks and signs of discharge ablation, and the static contact terminal for any oxydation and ablation; 3. Check the valve interlocking mechanism for reliability, the nylon roller for deformation or damage, and the shaft pin of each joint part for completeness or damage; 4. Check the flatness of the handcart slideway for deformation; 5. Check whether the handcart grounding contact terminal is well grounded; 6. Check whether the position of the handcart and the interlocking of the aviation plug are normal	1. Clean the contact box and the surface of the static contact terminal using a soft cloth with a mildly alkaline detergent, then wipe it with water, dry it and apply vaseline to the surface of the contact terminal. Wipe the silver-coated surface of the contact terminal lightly with a cloth dampened with alcohol. Replace the contact terminal if the ablation depth of the ablative surface on the contact terminal exceeds 1 mm. 2. Wipe the insulating part with a clean soft cloth to ensure the surface cleanliness. 3. Lubricate parts requiring flexible rotation with molybdenum disulfide. 4. Remove the internal foreign bodies, and use a file to remove from the slideway any burrs that hinder the normal movement of the handcart; 5. Replace damaged components and keep a record of non-critical damaged components that cannot be immediately replaced for maintenance
6	Inspection and maintenance of cable compartment	1. Check the inside for foreign bodies; 2. Check whether the connecting bolts and equipment retaining bolts are missing, loose or falling off; 3. Check whether the closing of the grounding knife switch is flexible and in place, whether the contact terminal and the contact finger are ablative, and whether the spring elasticity is good; 4. Check whether the connecting pins of each rotating joint are complete, and whether the cotter pins and circlip pins of the fixed pins are falling off or lost; 5. Check the support insulator, insulating sheath, and heat-shrinkable sleeve for condensation, discharge ablation, cracking, and falling off; 6. Check whether the heater works properly; 7. Check the appearance of electrical equipment such as current transformer and lightning arrester for any signs of discoloring, crack and ablation, and the wiring for secure connection; 8. Check whether the cable plugging mud is intact; 9. Check whether the phase sequence identification is complete	1. Tighten loose bolts, and make up missing bolts and pins; 2. Clean the surface of contact terminal and contact finger of the grounding knife switch using a soft cloth with mildly alkaline detergent, then wipe it with water, dry it, and apply vaseline to the surface; 3. 4. Clear out the debris in the cabinet, check the source of the left items such as bolts, take out the useless, and reset the useful; 5. Identify the causes for any traces of discharge or flashover; 6. Replace damaged components and keep a record of non-critical damaged components that cannot be immediately replaced for maintenance

Continued

S/N	Procedure	Inspection items	Maintenance content
7	Inspection and maintenance of instrument compartment	1. Check whether the LV circuit breakers of energy storage, control, heating and lighting power supply works normally; 2. Check whether the terminal strip is loose and whether there is foreign matter in the cabinet; 2. Change the position of the handcart, circuit breaker status, and grounding knife switch status, and check the signal, position indicators for correct indication and damage; 3. Check whether the secondary wire residue is loose or peeling; 4. Check whether the temperature and humidity controller works properly. 5. Check if the temperature and humidity controller is functioning properly	1. Clean the LV compartment and keep it clean; 2. Tighten the terminal strip and secondary wire. Note: If there is "Disconnect the power supply for control circuit and energy storage circuit of the equipment to be overhauled" and hang "No Closing, Work in Progress" in the work ticket, the power supply of control circuit and energy storage circuit can be switched on only with the consent of the licensor
8	Inspection and maintenance of handcart	1. Check the appearance of the handcart for any anomaly; 2. Check the conductive circuit for secure connection, and the connecting bolt between the contact terminal and the contact arm for fastening; 3. Check the chassis roller for any deformation; 4. Check whether the insulating parts are clean and in good condition to see if there are traces of discharge and flashover; 5. Check the plum contact terminal for any oxidation and ablation marks; 6. Check the clamping spring of the plum contact terminal for good elasticity; 7. Check whether the circuit breaker nameplate is intact, whether the opening/closing and energy storage identification are clear, and whether the panel is not deformed or damaged	1. Clean the surface of contact terminal using a soft cloth with mildly alkaline detergent, then wipe it with water, dry it, and apply medical vaseline to the surface of contact terminal; 2. Use a clean, soft cloth to wipe the insulation components, ensuring the cleanliness of the surface. 3. Lubricate the contact surfaces and equipment that produce rolling and sliding friction
9	Restore the site to its initial state	1. Close all the cabinet doors; 2. Cut off the power supply for control circuit and closing circuit; 3. Hang the sign board "No Closing, Work in Progress" back to the control circuit and closing circuit	
10	End of work	1. Clean up the work site, and make sure that elements and materials are fully utilized and the site is cleaned; 2. Convene a post-shift meeting; 3. Terminate the work ticket	It is strictly forbidden for the person in charge to clean up the site while terminating the work ticket

任务二　KYN28-12 高压开关柜"五防"联锁检查

一、工作任务

按照规范的作业流程以及工艺要求，对 KYN28-12 高压开关柜手车进行机械连锁检查，掌握"五防"闭锁装置的基本结构、修前准备、危险点预控、作业步骤、工艺要求及质量标准等操作技能。

二、引用标准

（1）《电气装置安装工程高压电器施工及验收规范》（GB 50147—2010）。
（2）《国家电网公司五项通用制度变电检修管理规定》（第 5 分册开关柜检修细则）。
（3）《国家电网公司变电运维管理规定》。
（4）《国家电网公司变电检修管理规定》。
（5）《国家电网公司电力安全工作规程》（变电部分）。
（6）《四川省电力公司高压开关柜现场维护导则》。

三、工作要求

（1）作业为室内停电作业，无天气要求。
（2）被检修设备与其他带电设备均应使用围栏隔离，面向通道处设置唯一出入口。
（3）被检修间隔要求母线与出线电缆均停电并且接地。
（4）作业人员精神状态良好，熟悉工作中安全措施、技术措施以及现场工作危险点。
（5）实训现场要求按生产现场规范布置安全措施，并严格执行标准化作业。
（6）作业人员应规范穿戴劳动保护用品，做好安全防护。

四、工作准备

被检修设备一次接线图见图 4-20。

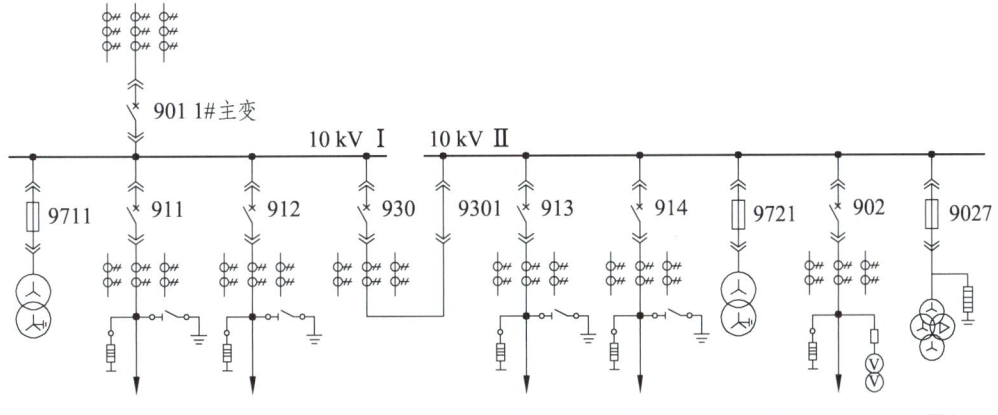

图 4-20　被检修设备一次接线图

项目四　高压开关柜检修　375

（一）危险点及预控措施

1. 高压触电

危险点：10 kV Ⅱ母停电，10 kV Ⅰ母整体视为带电，902 进线柜电缆室视为带电，930 分段柜视为带电，误入带电间隔可能导致高压触电。

预控措施：用围栏将被检修间隔与相邻带电设备（间隔）隔离，并且向作业现场内悬挂适量"止步，高压危险"标识牌，在通道处设置唯一出入口，悬挂"从此进出"标识牌；被检修间隔母线侧与出线电缆需各装设一组三相短路接地线；902 开关柜手车、930 开关柜手车需拉出至"检修"位置；902、930 开关柜门上悬挂"止步，高压危险"标识牌；工作时至少需要两人，一人监护一人操作，听工作负责指挥。

2. 低压触电

危险点：控制回路、合闸回路、照明回路低压断路器进线侧（上端）均低压电；合上空开后，断路器和开关柜二次回路均视为带电状态，触摸可能造成低压触电。

预控措施：若控制、储能低压断路器上悬挂了"禁止合闸，有人工作"标识牌时，工作负责人需要向运行人员申请更变安全措施得到许可后，方能合闸；勿将手指伸入低压断路器上端头；拔出航空插头前，需先将低压电源断开，拔出航空插头后不允许合上低压断路器；未得到负责人的许可，不允许打开断路器面板；工作时至少需要两人，一人监护一人操作，听工作负责指挥。

3. 机械伤害

危险点：断路器分合、接地刀闸分合时未通知其他工作班成员，可能会对工作班成员造成机械伤害；将断路器从"试验"位置转移至"检修"位置或将断路器从"检修"位置转移至"试验"位置时，手车可能发生倾倒，对操作者造成机械伤害。

预控措施：断路器分合、接地刀闸分合时，需要大声呼唱，得到工作班所有人员大声回应之后，方可操作；操作接地刀闸前，需要特别注意开关柜下柜门内是否有人员正在操作；未得到负责人的许可，不允许打开断路器面板；将断路器从"试验"位置转移至"检修"位置或将断路器从"检修"位置转移至"试验"位置时，需要用锁扣将转运小车锁在开关柜上，防止小车倾倒；工作时至少需要两人，一人监护一人操作，听工作负责指挥。

4. 设备损坏

危险点：未按照要求分合断路器、分合接地刀闸、野蛮操作，或者操作时遗留工具在柜体内，均会对设备造成不同程度的损坏。

预控措施：禁止野蛮操作，操作过程中，禁止遗留任何工具于柜内，若出现不能操作的现象，应当立即停止，不得使用蛮力，立即汇报负责人；工作时至少需要两人，一人监护一人操作，听工作负责指挥。

（二）危险点及预控措施

KYN28-12 高压开关柜手车机械联锁检查及维护工器具及材料见表 4-8。

表 4-8 KYN28-12 高压开关柜手车机械联锁检查工器具及材料

类别	名称	规格型号	数量	备注
专用工具	地刀操作把手		1 把	
	储能把手		1 把	
	手车把手		1 把	
通用工具	内六角扳手		1 套	
	平口钳		1 把	
	尖嘴钳		1 把	
	一字螺丝刀	150 mm	1 把	
	十字螺丝刀	150 mm	1 把	
	活动扳手	200 mm	1 把	
材料	二硫化钼润滑脂		0.5 kg	

（三）作业人员分工

KYN28-12 高压开关柜手车机械联锁检查人员分工见表 4-9。

表 4-9 KYN28-12 高压开关柜手车机械联锁检查人员分工

序号	工作岗位	数量	职责
1	工作负责人	1	负责本次工作的人员分工、现场查勘、作业方案制定、召开班前会、作业过程中安全监督、工作中突发状况的处理、工作质量的监督、班后会总结
2	操作人员	1	负责机械连锁检查的主要操作
3	辅助操作人员（可无）	1	辅助操作人员进行机械连锁的检查

五、作业程序

KYN28 高压开关柜手车机械联锁检查作业流程见表 4-10。

表 4-10 KYN28-12 高压开关柜手车机械联锁检查作业流程

序号	作业内容	作业步骤及标准	安全措施及注意事项
1	工作前准备工作	1. 检查工器具是否齐全，检查工器具外观和试验合格； 2. 工作负责人同工作许可人巡视待检修设备，确认工作票所列安全措施已经正确执行，安全措施是否完备，现场是否具备开工条件，必要时进行补充； 3. 执行工作许可手续； 4. 对工作班成员召开班前会； 5. 抄写设备铭牌参数	1. 工器具无损伤、变形、失灵现象，需要试验的工器具合格证在有效期内； 2. 巡视现场时禁止无关人员进入现场； 3. 班前会应包含工作地点双重名称；工作时间与内容；工作分工；工作危险点及预控措施；停电范围及工作现场安全措施； 4. 全体工作成员应当正确穿戴安全帽、工作服、工作鞋、劳保手套等劳动保护用品

续表

序号	作业内容	作业步骤及标准	安全措施及注意事项
2	确认设备状态	负责人带领工作班成员确认设备处于分闸未储能，接地刀闸处于合位	
3	检查闭锁电磁铁	1. 合上控制回路、储能回路电源； 2. 对断路器进行储能； 3. 断开控制回路、储能回路电源； 4. 打开断路器室柜门，按下"手动合闸"按钮	1. 工作票中若有"断开待检修设备控制回路、储能回路电源并悬挂'禁止合闸，有人工作'"这一项，则必须经许可人同一后，方可合控制回路、储能回路电源； 2. 手动合闸应当闭锁
4	检查地刀对手车闭锁	1. 关闭断路器室柜门； 2. 将手车从"试验位置"摇入至"工作位置"	1. 摇入时应当闭锁，闭锁时切忌大力摇动把手； 2. 检查完毕时，应摇回初始位置
5	检查断路器对手车闭锁（"试验位置"）	1. 分开接地刀闸； 2. 电动合闸断路器； 3. 将手车从"试验位置"摇入至"工作位置"	1. 摇入时应当闭锁，闭锁时切忌大力摇动把手； 2. 检查完毕时，应摇回初始位置
6	检查断路器室柜门对手车闭锁	1. 打开断路器室柜门； 2. 分开断路器； 3. 将手车从"试验位置"摇入至"工作位置"	1. 摇入时应当闭锁，闭锁时切忌大力摇动把手； 2. 检查完毕时，应摇回初始位置
7	检查"中间位置"对合闸闭锁	1. 手动解除柜门对手车闭锁； 2. 将手车从"试验位置"摇入至"中间位置"； 3. 储能； 4. 手动合闸断路器	手动合闸应当闭锁
8	检查"中间位置"对地刀闭锁	拉下接地刀闸挡板	接地刀闸挡板应当闭锁
9	检查"中间位置"对手车	解锁并拉动手车把手	手车把手应当闭锁
10	检查断路器对手车闭锁（"工作位置"）	1. 将手车从"中间位置"摇入至"工作位置"； 2. 手动/电动合闸断路器； 3. 将手车从"工作位置"摇出至"试验位置"	1. 摇出时应当闭锁，闭锁时切忌大力摇动把手； 2. 检查完毕时，应摇回初始位置
11	检查"航空插头"闭锁	1. 手动/电动分闸断路器； 2. 断开控制回路、储能回路电源； 3. 拔下"航空插头"	航空插头应当闭锁无法拔下

续表

序号	作业内容	作业步骤及标准	安全措施及注意事项
12	检查手车对断路器室柜门闭锁	1. 合上控制回路、储能回路电源； 2. 将手车从"工作位置"摇出至"试验位置"； 3. 关闭断路器室柜门； 4. 将手车从"工作位置"摇入至"中间位置"； 5. 打开断路器室柜门	断路器室柜门应当闭锁
13	检查接地刀闸对电缆室柜门闭锁	1. 将手车从"中间位置"摇出至"试验位置"； 2. 打开前下跪门、后柜门	前下柜门、后柜门应当闭锁
14	检查后柜门对接地刀闸闭锁	1. 合上接地刀闸； 2. 打开后柜门； 3. 分开接地刀闸	接地刀闸应当闭锁，闭锁时切忌大力操作接地刀闸
15	现场恢复至初始状态	1. 关闭所有柜门； 2. 断开控制回路、合闸回路电源； 3. 将"禁止合闸，有人工作"标识牌挂回控制回路、合闸回路处	
16	工作终结	1. 清理作业现场，做到"工完料尽场地清"； 2. 召开班后会； 3. 终结工作票	严禁负责人前去终结工作票的同时清理场地

Task 2 "Five-prevention" Interlocking Inspection of KYN28-12 HV Switch Cabinet

2.1 Work Tasks

Following the standard operation flow and process requirements, students are expected to carry out mechanical interlocking inspection of KYN28-12 HV switch cabinet handcart, and master the basic structure, repair preparation, hazard pre-control, operation procedures, process requirements and quality standards of the "five-prevention" locking device.

2.2 References

(1) *Code for Construction and Acceptance of High-voltage Electrical Apparatus of Electric Equipment Installation Engineering* (GB 50147-2010).

(2) *Five General Regulations of State Grid Corporation of China on Substation Maintenance Management* (*Volume 5: Rules for Maintenance of Switch Cabinets*).

(3) *Regulations of State Grid Corporation of China on Management of Substation Operation and Maintenance.*

(4) *Substation Maintenance Management Regulations of State Grid Corporation of China.*

(5) *Electric Power Safety Working Regulations (Power Transformation) of State Grid Corporation.*

(6) *Field Maintenance Guidelines for High-voltage Switch Cabinets of Sichuan Electric Power Company.*

2.3 Work Requirements

(1) The operation belongs to indoor power interruption operation, without weather requirements.

(2) Equipments under maintenance are to be isolated from other electrified equipments by a fence. The only exit is to be provided at the place oriented towards the passage.

(3) For the maintenance bay, the bus and outlet cable are required to be powered off and grounded.

(4) Operators are to be in good mental state, and are aware of safety measures, technical measures and hazards to site operation during operation.

(5) Take safety measures on practical training site as per regulations on production site, and strictly implement standard operation.

(6) Operators are requested to wear labor protection appliances to ensure safety protection.

2.4 Preparation for Work

Primary wiring diagram of the equipment under maintenance is shown in Fig. 4-20.

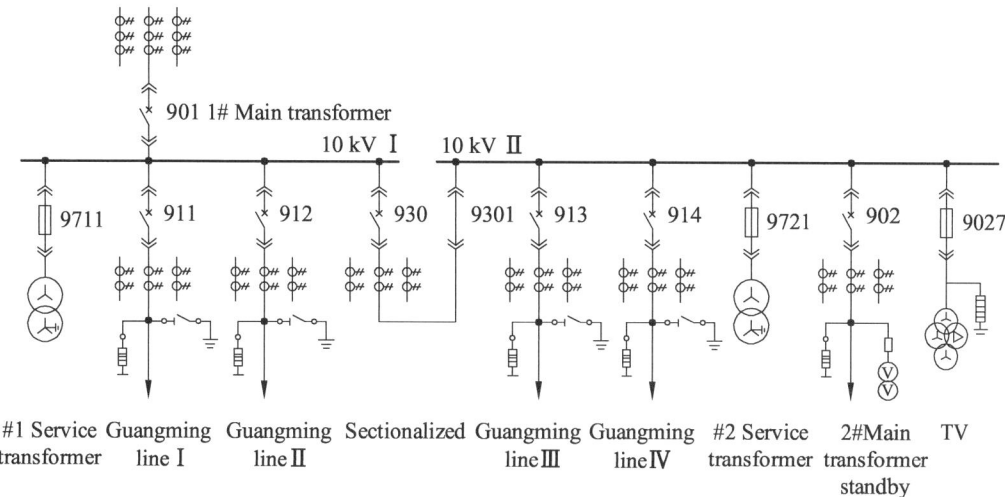

Fig. 4-20   Primary Wiring Diagram of the Equipment under Maintenance

2.4.1   Hazards and Preventive and Control Measures

1. High-voltage electric shock

Hazard: When 10kV Bus II is de-energized, 10 kV Bus I is considered energized as a whole, 902 incoming cable room is considered energized, and 930 sectionalizer is considered energized. Entering the energized area by mistake may result in high-voltage electric shock.

Preventive and control measures: Use fence to isolate maintenance interval from adjacent live equipments, and provide such sign boards as "Stop! High Voltage, Danger!" on operation site. The only exit is to be set at the passage, which is to be provided with the sign board indicating "Entrance/Exit"; At least two persons are required during work, one for supervision and the other for operation under instructions of person in charge of work.

2. Low-voltage electric shock

Hazard: There is low-voltage current in the incoming side (upper end) of low-voltage circuit breaker of control circuit, closing circuit and lighting circuit. After closing the air switch, the secondary circuits of circuit breaker and switch cabinet are all considered as live circuits, and touching the circuits may cause low-voltage electric shock.

Preventive andcontrolmeasures: If there is a "Do Not Close, Work in Progress" sign hanging on the control and energy storage low-voltage circuit breaker, the work supervisor must obtain permission from the operating personnel to change safety measures before closing the breaker. Do not insert fingers into the upper terminals of the low-voltage circuit breaker. Before unplugging the aviation plug, the low-voltage power supply must be disconnected first, and after unplugging the aviation plug, it is not allowed to close the low-voltage circuit breaker. Without the permission of the supervisor, do not open the circuit breaker panel.

At least two persons are required during work, one for supervision and the other for operation under instructions of person in charge of work.

3. Mechanical injury

Hazard: During opening and closing of circuit breaker and grounding knife switch, if notification is not given to other members of the work team, the members may suffer mechanical injury; When the circuit breaker is moved from the "Test" position to the "Maintenance" position or the circuit breaker is moved from the "Maintenance" position to the "Test" position, the handcart may be tripped, causing mechanical injury to the operator.

Preventive andcontrolmeasures: When operating the circuit breaker to open or close and the grounding knife switch to open or close, it is necessary to call out loudly and only proceed with the operation after receiving a loud response from all personnel in the work shift. Before operating the grounding knife switch, special attention must be paid to whether there are personnel inside the cabinet door of the switchgear. Without the permission of the supervisor, do not open the circuit breaker panel. When moving the circuit breaker from the "Test" position to the "Maintenance" position or from the "Maintenance" position to the "Test" position, use a lock to secure the transport trolley to the switchgear to prevent the trolley from tipping over.

At least two persons are required during work, one for supervision and the other for operation under instructions of person in charge of work.

4. Equipment damage

Hazard: The equipment may be damaged to varying degrees if opening and closing of circuit breaker and grounding knife switch are not in accordance with requirements, or rough handling is adopted, or tools are left in the cabinet during operation.

Preventive and control measures: Rough handling is prohibited. It is forbidden to leave any tool in the cabinet during operation. If there is a phenomenon that operation cannot be given during the operation, the operation shall be stopped immediately, no brute force shall be used, and report shall be sent to the person in charge immediately; At least two persons are required during work, one for supervision and the other for operation under instructions of person in charge of work.

2.4.2 Hazards and Preventive and Control Measures

Tools and materials for mechanical interlocking inspection of kyn28-12 HV switch cabinet handcart see Table 4-8.

Table 4-8 Tools and Materials for Mechanical Interlocking Inspection of KYN28-12 HV Switch Cabinet Handcart

Category	Name	Specification and model	Quantity	Remarks
Specialized tools	Grounding knife-switch operating handle		1	
	Energy storage handle		1	
	Handcart handlebar		1	
General tools	Allen wrench		1 set	
	Flat-nose pliers		1	
	Long-nose pliers		1	

Continued

Category	Name	Specification and model	Quantity	Remarks
General tools	Slotted screwdriver	150 mm	1	
	Cross screwdriver	150 mm	1	
	Monkey wrench	200 mm	1	
Material	Molybdenum disulfide grease		0.5 kg	

2.4.3 Division of labor among operators

See Table 4-9 for the personnel allocation for mechanical interlocking inspection of KYN28-12 HV switch cabinet handcart.

Table 4-9 Personnel Allocation for Mechanical Interlocking Inspection of KYN28-12 HV Switch Cabinet Handcart

S/N	Job	Quantity	Responsibilities
1	Person in charge of work	1	Responsible for the personnel allocation, site survey, stipulation of operation scheme, pre-shift meeting, safety supervision during operation, handling of emergencies during work, supervision of work quality and summary of post-shift meeting
2	Operator	1	Responsible for main operation of mechanical interlocking inspection
3	Auxiliary operators (optional)	1	Responsible for mechanical interlocking inspection

2.5 Operation Procedures

See Table 4-10 for the procedures for mechanical interlocking inspection of KYN28 centrally installed switchgear handcart.

Table 4-10 Procedures for Mechanical Interlocking Inspection of KYN28-12 HV Switch Cabinet Handcart

S/N	Scope of work	Operational steps and standards	Safety measures and precautions
1	Preparations before work	1. Check if all instruments and tools are complete, and if their appearance and test are acceptable; 2. The person in charge of work and work permitter shall make an tour inspection for equipments under maintenance, and confirm all safety measures as listed by work ticket have been properly implemented. It is also necessary to check if safety measures are complete, and if the site is provided with conditions for commencement of work, and make supplements as required; 3. Implement work permit procedures; 4. Call in toolbox meeting participated by members of work team; 5. Record parameters on the equipment nameplate	1. Instruments and tools are free of damage, deformation and malfunction. Qualification certificates for instruments and tools to be tested are within the term of validity; 2. Prevent other persons from entering the site during tour inspection; 3. Toolbox meeting shall cover dual designations of work place; Working hours and contents; Work division; Working hazards as well as preventive and control measures Power-cut scope and safety measures on work site; 4. All work members shall properly wear such labor protection appliances as safety helmet, working clothes, working shoes and protective gloves

Continued

S/N	Scope of work	Operational steps and standards	Safety measures and precautions
2	Confirm equipment status	The supervisor leads the team members to confirm that the equipment is in the "open" position without energy storage, and the grounding knife switch is in the "closed" position	
3	Check the latching electromagnet	1. Switch on the power supply for control circuit and energy storage circuit; 2. Carry out energy storage for the circuit breaker; 3. Cut off the power supply for control circuit and energy storage circuit; 4. Open the cabinet door of circuit breaker compartment and press the "Manual Closing" button	1. If there is "Disconnect the power supply for control circuit and energy storage circuit of the equipment to be overhauled" and hang "No Closing, Work in Progress" in the work ticket, the power supply of control circuit and energy storage circuit can be switched on only with the consent of the licensor; 2. Manual closing shall be latched
4	Check the latching of handcart by the grounding knife-switch	1. Close the cabinet doors of the circuit breaker compartment; 2. Joggle the handcart from the "Test" to the "Work" position	1. It should be latched when joggled, and do not shake the handle vigorously when latched; 2. After inspection, joggle back to the initial position
5	Check the latching of circuit breaker to the handcart ("Test" position)	1. Switch off the grounding knife switch; 2. Close the circuit breaker electrically; 3. Joggle the handcart from the "Test" to the "Work" position	1. When rocking in, it should be locked, and when locked, avoid vigorously shaking the handle; 2. After completing the inspection, rock it back to the initial position
6	Check the latching of the cabinet door of circuit breaker compartment to the handcart	1. Open the cabinet doors of the circuit breaker compartment; 2. Separate the circuit breakers; 3. Rock the handcart from the "Test Position" to the "Operating Position"	1. When rocking in, it should be locked, and avoid vigorously shaking the handle when locked; 2. After the inspection is complete, it should be rocked back to the initial position
7	Check the latching of "Middle" position to closing	1. Manually remove the latching of the cabinet door to the handcart; 2. Joggle the handcart from the "Test" to the "Middle" position; 3. Energy storage; 4. Close the circuit breaker manually	Manual closing should be locked
8	Check the latching of "Middle" position to the grounding knife switch	Switch on the grounding knife switch baffle	The grounding knife switch baffle shall be latched
9	Check the "Middle" position to the handcart	Unlock and pull the handlebars	Handlebars should be latched
10	Check the latching of circuit breaker to the handcart ("Work" position)	1. Joggle the handcart from the "Middle" to the "Work" position; 2. Do manual/electric closing of the circuit breaker; 3. Joggle the handcart from the "Work" to the "Test" position	1. It should be latched when joggled, and do not shake the handle vigorously when latched; 2. After the inspection is complete, it should be rocked back to the initial position

Continued

S/N	Scope of work	Operational steps and standards	Safety measures and precautions
11	Check the latching of "aviation plug"	1. Do manual/electric opening of the circuit breaker; 2. Cut off the power supply for control circuit and energy storage circuit; 3. Unplug the "aviation plug"	The aviation plug should be latched and cannot be unplugged
12	Check the latching of the handcart to the cabinet door of circuit breaker compartment	1. Switch on the power supply for control circuit and energy storage circuit; 2. Joggle the handcart from the "Work" to the "Test" position; 3. Close the cabinet door of circuit breaker; 4. Joggle the handcart from the "Work" to the "Middle" position; 5. Open the cabinet doors of the circuit breaker compartment	The cabinet doors of the circuit breaker compartment should be latched
13	Check the latching of the grounding knife switch to the cabinet doors of the cable compartment	1. Joggle the handcart from the "Middle" to the "Test" position; 2. Open the front lower and the rear cabinet doors	The front lower and the rear cabinet doors should be latched
14	Check the latching of the rear cabinet door to the grounding knife switch	1. Close the grounding knife switch; 2. Open the rear cabinet door; 3. Switch off the grounding knife switch	The grounding knife switch should be latched, and do not operate the grounding knife switch vigorously when latched
15	Restore the site to its initial state	1. Close all the cabinet doors; 2. Cut off the power supply for control circuit and closing circuit; 3. Hang the "No Closing, Work in Progress" sign board back to the control circuit and closing circuit	
16	Work completion	1. Clean up the work site, and make sure that elements and materials are fully utilized and the site is cleaned; 2. Convene a post-shife meeting; 3. Terminate the work ticket	It is strictly prohibited for the responsible person to clean the site while closing out the work permit.

项目 五 组合电器检修

模块一 组合电器基础知识

一、六氟化硫组合电器的概念

六氟化硫组合电器又被称为气体绝缘全封闭组合电器（Gas Insulated Substation, GIS）。它将断路器、隔离开关、母线、接地开关、互感器、出线套管或电缆终端头等分别装在各自密封间中，以金属筒为外壳，集中组成一个整体，内部充以一定压力的六氟化硫气体作为绝缘介质。

GIS 根据安装地点可分为户外式和户内式两种（见图 5-1、5-2）。

图 5-1　户外 GIS 设备

图 5-2　户内 GIS 设备

二、六氟化硫组合电器的特点及适用范围

1. GIS 主要优点

（1）可靠性高。

由于带电部分全部封闭在 SF_6 气体中，不会受到外界环境的影响。

（2）安全性高。

由于 SF_6 气体具有很高的绝缘强度，并为惰性气体，不会产生火灾；带电部分全部封闭在接地的金属壳体内，实现了屏蔽作用，也不存在触电的危险。

（3）占地面积小。

由于采用具有很高的绝缘强度 SF_6 气体作为绝缘和灭弧介质，使得各电气设备之间、设备对地之间的最小安全净距减小，从而大大缩小了占地面积。

（4）安装方便。

组合电器可在制造厂家装配和试验合格后，再以间隔的形式运到现场进行安装，工期大大缩短。

（5）维护方便，检修周期长。

因其结构布局合理，灭弧系统先进，大大提高了产品的使用寿命，可长达 30 年，因此检修周期长，维修工作量小，而且由于小型化，离地面低，因此日常维护方便。

2. GIS 主要缺点

（1）密封性能要求高。

装置内 SF_6 气体压力的大小和水分的多少会直接影响整个装置运行的性能和人员的安全性，因此，GIS 对加工的精度有严格的要求。

（2）价格较昂贵。

GIS 将除变压器以外的所有电气设备安装与铝合金材料的壳体内，金属消耗量大，造价高。

（3）故障后危害较大。

首先，故障发生后造成的损坏程度较大，有可能使整个系统遭受破坏。其次，检修时有毒气体（SF_6 气体与水发生化学反应后产生）会对检修人员造成伤害。

3. 适用范围

近年来为了减少占地面积，六氟化硫全封闭组合电器得到了广泛应用，目前，我国的 GIS 使用的起始电压为 110 kV 及以上，主要在以下场合使用：

（1）占地面积较小的地区，如市区变电站。

（2）高海拔地区或高烈度地震区。

（3）外界环境较恶劣的地区。我国西北电网建设的 750 kV 工程，采用的 GIS 组合电器已在变电站投入运行。

三、六氟化硫组合电器的结构形式

1. GIS 设备罐体结构形式

（1）全三相共箱式。

不仅三相母线，而且三相断路器和其他电气元件采用共箱罐体。

（2）不完全三相共箱式。

母线采用三相共箱式，而断路器和其他电气元件采用分箱式。

（3）全分箱式。

包括母线在内的所有电气元件都采用分箱式罐体。

2. GIS 设备出线方式

（1）架空线引出方式。

GIS 设备通过充气套管与架空线相接。

（2）电缆引出方式。

GIS 设备通过电缆终端与电力电缆相连，密封垫的一侧为 SF_6 气体，另一侧为电缆油。

（3）GIS 设备出线端直接与主变压器对接。

GIS 设备经油气套管直接与主变压器连接，此时连接套管的一侧充有 SF_6 气体，另一侧则充有变压器油。

四、组合电器的结构

（一）GIS 总体结构

GIS 一般由实现各种不同功能的单元组成，称间隔，主要有进（出）线间隔、PT 间隔、母联间隔、母线计量保护间隔等；并根据用户的不同要求实现单母线、单母线分段、双母线、桥形接线等不同的接线方式（见图 5-3）。

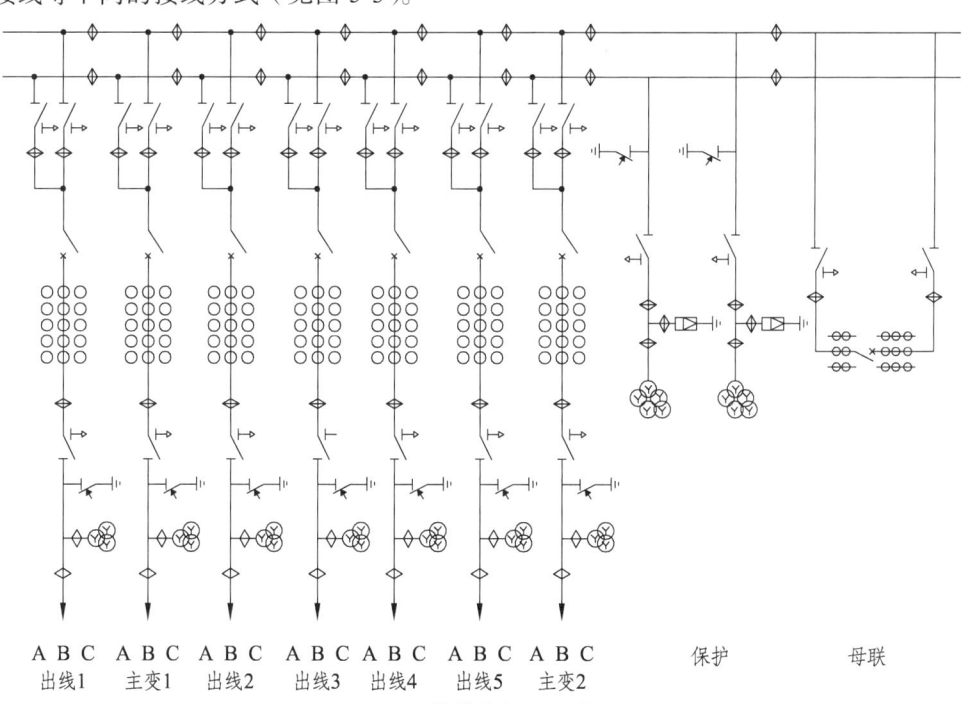

图 5-3　双母线接线间隔一次原理图

如图 5-4 所示，该 GIS 采用双母线接线，由 5 个出线间隔、2 个主变进线间隔、1 个母联间隔、2 个母线计量保护间隔组成。

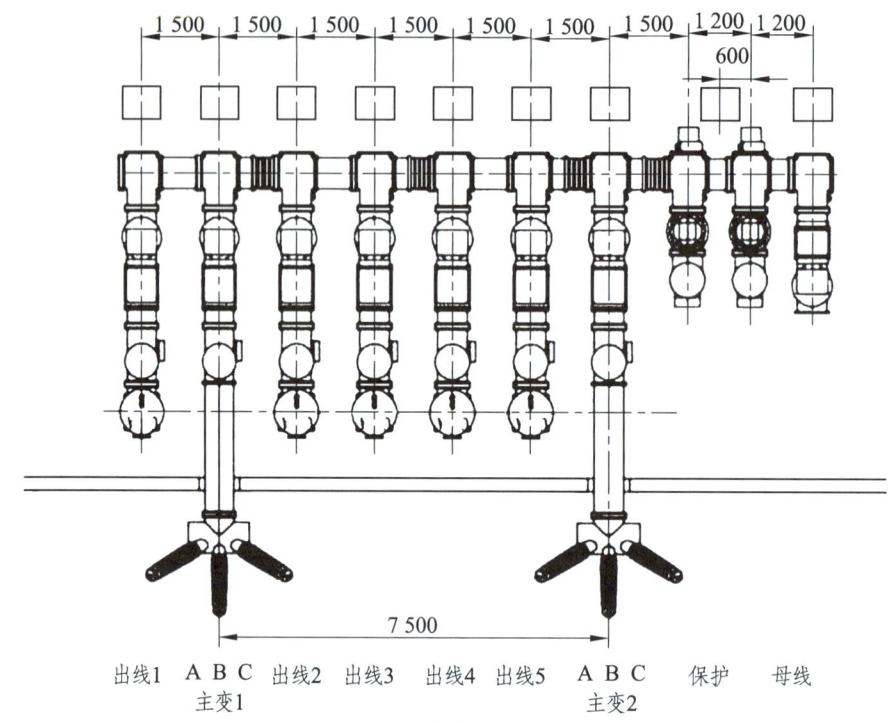

出线1　A B C　出线2　出线3　出线4　出线5　A B C　保护　母线
　　　　主变1　　　　　　　　　　　　　　主变2

图 5-4　双母线接线间隔断面图

GIS 的每一个间隔，用不通气的盆式绝缘子（气隔绝缘子）划分为若干个独立的 SF_6 气室，即气隔单元。各独立气室在电路上彼此相通，而在气路上则相互隔离。一般断路器压力高，它和电流互感器组成一个气室；主母线、电压互感器、避雷器分别为独立的气室，其他元件根据工程确定气室划分。各气室分别由相应的密度控制器监测气体压力。

GIS 一般每间隔设有一个就地控制柜，各元件控制、状态信号，各气室密度监测信号，以及电压、电流互感器二次出线全部引到就地控制柜，并通过就地控制柜与主控室相连。

如图 5-5 所示为单母线架空进（出）线间隔，该间隔部件组成包括：汇控柜、断路器、隔离开关、接地开关、母线筒组件、电流互感器、连接筒体、进出线套管、支架、机架等。

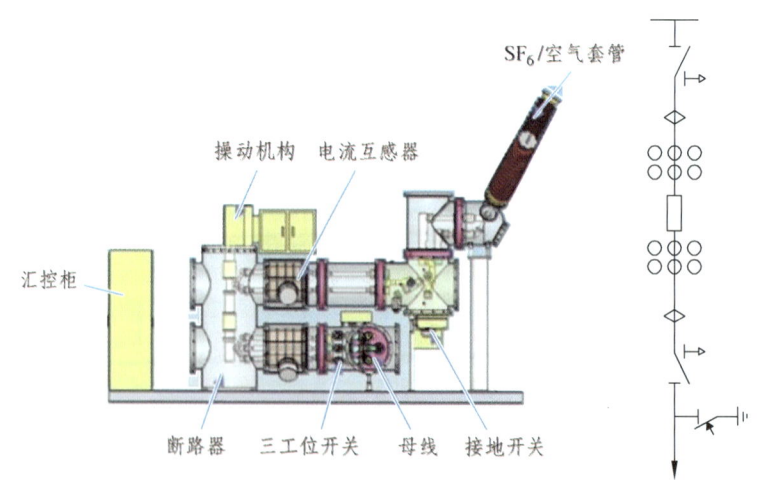

图 5-5　架空进（出）线间隔断面图与一次原理图

如图 5-6 所示为母联间隔，起到连接两组母线的作用，该间隔部件组成包括：汇控柜、断路器、隔离开关、母线筒组件、电流互感器、连接筒体、支架、机架等。

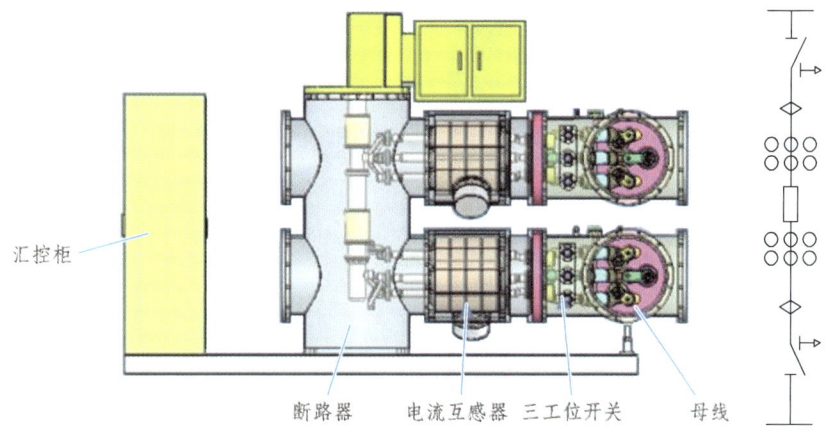

图 5-6　母联间隔断面图与一次原理图

如图 5-7 所示为母线 PT 间隔，起到计量和防雷保护的作用，该间隔部件组成包括：汇控柜、隔离开关、母线筒组件、电压互感器、连接筒体、支架、机架等。

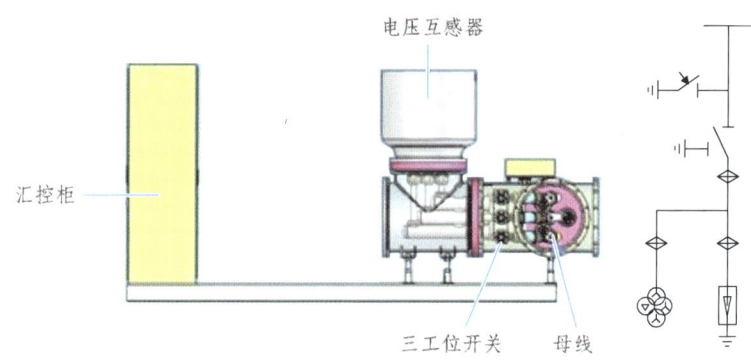

图 5-7　母线 PT 间隔断面图与一次原理图

（二）GIS 部件结构

1. 汇控柜

汇控柜（见图 5-8）又叫就地控制柜，既是 GIS 间隔内、外各元件之间进行电气联络的中继枢纽，也是对 GIS 设备进行现场控制、监视以及进行遥测、遥控、遥调、遥信的集中枢纽，对电设备的正常运行起着非常重要的作用。

GIS 的就地控制柜一般具有就地操作、信号传输、保护和中继、对 GIS 各间隔气室进行监视等功能，主要功能如下：

（1）对间隔内一次设备如断路器、三工位开关、快速接地开关等实施就地远方选择操作。既可实现在控制柜上对上述一次设备进行就地操作，又可在 GIS 正常运行时改为远方操作。

（2）监视断路器、隔离开关、接地开关的分合闸位置状态。

（3）监视各气室 SF_6 气体密度是否处于正常状态。

图 5-8 汇控柜内外结构图

（4）监视断路器储能弹簧的储能状态。

（5）监视控制回路电源是否正常。

（6）显示 GIS 一次电气设备的主接线形式及运行状态。

（7）实现断路器、隔离开关、接地开关之间的电气联锁及间隔与间隔之间的电气联锁。

（8）监测 GIS 设备机构箱及端子箱内的温湿度并自动投入加热除湿装置。

（9）作为 GIS 各元件间及 GIS 与主控室之间控制、信号的中继端子箱接收和发送信号。

2. 断路器

断路器是 GIS 中主要元件，用于输变电线路中，作为电力系统的控制和保护设备。

（1）断路器本体。

如图 5-9 所示，为三相共箱罐式结构，三相共用一台弹簧操动机构，机械联动。动触座通过绝缘台固定在罐顶，由动触座、中间触头、导向等组成，对动触头起支持、导向、导电的作用；动触头由喷口、动主触头、动弧触头、气缸、拉杆等组成，通过绝缘拉杆与机构相连，在弹簧机构的带动下实现分合闸操作。静触头通过绝缘子与动触头相连，包括静触座、静主触头、静弧触头、屏蔽罩等组成。主导电回路为：动触座上的梅花触头—动触座—中间触头—气缸—动主触头—静主触头—静触座—静触座上的梅花触头。

断路器外形及灭弧室装配见图 5-10。

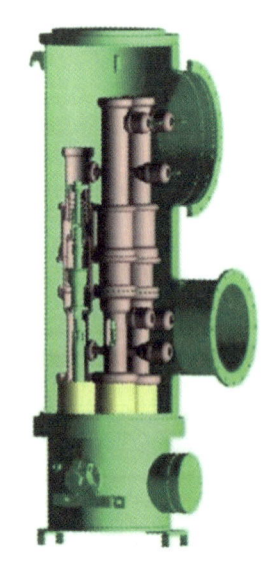

图 5-9 断路器内部剖面图

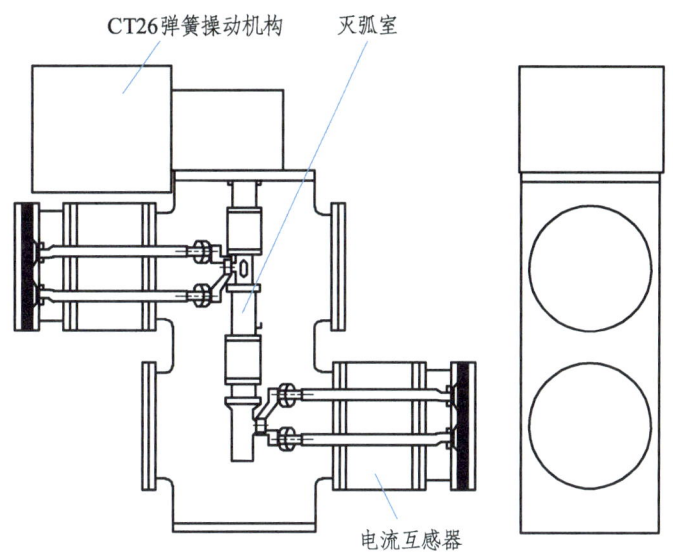

图 5-10 断路器外形及灭弧室装配示意图

如图 5-11 所示,断路器采用"热膨胀+助吹"的自能式灭弧结构。当开断短路电流时,电弧在动静弧触头间燃烧,巨大的能量加热膨胀室内的 SF_6 气体使温度升高,膨胀室内气体压力随之升高,产生内外压差;当动触头分闸达到一定位置,静弧触头拉出喷口,产生强烈气吹,在电流过零点时熄灭电弧。开断过程中,由于电弧能量大,膨胀室内压力高于辅助压气室内压力上升,膨胀室阀片闭合,压气室阀片打开,压气室压力释放。

当开断小电感、电容电流或负荷电流时,所开断电流小,电弧能量也较小,膨胀室内压力上升比辅助压气室压力上升慢,压气室阀片闭合,膨胀室阀片打开,压缩气体进入膨胀室,产生气吹,在电流过零点时熄灭电弧。

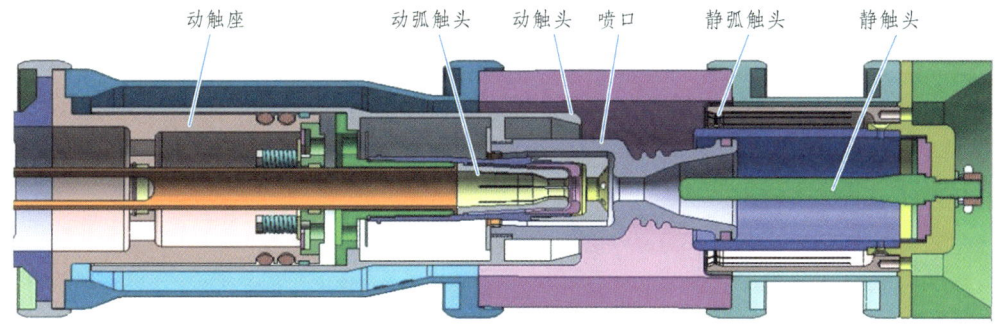

图 5-11 断路器灭弧室结构

(2) 断路器操动机构。

下面以断路器配用 CT26 型弹簧操动机构为例,分析操动机构工作原理。

① 合闸弹簧储能。

合闸弹簧处于预压缩状态,储能拐臂通过蜗轮、蜗杆与电机相连。储能时,电机通过蜗杆带动蜗轮转动,蜗轮通过轴销推动储能拐臂逆时针旋转,通过与储能拐臂键联的储能轴的传动,弹簧拐臂通过与其相连接的销及拉杆拉动合闸弹簧压缩储能,当销越过左侧顶点位置时，合闸弹簧带动储能拐臂逆时针转过约 5º，储能拐臂上的储能保持销被保持挚子扣住，完成储能动作；此时电机电源被切断，棘爪与棘轮相脱离。

② 合闸操作。

断路器处于分闸位置，合闸弹簧已储能。

当机构得到合闸指令，合闸线圈受电，合闸电磁铁的动铁芯吸合带动合闸导杆撞击合闸挚子顺时针旋转，释放储能保持挚子，合闸弹簧带动棘轮逆时针快速旋转，与棘轮同轴的凸轮打击输出拐臂上的合闸滚子，使拐臂向上运动，通过连杆带动断路器本体实现合闸操作，此时输出拐臂上的合闸保持销被合闸保持挚子扣住实现合闸保持；同时与输出拐臂同轴的分闸弹簧拐臂压缩分闸弹簧储能，准备分闸操作。合闸操作也可通过手动撞击合闸电磁铁导杆实现。

合闸操作完成后，行程开关自动投入电机再次对合闸弹簧储能。见图 5-12、5-13。

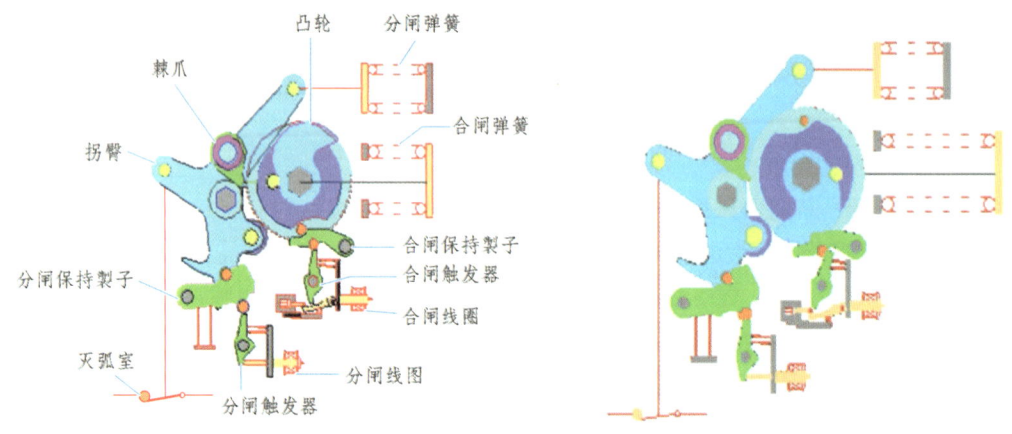

图 5-12 合闸位置（合闸弹簧储能）　　　图 5-13 合闸位置（合闸弹簧释放）

③ 分闸操作。

断路器处于合闸位置，合闸弹簧与分闸弹簧均已储能。机构接到分闸指令，分闸线圈受电，分闸电磁铁动铁芯吸合带动分闸导杆撞击分闸挚子顺时针旋转，释放合闸保持挚子，分闸弹簧释放能量通过拐臂、连杆带动断路器本体实现分闸操作。见图 5-14。

分闸操作也可通过手力撞击分闸电磁铁导杆实现。

图 5-14 分闸位置（合闸弹簧储能）

3. 三工位隔离开关

三工位隔离开关是将隔离开关和接地开关集成在同一模块内,可实现接通、隔离、接地三种工况。接地开关可与外壳隔离,当需要进行继电保护的调整和试验,电缆检查和电缆故障定位、直流电阻测量等工作时,可以通过接地开关的动触头,从外面与 GIS 主回路的导体进行电器连接,极大地方便了试验工作,提高了准确性。

三工位隔离开关配用电动操动机构,机构包括两台驱动电机,通过电机的正反转驱动丝杠转动,丝杠带动驱动螺母做直线运动,驱动螺母通过销轴推动输出轴转动,经齿轮、齿条的转换,实现动触头在接通⟷隔离⟷接地间的往复运动。见图 5-15、5-16、5-17。

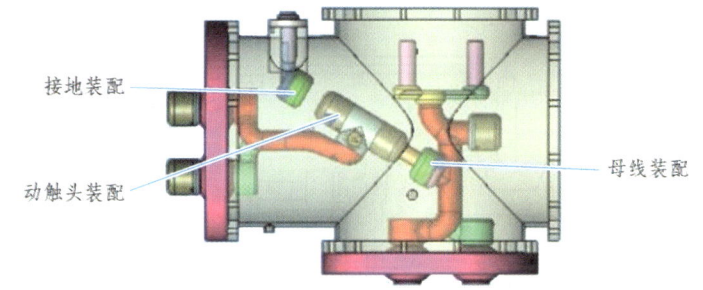

图 5-15　隔离开关处于"接通位置"

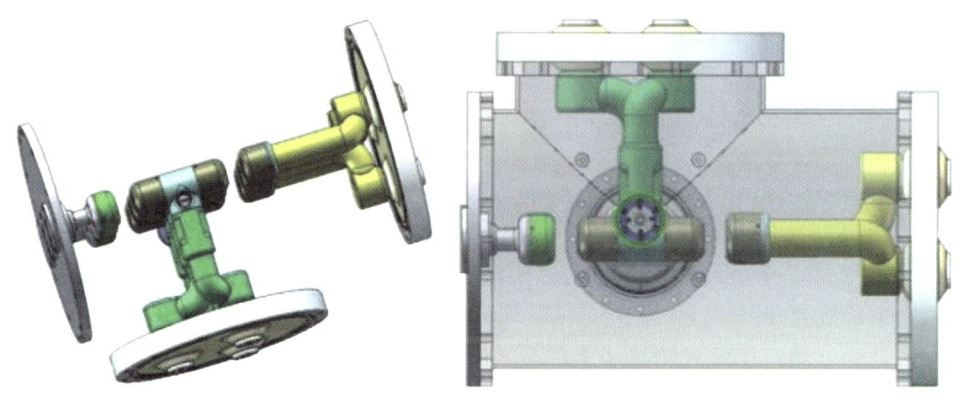

图 5-16　隔离开关处于"隔离位置"

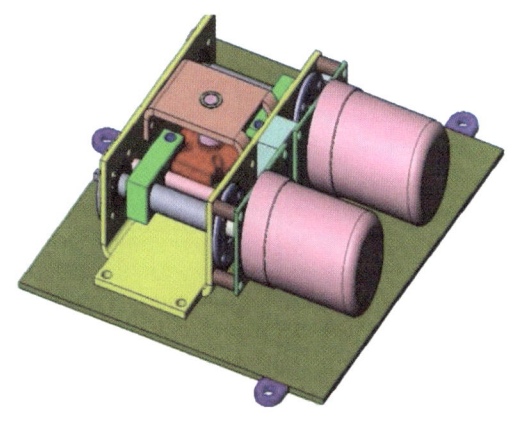

图 5-17　隔离开关电动操动机构

4. 接地开关

除三工位开关内的接地开关（用于在检修时保护安全的工作接地）外，也可单独安装，并有两种形式，即具有关合短路电流及开合感应电流能力的快速接地开关（FES 又称故障关合接地开关）和用于在检修时保护安全的工作接地开关（ES）；工作接地开关配用电动机机构，快速接地开关配用电动弹簧操动机构。接地开关可与工作接地的壳体绝缘断开，当需要进行回路电阻测量、机械特性等试验时，将与接地外壳相连的接地母线拆除，即可通过接地开关的动触头与主回路进行电气连接，极大方便了试验工作。快速接地开关与相关的三工位开关、断路器通过机构进行电气联锁，以防止误操作。见图 5-18。

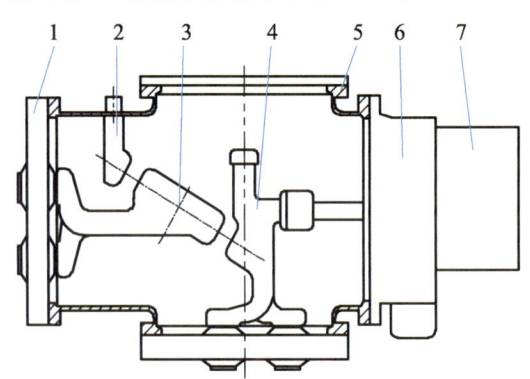

1—盘式绝缘子；2—接地装配；3—动触头装配；4—母线装配；
5—壳体；6—快速接地开关；7—机构。

图 5-18 三工位、快速接地开关示意图

5. 母线

SF_6 气体绝缘母线用于连接 GIS 各种元件，母线可连续通过额定电流及耐受动、热稳定电流。

在共箱 GIS 中，分支母线与主母线具有同样尺寸的外壳，所不同的是分支母线中的导体直径可能比主母线要小。

母线由外壳、固定于盆式绝缘子上的分支导体及三相导电杆组成。

为了吸收热胀冷缩变形和装配误差，在导体连接部分采用梅花触头，并在母线连接合适位置安装波纹管。见图 5-19、5-20。

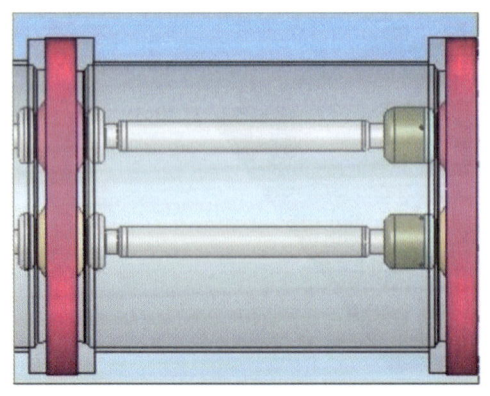

图 5-19 母线装配

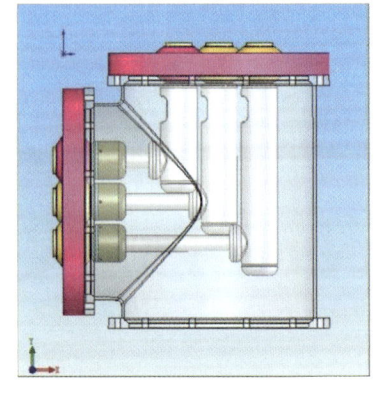

图 5-20 母线三通装配

6. 电流互感器

电流互感器是 GIS 中实现电流量的测量与过电流保护功能的元件。如图 5-21 所示，GIS 配用的电流互感器为三相封闭、穿心式结构，一次线圈为主回路导电杆，二次线圈缠绕在环形铁心上。导电杆与二次线圈间有屏蔽筒，二次线圈的引出线通过环氧浇注的密封端子板引到外部。

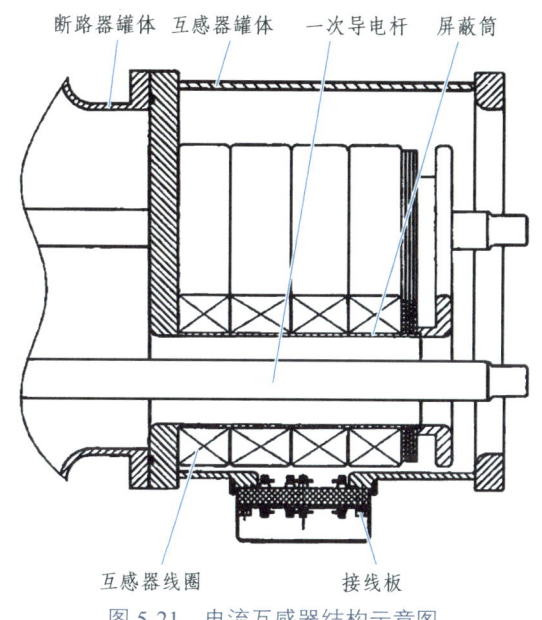

图 5-21　电流互感器结构示意图

7. 电压互感器

图 5-22 为 GIS 三相共箱式电压互感器，具有电气测量和电气保护作用。

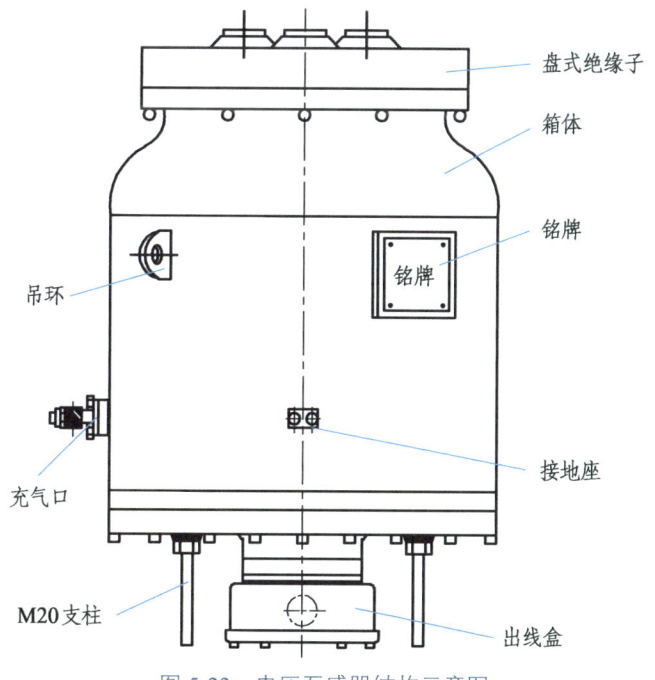

图 5-22　电压互感器结构示意图

电压互感器运行时二次侧严禁短路，否则二次侧产生的巨大电流将导致电压互感器损坏。

8. 避雷器

如图 5-23 所示避雷器采用三相共箱罐式结构。氧化锌电阻片具有良好的伏安特性和较大的通流容量，在正常运行电压下，氧化锌电阻片呈现出极高的电阻，使流过避雷器的电流只有微安级，当系统出现危害电器设备绝缘的大气过电压或操作过电压时，氧化锌电阻片呈现低电阻，使避雷器的残压被限制在允许值以下，并且吸收过电压能量，从而对电力设备提供可靠的保护。

在避雷器附属箱上部安装着在线监测装置，可在运行中记录避雷器的动作次数和泄露电流。避雷器外观见图 5-24。

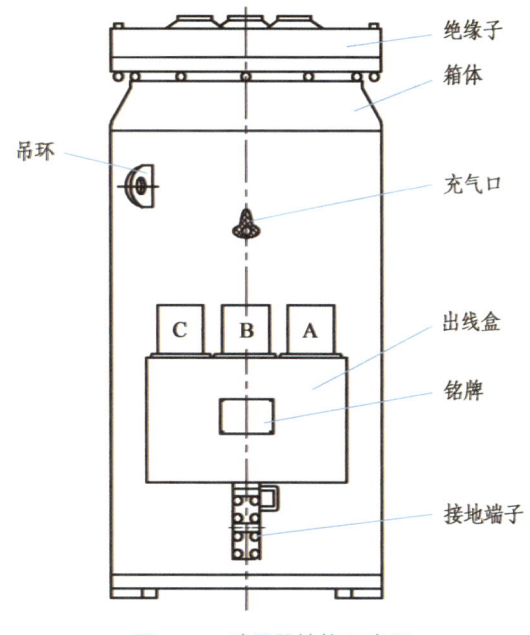

图 5-23 避雷器结构示意图

图 5-24 避雷器外观图

9. GIS 中的终端元件

在 GIS 中，作为 GIS 与进、出线或主变压器的连接元件，可有以下三种形式：SF_6/空气套管、SF_6/油套管及电缆终端。通常 SF_6 充气套管和电缆终端用来连接 GIS 与进、出线或变压器，而 SF_6/油套管仅用来连接 GIS 与变压器。

（1）SF_6/空气套管（见图 5-25）。

SF_6/空气套管为分相式结构，三相共箱的 GIS 本体，从套管处变成三相，然后与外部连接。

（2）SF_6/油套管。

SF_6/油套管专门用来连接 GIS 与变压器，通过 SF_6 管道母线和 SF_6/油套管，将 GIS 与变压器直接连接在一起。通常分为侧出线和顶出线两种形式。

通常油气套管的连接及制造分工应通过用户由 GIS 制造厂和变压器制造厂共同商定。

（3）电缆终端（见图 5-26）。

当 GIS 采用电缆进出线时，电缆通过电缆密封终端与 GIS 本体连在一起。

通常电缆终端的连接及制造分工应通过用户由 GIS 制造厂和电缆终端制造厂共同商定。

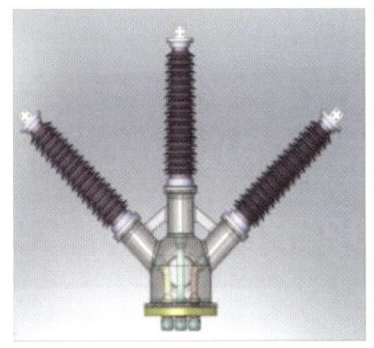

图 5-25　SF/空气套管

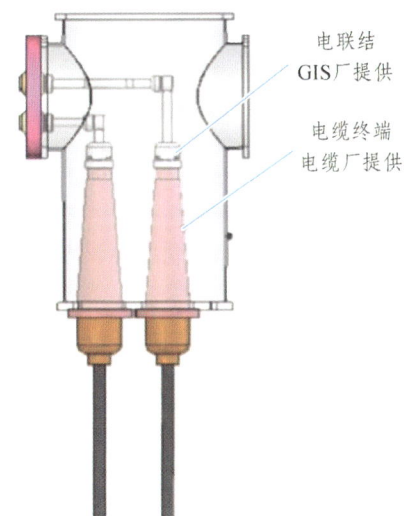

图 5-26　电缆终端

10. GIS 中的气体监视装置

为了监视 GIS 设备各气室 SF_6 气体是否泄漏,根据各厂家设计不同装有压力表或密度计,密度计装有温度补偿装置,一般不受环境温度的影响。为防止 SF_6 压力过高,超出正常压力,又装有防爆装置。

图 5-27 为 GIS 中的密度继电器,压力表中绿色区域为正常压力值,黄色区域为告警压力值,红色区域为闭锁压力值。

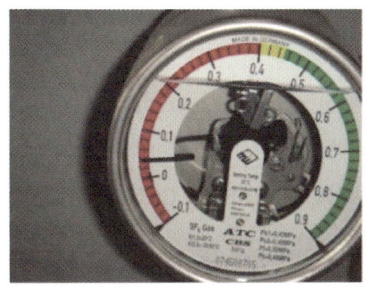

图 5-27　密度继电器

项目五　组合电器检修

Program 5 Maintenance of Gas Insulated Switchgear (GIS)

Module 1 Fundamentals of GIS

5.1.1 Concept of Sulfur Hexafluoride GIS

Sulfur hexafluoride GIS is also known as gas insulated switchgear (GIS). Circuit breakers, disconnectors, busbar, grounding switches, mutual inductors, outgoing line bushings or cable terminal heads are installed in their own enclosed spaces, and it has a metal cylinder as the housing, which make up an entirety in which sulfur hexafluoride gas at certain pressure is filled as insulation medium.

GIS can be classified into outdoor and indoor types considering installation location (See Fig. 5-1, 5-2).

Fig. 5-1   Outdoor GIS   Fig. 5-2   Indoor GIS

5.1.2 Characteristics and Scope of Application of Sulfur Hexafluoride GIS

1. Main advantages of GIS

(1) High reliability.

As the live parts are completely enclosed in SF_6 gas, they will not be affected by external environment.

(2) High safety.

As SF_6 gas has very high insulation strength and is an inert gas, no fire will occur. The live parts are completely enclosed in a grounded metal housing, providing shielding effect and eliminating risk of electric shock.

(3) Small footprint.

Due to use of SF_6 gas with high insulation strength as insulation and arc extinguishing medium, the minimum safe clearance between electrical equipment and between equipment and the ground is reduced, greatly reducing footprint.

(4) Easy installation.

A GIS can be assembled and tested by the manufacturer before delivering to the site for installation in the form of a bay, which greatly shortens construction period.

(5) Easy maintenance, long maintenance cycle.

Due to its reasonable structural layout and the advanced arc extinguishing system, service life of the product is greatly extended up to 30 years. Therefore, the maintenance cycle is long, maintenance workload is low, and as it is compact and has small ground clearance, daily maintenance is easy.

2. Main disadvantages of GIS

(1) High sealing performance requirement.

SF_6 gas pressure and water content in the device will have direct impact on performance of the entire device and safety of personnel. Therefore, GIS has strict requirement for processing accuracy.

(2) High price.

All electrical equipment except for transformer are installed in aluminum alloy housing of GIS, resulting in high metal consumption and high cost.

(3) High risk after fault.

Firstly, a fault may cause severe damage and may cause damage to the whole system. Secondly, toxic gases (from chemical reaction between SF_6 gas and water) during maintenance can cause harm to maintainers.

3. Scope of application

In recent years, in order to reduce footprint, sulfur hexafluoride fully-enclosed switchgear are widely used. At present, starting voltage used for GIS in China is 110 kV and higher, and it is mainly for the following applications:

(1) Areas with smaller footprint, e.g., urban substations.

(2) Areas with high altitude or high seismic intensity.

(3) Areas with harsh external environments. The GIS for the 750 kV project built by Northwest Power Construction Engineering in China has been put into operation at substations.

5.1.3 Structure of Sulfur Hexafluoride GIS

1. GIS tank structure

(1) Three-phases common box type.

Common box tank is used for not only three-phase busbar but also three-phase circuit breakers and other electric elements.

(2) Non-complete three-phase common box type.

Three phases of busbar share a same box, and circuit breakers and other electric elements have their own box.

(3) Fully separated types.

Separate box-type tank is used for all electric elements including busbar.

2. GIS outgoing lines

(1) Overhead outgoing lines.

GIS is connected to overhead lines through gas filled bushing.

(2) Cable outgoing.

GIS is connected to power cables through cable terminals, and there is SF_6 gas on one side of the sealing gasket and cable oil on the other side.

(3) GIS outgoing terminal is directly connected to the main transformer.

GIS is directly connected to the main transformer through oil-filled bushing, and now one side of the connecting bushing is filled with SF_6 gas, and the other side is filled with transformer oil.

5.1.4 GIS Structure

1. GIS overall structure

GIS is usually composed of units providing different functions which are called bays, including incoming (outgoing) line bays, PT bays, busbar bay, bus metering protection bay, etc; Different types of wiring such as single busbar, segmented single busbar, double busbar, bridge connection, etc. are used depending on different demands of users (See Fig. 5-3).

As shown in Fig. 5-4, dual-bus wiring is used for GIS, and there are 5 outgoing line bays, 2 main transformer incoming line bays, 1 bus bay and 2 bus metering protection bays.

Every bay of GIS is divided into a few independent SF_6 gas chambers which are known as gas separating units using non-ventilated basin insulators (gas separated insulators). The gas chambers on the circuit are interconnected, while those on a gas circuit are isolated from one another. Generally, circuit breakers have high pressure and form an gas chamber along with current transformers; Main bus, voltage transformer and lightning arrester are independent gas chambers, and other elements are classified on project basis. The gas chambers are monitored for gas pressure by corresponding density controllers.

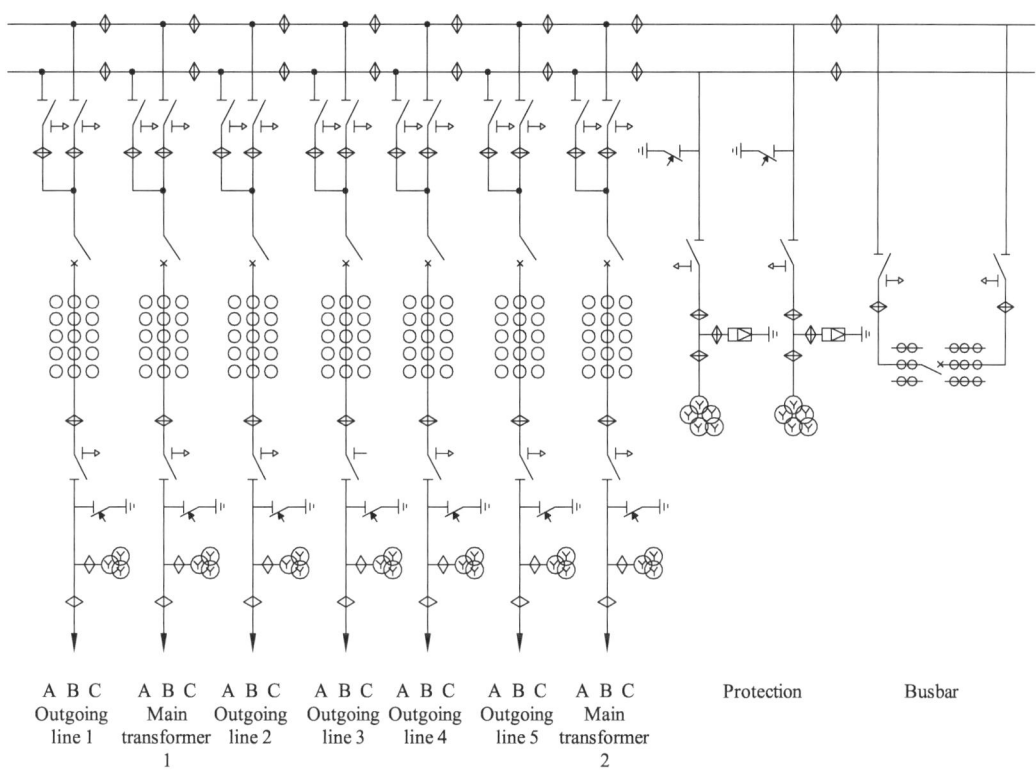

Fig. 5-3　Primary Schematic Diagram of Double-bus Wiring

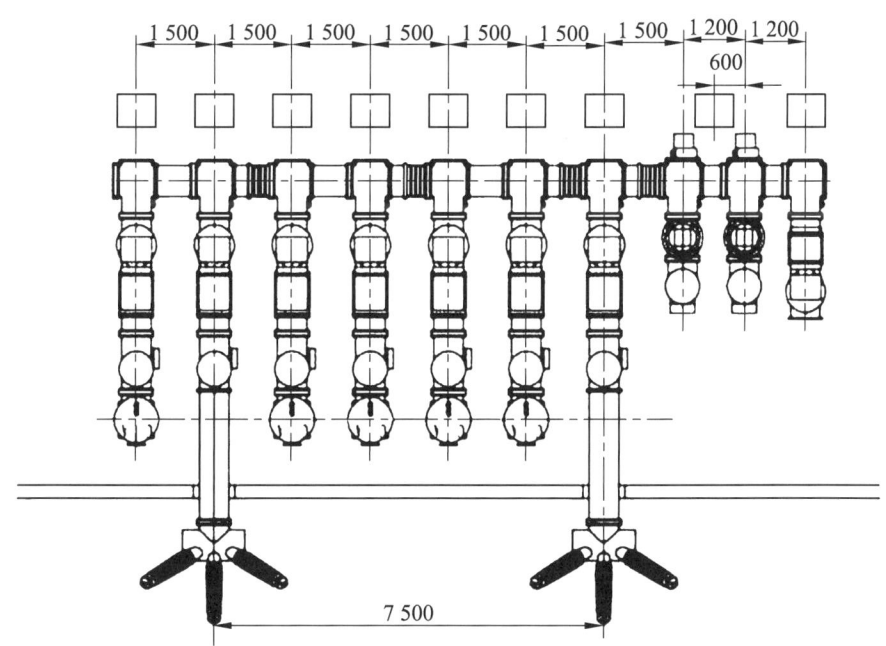

Fig. 5-4　Cross Section Diagram of Double Bus Connection Bay

GIS generally has a local control cabinet for each bay, and element control and status signals, density monitoring signals from the gas chambers and secondary outgoing lines of voltage and current transformers are all led into local control cabinet and connected into the main control room through local control cabinet.

Fig. 5-5 shows a single-bus overhead incoming (outgoing) line bay composed of: control cabinet, circuit breaker, disconnector, grounding switch, bus tube assembly, current transformer, connecting tube, incoming and outgoing line bushings, bracket, frame, etc.

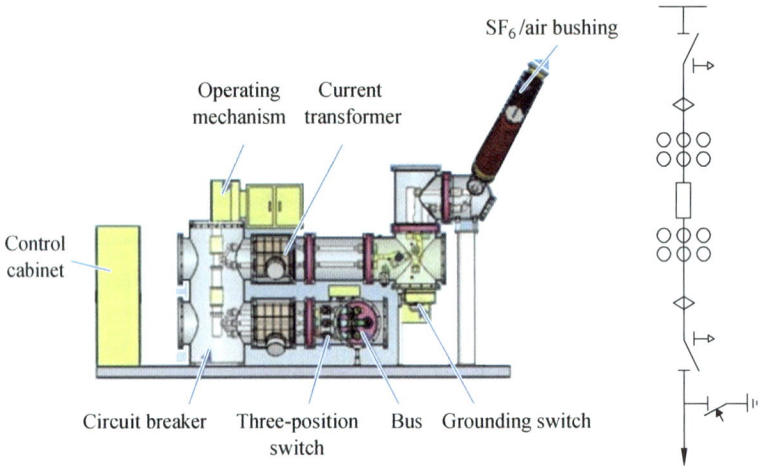

Fig. 5-5 Cross Section and Primary Schematic Diagram of Overhead Incoming (Outgoing) Line Bays

Fig. 5-6 shows a bus bay which connects two buses and consists of: control cabinet, circuit breaker, disconnector, bus tube assembly, current transformer, connecting tube, bracket, frame, etc.

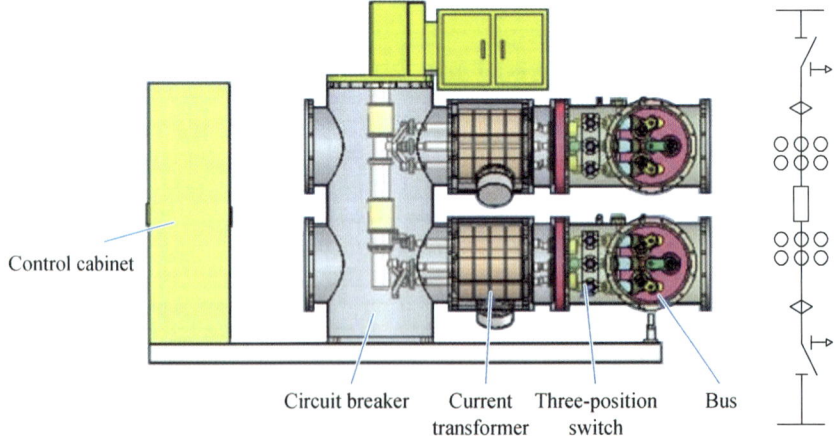

Fig. 5-6 Cross Section and Primary Schematic Diagram of Bus Bay

Fig. 5-7 shows a bus PT bay which is for metering and lightening protection and consists of control cabinet, disconnector, bus tube assembly, voltage transformer, connecting tube, bracket, frame, etc.

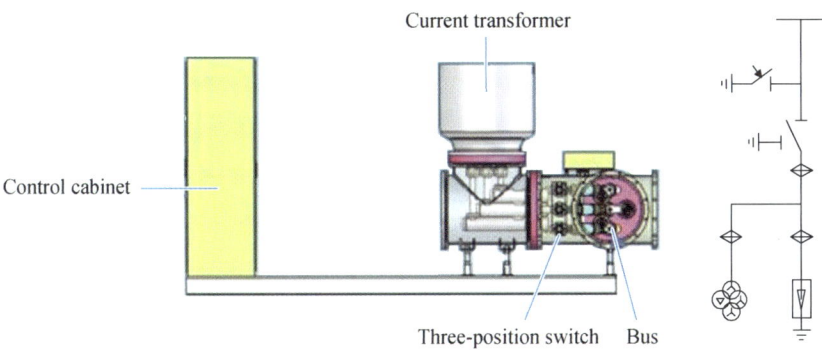

Fig. 5-7　Cross Section and Primary Schematic Diagram of Bus PT Bay

2. GIS component structure

(1) Control cabinet.

Control cabinet (See Fig. 5-8) is also known as local control cabinet. It is not only a relay hub for electric connection between elements inside and outside GIS bays, but also a centralized hub for on-site control, monitoring, telemetry, remote control, remote adjustment and remote signaling by GIS. It plays a very important role in normal operation of electric equipment.

Fig. 5-8　Inner and Outer Structural Diagrams of Control Cabinet

Local control cabinet of GIS generally allow local operation, signal transmission, protection and relay and GIS bays monitoring. The main functions include:

① Provide selection of local/remote operation of primary equipment such as circuit breakers, three-position switches and fast grounding switches in the bays. It allows local operation of the above-mentioned equipment on control cabinet and switchover to remote operation when GIS is in normal operation.

② Monitor open position and closed position status of circuit breakers, disconnectors and grounding switches.

③ Monitor SF_6 gas density in the gas chambers.

④ Monitor energy storage state of circuit breaker energy storage spring.

⑤ Monitor control loop power supply.

⑥ Display main wiring modes and operating status of primary electrical equipment of GIS.

⑦ Implement electric interlocking between circuit breakers, disconnector, grounding switches and electric interlocking between bays.

⑧ Monitor temperature and humidity in GIS mechanism box and terminal box and automatically switch on heating and dehumidification device.

⑨ Serve as control and signal relay terminal box between GIS elements and between GIS and main control room to receive and send signals.

(2) Circuit breaker.

Circuit breaker is the main element in a GIS which is used in transmission and transformation lines as power system control and protection equipment.

① Circuit breaker body.

Fig. 5-9 shows a tank structure in which three phases share a same box, where three phases share a spring operating mechanism and are mechanically interlocked. The moving contact base is fixed on tank top through an insulation platform and consists of moving contact base, intermediate contact terminal, guide, etc., which supports, guides and conducts electricity to moving contact terminals; A moving contact terminal is composed of nozzle, main moving contact, moving arcing contact, cylinder, tie rod, etc. It is connected to the mechanism through insulated tie rod and allows opening and closing operations by the driving force from the spring mechanism. Static contact terminal is connected to moving contact terminal through insulator and consists of static contact base, main static contact, static arcing contact, shield, etc. The main conductive circuit consists of: quincunx contact terminal on moving contact base–moving contact base–intermediate contact terminal–cylinder–main moving contact–main static contact terminal–static contact base–quincunx contact on the static contact base.

Fig. 5-10 shows the circuit breaker appearance ande arc extinguishing chamber assembly. As shown in Fig. 5-11, the circuit breaker has self-energy arc extinguishing structure of "thermal expansion+auxiliary blowing". In short-circuit current cut-off, the arc burns between moving and stationary arc contacts, and huge energy heats the SF_6 gas in the expansion chamber, resulting in temperature rise. Gas pressure in the expansion chamber also rises, giving rise to differential pressure between inside and outside; When the moving contact terminal opens up to a certain position, the static arc contact will separate itself from the nozzle, generating violent air blowing which extinguishes the arc when current flows across zero. During opening, due to high arc energy, pressure in expansion chamber goes beyond that in auxiliary compressor chamber, expansion chamber valve plate closes, compression chamber valve plate opens, and compression chamber pressure is released.

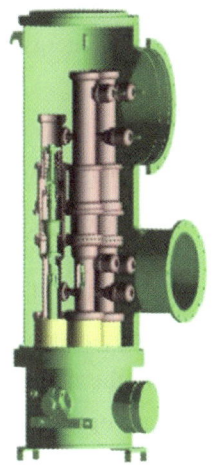

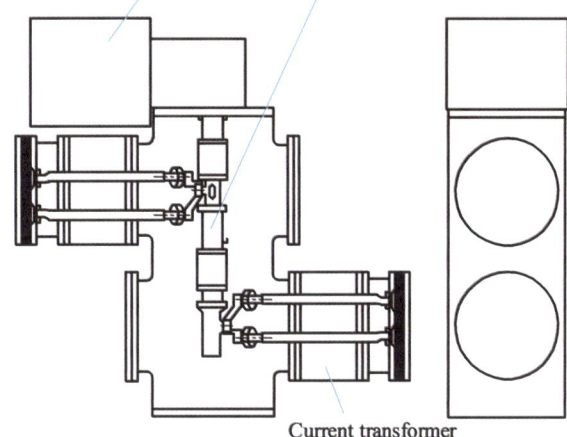

Fig. 5-9　Inside Section of Circuit Breaker

Fig. 5-10　Diagram of Circuit Breaker Appearance and Arc Extinguishing Chamber Assembly

In breaking low inductance, capacitance current or load current, breaking current is low, and arc energy is also low, pressure rise in expansion chamber is slower than that in auxiliary compression chamber, and valve plate of the compression chamber is closed, valve plate of expansion chamber is opened, and compressed gas flows into expansion chamber to generate air blowing, and arc is extinguished when current flows across zero point.

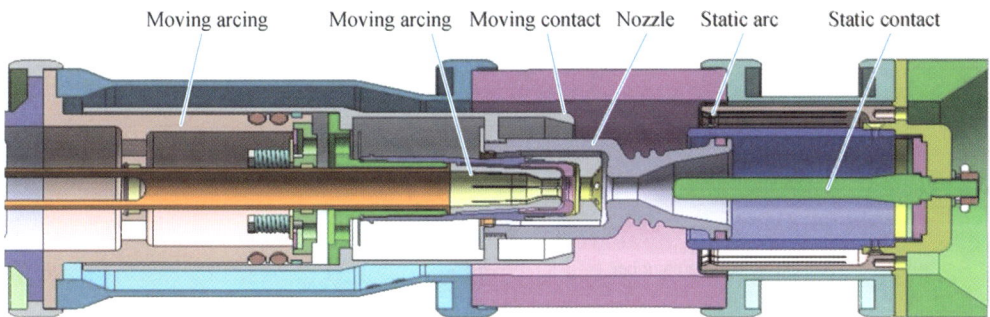

Fig. 5-11　Circuit Breaker Arc Extinguishing Chamber Structure

② Circuit breaker operating mechanism.

In the following section, work principle of the operating mechanism will be analyzed using CT26 spring operating mechanism for circuit breakers as example.

A. Closing spring energy storage.

The closing spring is at pre-compressed state, and the energy storage crank arm is connected to the motor through worm gear and worm. In energy storage, the motor drives the worm wheel through the worm, and the worm wheel drives the energy storage crank arm through the shaft pin to rotate counterclockwise; by the driving act of the energy storage shaft binding with the energy storage crank arm, the spring crank arm pulls the closing spring for compression and energy storage through the pins and tie rod in connection with it; when the pin gets across the left vertex position, the closing spring has the energy storage crank arm rotate by about 5°, and the energy storage holding pin on the energy storage crank arm is held back by the latch and energy storage is completed; now, motor power is cut off, and pawls disengage from ratchet wheel.

B. Closing operation.

The circuit breaker is at open position, and closing spring has completed energy storage.

When the mechanism receives a closing command, the closing coil is energized, and moving iron core of the closing electromagnet actuates and drives the closing guide rod to hit the closing latch and have it rotate clockwise; when energy storage holding latch is released, the closing spring drives the ratchet wheel to rotate quickly counterclockwise, and the cam sharing a same shaft with the ratchet wheel strikes the closing roller on the output crank arm to have the crank arm move upwards, and the connecting rod drives circuit breaker body to allow closing operation; now, the closing holding pin on the crank arm is held down by the closing holding latch to allow closing holding; And the closing spring crank arm sharing a same shaft with the output crank arm compresses the opening spring to allow energy storage and opening operation. Closing operation can also be completed by striking closing electromagnet with hand.

After closing operation, the travel switch automatically switches on the motor to store energy for the closing spring once again. See Fig. 5-12, 5-13.

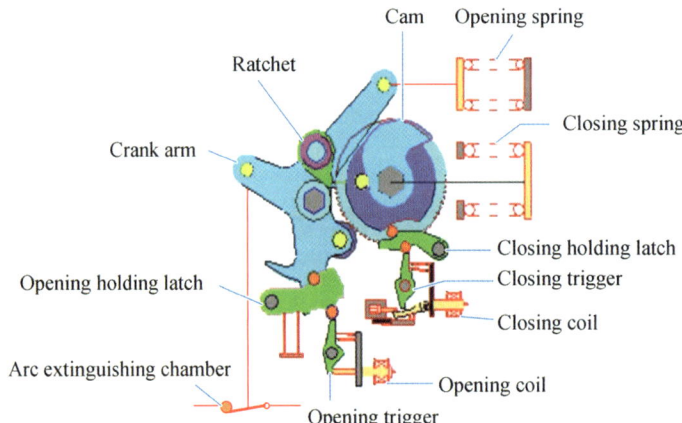

Fig. 5-12　Closed Position (Closing Spring Energy Storage)

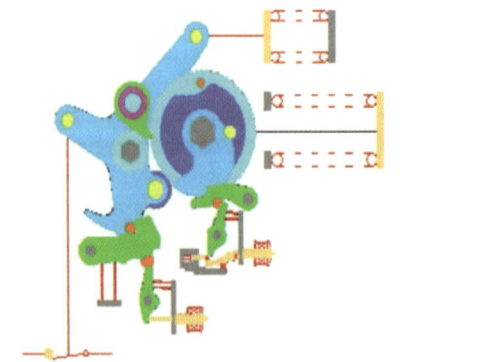

Fig. 5-13　Closed Position (Closing Spring Release)

C. Opening operation.

The circuit breaker is at closed position, and the closing spring and opening spring have completed energy storage. When the mechanism receives an opening command, the opening coil is energized, moving core of the opening electromagnet actuates the opening guide rod to strike the opening latch to allow it rotate clockwise; when the closing holding latch is released, the opening spring releases energy and allow opening operation through crank arm and connecting rod which drives circuit breaker body. See Fig. 5-14.

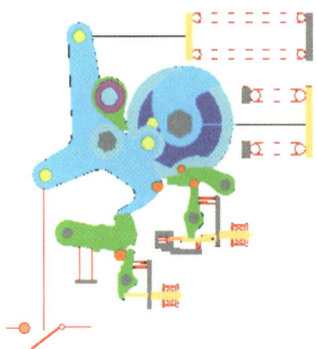

Figu. 5-14 Open Position (Closing Spring Energy Storage)

Opening operation can also be completed by manually striking the opening electromagnet guide rod.

(3) Three-position disconnector.

Three-position disconnector integrates a disconnector and grounding switch in a same module to allow three operating states: ON, disconnected and grounded. Grounding switch can be isolated from the housing. When relay protection adjustment and test, cable inspection, cable fault locating and DC resistance measurement are required, it can be electrically connected to GIS main circuit conductor externally through moving contact terminal of grounding switch to allow much more easier test and higher accuracy.

Three-position disconnector has an electric operating mechanism composed of two driving motors. Forward and reverse rotations of the motors drive the screw to rotate, and the screw drives the driving nut to move linearly. The driving nut drives the output shaft to rotate through a pin shaft to allow reciprocating motion of moving contact terminal among ON state ⟷ disconnected ⟷ grounded states through switchover by gears and racks. See Fig. 5-15, 5-16, 5-17.

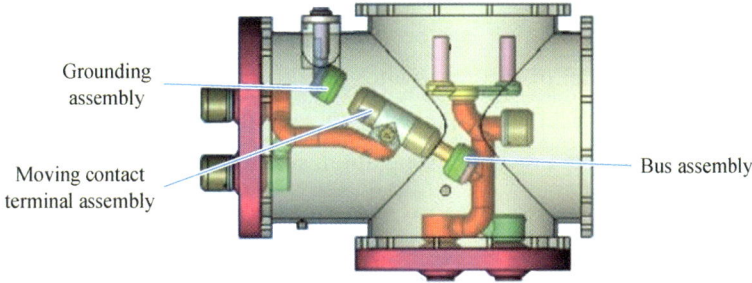

Fig. 5-15 Disconnector at "ON Position"

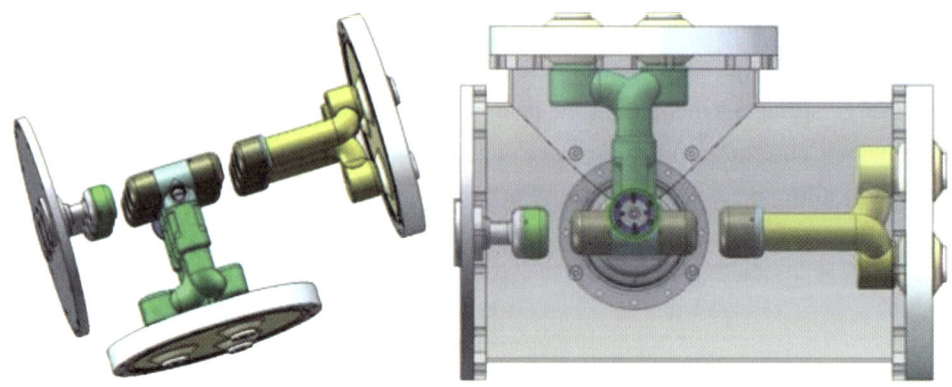

Fig. 5-16　Disconnector at "OFF Position"

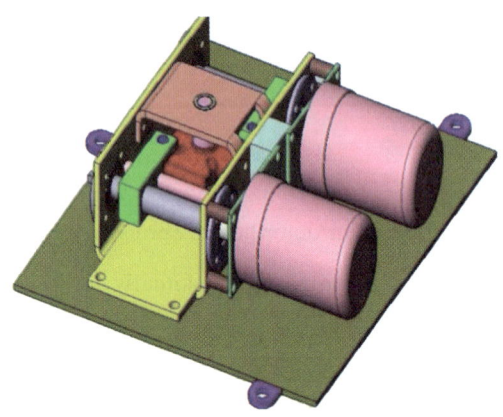

Fig. 5-17　Electric Operating Mechanism of Disconnector

(4) Grounding switch.

In addition to the grounding switch in the three-position switch (operating grounding for protecting safety during maintenance), it can also be installed separately and has two forms: fast earthing switch (FES, also known as fault closing grounding switch) capable of closing short-circuit current and opening/closing induced current and operating earthing switch (ES) for protecting safety during maintenance; The working grounding switch is equipped with a motor mechanism, and the fast grounding switch is equipped with an electric spring operating mechanism. Grounding switch can be insulated and disconnected from operating grounding housing. When circuit resistance measurement, mechanical characteristics test and other tests are required, the earthing bus connected to the earthing housing can be removed, thus electric connection to main circuit through moving contact terminal of earthing switch is allowed, making test much more easier. The fast grounding switch is electrically interlocked with relevant three-position switches and circuit breakers through a mechanism to prevent misoperation. see Fig. 5-18.

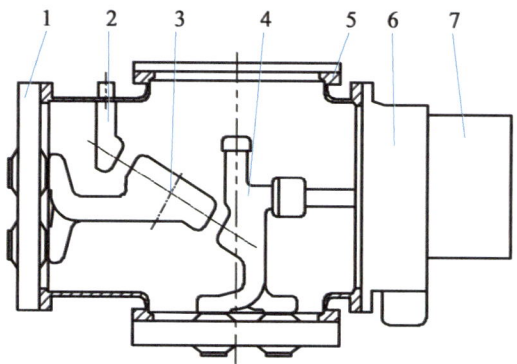

1-Disc insulator; 2-Grounding assembly; 3-Moving contact terminal assembly;
4-Bus assembly; 5-Housing; 6-Quick grounding switch; 7-Mechanism.

Fig. 5-18　Schematic Diagram of Three-position Fast Grounding Switch

(5) Bus.

SF_6 gas insulated bus is used to connect GIS elements, and the bus allows continuous flow of rated current and withstands dynamic and thermal stability current.

In a common box-type GIS, branch bus has the same housing size as the main bus, but diameter of conductor in the branch bus may be smaller than diameter of the main bus.

A bus consists of housing, branch conductors fixed on basin insulator and three-phase conductive rod.

To absorb deformation due to thermal expansion and assembly error, quincunx contact terminals are used in conductor connection, and bellows are installed at proper position of bus connection. See Fig. 5-19, 5-20.

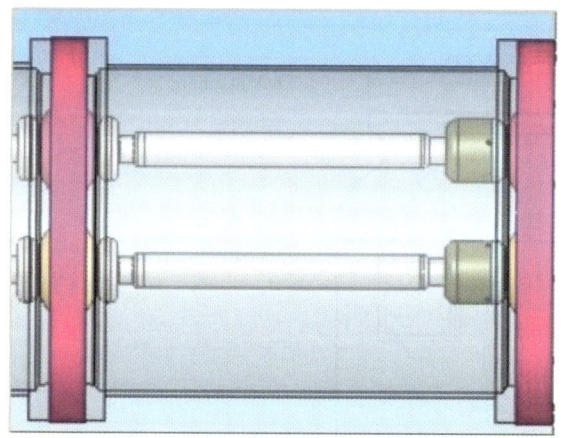

Fig. 5-19　Bus Assembly　　　　　　Fig. 5-20　Bus Tee Assembly

(6) Current transformer.

Current transformer is an element in GIS for measuring magnitude of current and overcurrent protection. As shown in Fig. 5-21, the current transformer for GIS has a three-phase enclosed cross-core structure, with primary coil as the main circuit conductive rod and secondary coil winding around the annular core. There is a shielding tube between the conductive rod and the secondary coil, and outgoing line of the secondary coil is led to outside through epoxy cast sealed terminal board.

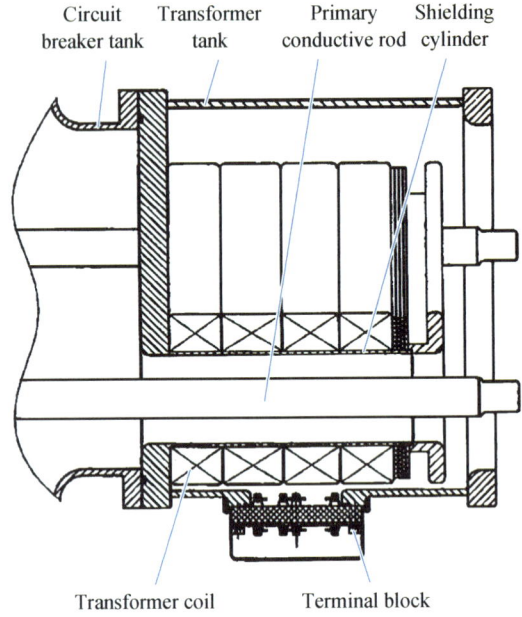

Fig. 5-21　Structural Representation of Current Transformer

(7) Voltage transformer.

What is shown in Fig. 5-22 is a GIS voltage transformer with three phases sharing a same box, which provides electrical measurement and protection.

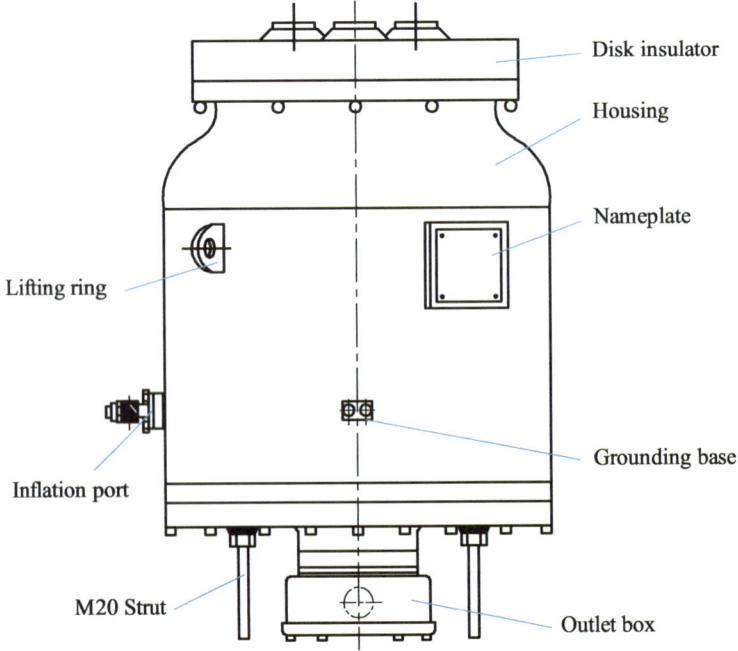

Fig. 5-22　Structural Representation of Voltage Transformer

When voltage transformer is running, secondary side short circuit is strictly forbidden, otherwise, huge current generated at secondary side will cause damage to voltage transformer.

(8) Lightning arrester.

Fig. 5-23 shows a lightning arrester with tank structure in which three phases share a same box. Zinc oxide resistors have good volt-ampere characteristics and high current capacity. At normal operating voltage, zinc oxide resistors have extremely high resistance, thus current flowing through lightning arrester is at microampere level. When the system has atmospheric overvoltage or operational overvoltage that endangers electrical equipment insulation, the zinc oxide resistor has low resistance, keeping residual voltage in lightning arrester below allowable value and allowing overvoltage energy absorption, providing reliable protection for electric equipment.

An online monitoring device is installed on upper part of the lightning arrester accessory box which will record number of actions and leakage current from lightning arrester during operation.

Fig. 5-24 shows the outside view of lightning arrester.

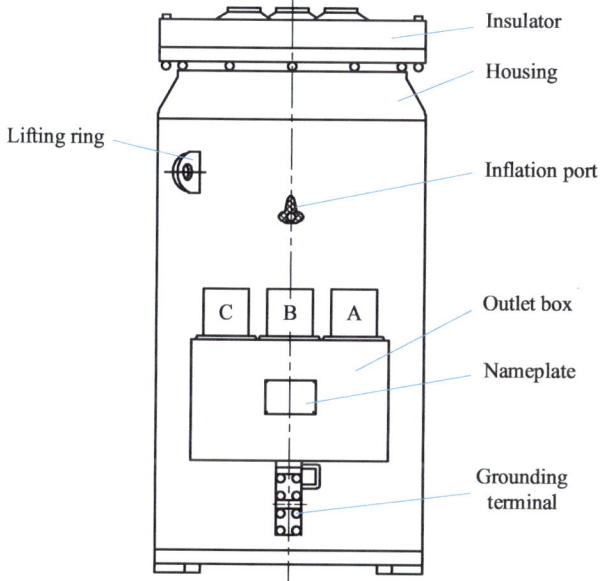 

Fig. 5-23　Structural Representation of Lightning Arrester　　Fig. 5-24　Outside View of Lightning Arrester

(9) Terminal elements in GIS.

In a GIS, as connecting elements between GIS and incoming and outgoing lines or main transformers, there are SF_6/air bushing, SF6/oil bushing and cable terminal. Usually, SF_6 gas filled bushing and cable terminal are for connecting GIS to incoming/outgoing lines or transformers, while SF_6/oil-filled bushing are only used to connect GIS to transformers.

① SF_6/air bushing (see Fig. 5-25).

SF_6/air bushing has a split-phase structure. GIS body in which three phases share a same box becomes three phases at the bushing and then connected to the outside.

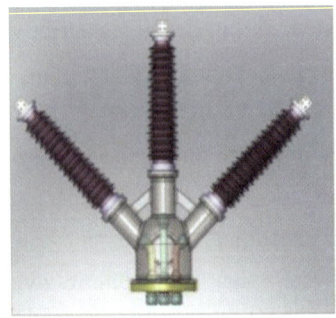 

Fig. 5-25　SF6/air bushing

② SF_6/oil bushing.

SF_6/oil bushing is exclusively used to connect GIS and transformer and direct connect GIS and transformer through SF_6 pipeline bus and SF_6/oil bushing. There are usually side outlet and top outlet types.

In general, oil and gas bushings connection and manufacturing shall be agreed upon between GIS manufacturer and transformer manufacturer through the user.

③ Cable terminal (see Fig. 5-26).

When GIS has cable inlet and outlet lines, the cables are connected to GIS body through cable sealing terminals.

Usually, cable terminals connection and manufacturing shall be agreed on between GIS manufacturer and cable terminal manufacturer through the user.

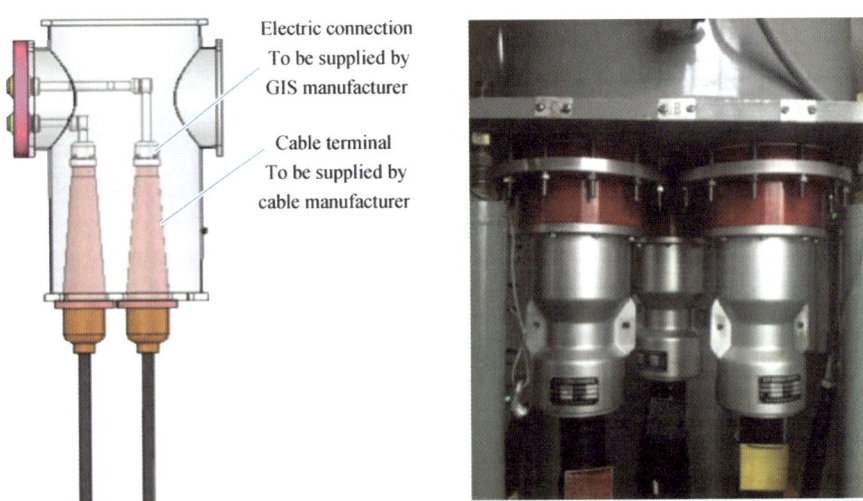

Fig. 5-26　Cable Terminal

(10) Gas monitoring devices in GIS.

In order to monitor SF_6 gas leakage in gas chambers of GIS equipment, pressure gauge or density meter is installed considering design of different manufacturers. Density meter has temperature compensation device and is generally free from impact by ambient temperature. To avoid SF_6 overpressure going beyond normal pressure, explosion-proof devices are installed.

Fig. 5-27 shows a density relay in GIS. The green area in pressure gauge represents normal pressure, the yellow area represents warning pressure, and the red area represents locking pressure.

Fig. 5-27　Density Relay

模块二 组合电器的检修

一、组合电器检修分类

GIS 设备检修工作分为四类：A 类检修、B 类检修、C 类检修、D 类检修。

1. A 类检修

A 类检修是指整体性检修，包括整体更换、解体检修。

2. B 类检修

B 类检修指维持气室密封情况下实施的局部性检修，包括部件解体检查、维修及更换。

3. C 类检修

C 类检修指例行检查及试验，包含本体检查维护、操动机构检查维护及整体调试。

4. D 类检修

D 类检修指在不停电状态下进行的检修，包含专业巡视、SF_6 气体补充、空压机润滑油更换、部分辅助二次元器件更换、金属部件防腐处理、传动部件润滑处理、箱体维护、互感器二次接线检查维护、避雷器泄漏电流监视器（放电计数器）检查维护、带电检漏及堵漏处理等不停电工作。

二、组合电器专业巡视要求

1. 组合电器外观巡视

（1）外壳、支架等无锈蚀、松动、损坏，外壳漆膜无局部颜色加深或烧焦、起皮。

（2）外观清洁，标志清晰、完善。

（3）压力释放装置无异常，其释放出口无障碍物。

（4）接地端子无过热，接触完好。

（5）各类管道及阀门无损伤、锈蚀，阀门的开闭位置正确，管道的绝缘法兰与绝缘支架良好。

（6）盆式绝缘子外观良好，无龟裂、起皮，颜色标示正确。

（7）二次电缆护管无破损、锈蚀，内部无积水。

2. 断路器巡视

（1）SF_6 气体密度值正常，无泄漏。

（2）无异常声响或气味，防松螺母无松动。

（3）分、合闸到位，指示正确。

（4）对于三相机械联动断路器检查相间连杆与拐臂所处位置无异常，连杆接头和连板无裂纹、锈蚀；对于分相操作断路器检查各相连杆与拐臂相对位置一致。

（5）拐臂箱无裂纹。

（6）机构内金属部分及二次元器件无腐蚀。

（7）机构箱密封良好，无进水受潮、无凝露，加热驱潮装置功能正常。

（8）对于液压、气动机构，分析后台打压频度及打压时长记录，无异常。

（9）对于液压机构，机构内管道、阀门无渗漏油，液压压力指示正常，各功能微动开关触点与行程杆间隙调整无逻辑错误，液压油油位、油色正常。

（10）对于气动机构，气压压力指示正常，空压机油无乳化。

（11）对于弹簧机构，分、合闸脱扣器和动铁心无锈蚀，机芯固定螺栓无松动，齿轮无破损，咬合深度不少于三分之一，挡圈无脱落，轴销无开裂、变形、锈蚀。

（12）加热装置功能正常，按要求投入。

（13）分合闸缓冲器完好，无渗漏油等情况发生。

（14）检查储能电机无异常。

3. 隔离开关巡视

（1）SF_6气体密度值正常，无泄漏。

（2）无异常声响或气味。

（3）分、合闸到位，指示正确。

（4）传动连杆无变形、锈蚀，连接螺栓紧固。

（5）卡、销、螺栓等附件齐全，无锈蚀、变形、缺损。

（6）机构箱密封良好。

（7）机械限位螺钉无变位，无松动，符合厂家标准要求。

4. 接地开关巡视

（1）SF_6气体密度值正常，无泄漏。

（2）无异常声响或气味。

（3）分、合闸到位，指示正确。

（4）传动连杆无变形、锈蚀，连接螺栓紧固。

（5）卡、销、螺栓等附件齐全，无锈蚀、变形、缺损。

（6）机构箱密封情况良好。

（7）接地连接良好。

（8）机械限位螺钉无变位，无松动，符合厂家标准要求。

5. 电流互感器巡视

（1）SF_6气体密度值正常，无泄漏。

（2）无异常声响或气味。

（3）二次电缆接头盒密封良好。

6. 电压互感器巡视

（1）SF_6气体密度值正常，无泄漏。

（2）无异常声响或气味。

（3）二次电缆接头盒密封良好。

7. 避雷器单元巡视

（1）SF_6气体密度值正常，无泄漏。

（2）无异常声响或气味。

（3）放电计数器（在线监测装置）无锈蚀、破损，密封良好，内部无积水，固定螺栓（计数器接地端）紧固，无松动、锈蚀。

（4）泄漏电流不超过规定值的10%，三相泄漏电流无明显差异。

（5）计数器（在线监测装置）二次电缆封堵可靠，无破损，电缆保护管固定可靠、无锈蚀、开裂。

（6）避雷器与放电计数器（在线监测装置）连接线连接良好，截面积满足要求。

8. 母线巡视

（1）SF_6气体密度值正常，无泄漏。

（2）无异常声响或气味。

（3）波纹管外观无损伤、变形等异常情况。

（4）波纹管螺柱紧固符合厂家技术要求。

（5）波纹管波纹尺寸符合厂家技术要求。

（6）波纹管伸缩长度裕量符合厂家技术要求。

（7）波纹管焊接处完好、无锈蚀。固定支撑检查无变形和裂纹，滑动支撑位移在合格范围内。

9. 进出线套管、电缆终端单元巡视

（1）SF_6气体密度值正常，无泄漏。

（2）无异常声响或气味。

（3）高压引线连接正常，设备线夹无裂纹、无过热。

（4）外绝缘无异常放电、无闪络痕迹。

（5）外绝缘无破损或裂纹，无异物附着，辅助伞裙无脱胶、破损。

（6）均压环无变形、倾斜、破损、锈蚀。

（7）充油部分无渗漏油。

（8）电缆终端与组合电器连接牢固，螺栓无松动。

（9）电缆终端屏蔽线连接良好。

10. 汇控柜巡视

（1）汇控柜外壳接地良好，柜内封堵良好。

（2）汇控柜密封良好，无进水受潮、无凝露，加热驱潮装置功能正常。

（3）汇控柜内干净整洁，无变形和锈蚀。

（4）钢化玻璃无裂纹、损伤。

（5）柜内二次元件安装牢固，元件无锈蚀，无烧伤过热痕迹。

（6）柜内二次线缆排列整齐美观，接线牢固无松动，备用线芯端部进行绝缘包封。

（7）智能终端装置运行正常，装置的闭锁告警功能和自诊断功能正常。

（8）空调运行正常，温度满足智能装置运行要求

（9）断路器、隔离开关及接地开关位置指示正确，无异常信号。
（10）带电显示器安装牢固，指示正确。

11. 集中供气系统巡视

（1）空气压缩机油位正常，油位应在油窗 1/2 左右，油质无乳化。
（2）压缩机风扇转动灵活，与储气罐及其压缩空气管道密封完好，传动皮带无开裂、松动等异常。
（3）高压储气罐压力指示正常。
（4）高压储气罐安全装置、阀门等清洁、完好。
（5）空压屏阀门开闭状态满足运行要求。
（6）气水分离器及自动排污装置外观完好，管道连接牢固，接线正确。

三、组合电器检修内容

（一）检修要求

GIS 设备的检修必须严格执行制造厂相关导则和工艺要求，拟订专项检修方案和作业指导书，作业指导书中应明确检修环境、"三措"（组织措施、技术措施、安全措施）、工序、工艺质量标准等的要求，检修实施严格执行作业指导书。

（二）检修项目

（1）断路器的检查和检修。
（2）隔离开关、接地开关和快速接地开关的检查和检修。
（3）母线的检查和检修。
（4）GIS 与电缆直接连接的检查和检修。
（5）GIS 与电力变压器直接连接的检查和检修。
（6）SF_6 气体系统的检查和检修。
（7）汇控柜箱和二次元器件的检查和检修。
（8）外壳、构支架、基础及接地连接的检查和检修。
（9）辅助系统的检查和检修。
（10）其他元件的检查和检修。

（三）主要检修内容

1. 断路器检修

（1）断路器本体。
① 检查引弧触头烧损程度。
② 检查喷口烧损程度。
③ 检查触指磨损程度。
④ 检查并清洁灭弧室及其绝缘件。
⑤ 更换吸附剂及密封圈。

⑥检查调整相关尺寸。

⑦检查合闸电阻及其传动部件（如有）。

⑧检查并联电容器（如有）。

（2）弹簧操动机构。

①检查分合闸线圈和脱扣打开尺寸及磨损情况。

②检查辅助开关切换情况。

③检查弹簧疲劳程度。

④检查轴、销、锁扣等易损部位，复核机构相关尺寸。

⑤检查缓冲器，更换缓冲器油（垫）及密封件。

⑥检查电机工作情况及储能时间。

（3）气动操动机构。

①检查分合闸线圈。

②检查辅助开关切换情况。

③检查并清洗操作阀、信号缸，更换密封圈。

④检查压力开关并校核各级压力接点设定值。

⑤检查建立压力时间（零表压起至额定压力）。

⑥检查缓冲器，更换缓冲器油及密封件。

⑦检查管道密封情况。

⑧气动弹簧操动机构应检查轴、销、锁扣等易损部位，复核机构相关尺寸。

⑨检查转动、传动部位润滑情况。

⑩检查本间隔储气罐及相关阀门。

（4）液压操动机构。

①检查分合闸线圈。

②检查辅助开关切换情况。

③清洗并检查操作阀，更换密封圈。

④校核各级压力接点设定值并检查压力开关。

⑤检查打压时间（零表压起至额定压力）。

⑥检查油泵、安全阀是否正常工作。

⑦检查预充氮气压力，对活塞杆结构储压器应检查微动开关，若有漏氮及微动开关损坏应处理或更换。

⑧液压弹簧机构应检查弹簧储能前后尺寸。

⑨清洗油箱、更换液压油后排气。

⑩检查防慢分装置功能正常。

2. 隔离开关、接地开关和快速接地开关检修

（1）检查实际分合位置和触头磨损情况。

（2）更换吸附剂及密封圈。

（3）操动机构。

①检查联锁线圈、电机工作情况。

② 检查辅助开关、微动开关切换情况。
③ 气动操动机构检查清洗电磁阀、清洗并检查操作阀，更换密封圈。
④ 检查轴、销、锁扣等易损部位，复核机构相关尺寸。
⑤ 检查并补充转动、传动部位的润滑油脂。
⑥ 检查电机转子轴承及碳刷磨损情况（直流电机）。
⑦ 检查机械限位尺寸。
⑧ 检查快速接地开关操动机构弹簧、缓冲器。

3. SF_6 气体系统

（1）校验 SF_6 密度继电器、压力表或密度表（条件允许可不停电校验）。
（2）检测 GIS 气室及管道的泄漏。
（3）测量 SF_6 气体湿度和纯度。
（4）对打开的气室更换吸附剂或根据制造厂要求定期更换。

4. 母线的检查和检修

（1）检查母线筒内绝缘件。
（2）更换吸附剂及密封圈。
（3）检查母线筒内导体连接及固定情况。

5. GIS 设备与电缆直接连接的检查和检修

（1）检查绝缘件。
（2）更换吸附剂及密封圈。
（3）检查电缆筒内导体连接及固定情况。
（4）检查屏蔽罩。
（5）检查放电间隙。

6. GIS 设备与电力变压器直接连接的检查和检修

（1）检查绝缘件。
（2）更换吸附剂及密封圈。
（3）检查内部导体连接及固定情况。
（4）检查屏蔽罩。
（5）检查放电间隙。
（6）检查连接伸缩节。

7. 汇控柜(箱)和二次元器件的检查和检修

（1）检查江控柜（箱）密封情况，更换老化的箱门密封圈。
（2）检查二次电缆封堵情况，更换老化开裂或脱落的防火封堵泥。
（3）检查箱内二次元件接地情况。
（4）检查切换开关、继电器、接触器、空气断路器、温湿度控制器、加热器、限位开关、端子排、信号指示灯、整流模块等二次元器件，酌情更换或按制造厂要求定期更换。

8. 外壳、构支架、基础及接地连接

（1）检查外壳漆层。

（2）检查伸缩节有无扭曲，拉伸或压缩尺寸在允许范围内。

（3）检查各气室防爆膜有无锈蚀、未堵塞。

（4）检查三相汇流排连接情况。

（5）检查支架、构架及接地连接有无锈蚀、变形或损坏。

（6）检查设备基础有无沉降。

9. 辅助系统

（1）气动机构的集中供气系统。

① 检查空压机阀板、活塞环和曲轴箱，更换密封件、滤芯及易损件，清洗进气滤网，有油压缩机应调换压缩机油；压缩机运转时间较长须更换部件较多时建议调换整台压缩机。

② 检查油气分离装置和自动（手动）排污阀工作情况。

③ 检查电动机及调换传动皮带。

④ 检查安全阀开启/关闭压力；检查减压阀、逆止阀和其他阀门，必要时应更换损坏部件。

⑤ 测量打压时间；检查压缩空气管道泄漏情况；有减压装置的应检查一/二级压力；校验各级压力开关设定值。

⑥ 检查气站控制系统二次元器件。

（2）在线监测系统的维护检查。

① 传感器的检查维护。

② 后台信号处理系统检查维护。

③ 二次元件等检查维护。

④ 根据制造厂建议对于消耗品及易损件进行更换。

10. 其他部件

（1）电流互感器、电压互感器、避雷器、带电显示器等其他可根据相关规定进行检查。

（2）检查出线套管外绝缘情况（复合套管应定期检查憎水性）。

（3）检查与其他一次设备的连接情况（与变压器连接的油气套管及与电缆连接的电缆终端）。

Module 2　GIS Maintenance

5.2.1　Classification of GIS maintenance

GIS equipment maintenance is classified into four types: Class A maintenance, Class B maintenance, Class C maintenance and Class D maintenance.

1. Class A maintenance

Class A maintenance refers to overall maintenance, including overall replacement and disassembly maintenance.

2. Class B maintenance

Class B maintenance refers to local maintenance during which gas chamber is kept sealed, including disassembly inspection, repair and replacement.

3. Class C maintenance

Class C maintenance refers to routine inspection and testing, including body inspection and maintenance, operating mechanism inspection and maintenance and overall commissioning.

4. Class D maintenance

Class D maintenance refers to maintenance carried out without power cut, including special patrol, SF_6 gas replenishment, air compressor lubricating oil replacement, replacement of some auxiliary and secondary components, metal parts antiseptic treatment, transmission parts lubrication, box maintenance, secondary wiring inspection and maintenance of transformer, lightning arrester leakage current monitor (discharge counter) inspection and maintenance, leakage detecting at charged state and leaking stoppage.

5.2.2　Professional Inspection of GIS

1. GIS appearance inspection

(1) Make sure that housings, brackets, etc. are free of rust, loosening or damage, and film on the housing has no local color burn or charring or peeling.

(2) The appearance is clean, and the markings a clear and complete.

(3) The pressure release device is normal, and the release outlet is unobstructed.

(4) The grounding terminal is free of overheat and has good contact.

(5) The pipelines and valves are free of damage and rust, opening and closing positions of the valves are correct, and piping insulation flanges and insulation supports are at good condition.

(6) Basin insulators have intact appearance and have no cracks or peeling, and color markings are correct.

(7) The secondary cable conduits are free from damage or rust and water.

2. Circuit breaker inspection

(1) SF_6 gas intensity is normal and there is no leakage.

(2) There is no abnormal sound or odor, and the locking nut is tight.

(3) Opening and closing are adequate, and indication is correct.

(4) For three-phase mechanically-linked circuit breakers, check that interphase connecting rod and crank arm positions are normal, connecting rod and plate are free of crack and rust. For split-phase operated circuit breakers, check that relative positions of connecting rods at the phases and crank arms are consistent.

(5) The crank arm box is free of crack.

(6) Metal parts in the mechanism and secondary elements are free of corrosion.

(7) The mechanism box is well sealed, there is no water ingress, moisture or condensation, and the heating and moisture removal devices are functioning normally.

(8) For hydraulic and pneumatic mechanisms, analyze records of background compression frequency and duration and make sure that there is no anomaly.

(9) For hydraulic mechanism, make sure that there is no oil leakage from pipelines and valves in the mechanism, hydraulic pressure indication is normal, adjustment of clearance between functional microswitch contact and travel rod has no logic error and hydraulic oil level and color are normal.

(10) Pneumatic mechanism has normal air pressure indication, and air compressor oil is not emulsified.

(11) For spring mechanism, make sure that there is no rust on opening and closing releases and moving core, core fixing bolts are tight, gears are intact, engagement depth is no less than one-third, retaining ring is not detached and shaft pins are free of crack, deformation or corrosion.

(12) Heating devices are functioning normally and has been put into operation as required.

(13) The opening and closing buffers are intact and there is no oil seepage or leakage.

(14) Check motor energy storage for abnormality.

3. Disconnector inspection

(1) SF_6 gas intensity is normal and there is no leakage.

(2) No abnormal noise or smell.

(3) Opening and closing are adequate, and indication is correct.

(4) Check that transmission connecting rod is free of deformation and rust, and connecting bolts are tightened.

(5) Check that accessories such as clamps, pins, bolts, etc. are complete and free of rust, deformation or defect.

(6) Mechanism box are well sealed.

(7) Mechanical limits screws are free of displacement or loosening and conform to standards of the manufacturer.

4. Grounding switch inspection

(1) SF_6 gas intensity is normal and there is no leakage.

(2) No abnormal noise or smell.

(3) Opening and closing are adequate, and indication is correct.

(4) Check that transmission connecting rod is free of deformation and rust, and connecting bolts are tightened.

(5) Check that accessories such as clamps, pins, bolts, etc. are complete and free of rust, deformation or defect.

(6) Mechanism box are well sealed.

(7) Grounding connection is at good condition.

(8) Mechanical limits screws are free of displacement or loosening and conform to standards of the manufacturer.

5. Current transformer inspection

(1) SF_6 gas intensity is normal and there is no leakage.

(2) No abnormal noise or smell.

(3) Secondary cable junction boxes are well sealed.

6. Voltage transformer inspection

(1) SF_6 gas intensity is normal and there is no leakage.

(2) No abnormal noise or smell.

(3) Secondary cable junction boxes are well sealed.

7. Lightning arrester unit inspection

(1) SF_6 gas intensity is normal and there is no leakage.

(2) No abnormal noise or smell.

(3) Discharge counter (online monitoring device) is free of rust or damage, well sealed and has water. Fixing bolts (counter grounding terminal) are fastened tightly and are not rust eaten.

(4) Leakage current is no more than 10% of the specified value, and there is no significant difference in three-phase leakage current.

(5) Secondary cable of the counter (online monitoring device) is reliably sealed and free of damage, and cable protection conduits are fixed reliably and there is no rust or crack.

(6) The lightning arrester and the discharge counter (online monitoring device) are well connected, and the cross-sectional area is as required.

8. Bus inspection

(1) SF_6 gas intensity is normal and there is no leakage.

(2) No abnormal noise or smell.

(3) Bellows are free from any anomaly such as damage or deformation.

(4) Bellows bolts fastening meets technical requirements of the manufacturer.

(5) Ripple size of bellows meets technical requirements of the manufacturer.

(6) Allowance of expansion length of bellows meets technical requirements of the manufacturer.

(7) Welds of bellows are intact and free of rust. Fixed supports are free of deformation or cracks, and sliding support displacement is within acceptable range.

9. Inspection of incoming and outgoing line bushings and cable terminal units

(1) SF_6 gas intensity is normal and there is no leakage.

(2) No abnormal noise or smell.

(3) High-voltage lead connection is correct and wire clamp is free of crack or overheating.

(4) There is no abnormal discharge or flashover marks on external insulation.

(5) External insulation is free of damage or crack or foreign object, and the auxiliary umbrella skirt is free of debonding and damage.

(6) The grading ring is free of deformation, inclination, damage or corrosion.

(7) There is no oil leakage in the oil filled part.

(8) The cable terminal is firmly connected to GIS and the bolts are tight.

(9) Cable terminal shield wire is well connected.

10. Control cabinet inspection

(1) The control cabinet housing is well grounded and the cabinet is well plugged.

(2) The control cabinet is well sealed, there is no water ingress, moisture or condensation, and the heating and moisture removal devices are functioning normally.

(3) The control cabinet is clean and tidy and free of deformation or rust.

(4) Tempered glass is free of crack or damage.

(5) Secondary elements in the cabinet are firmly installed, and the elements are free of rust, burn and overheating.

(6) Secondary cables in the cabinet are arranged neatly and aesthetically, the connections are firm, and standby core ends are insulated and sealed.

(7) The smart terminal devices operate normally, and locking alarm function and self-diagnosis function are normal.

(8) The air conditioner is operating normally and temperature satisfies the demand for equipment operation.

(9) Circuit breakers, disconnectors and grounding switch have correct position indications and there is no abnormal signal.

(10) The live display is securely installed and provide correct indication.

11. Centralized gas supply system inspection

(1) Air compressor has normal oil level which shall be at about 1/2 of the oil window, and oil is not emulsified.

(2) Compressor fan can rotate freely and is well sealed with air tank and its compressed air pipeline, and the driving belt is free of crack, loosening and other abnormalities.

(3) Pressure indication of the high-pressure air tank is normal.

(4) Safety devices and valves of the high-pressure air tank are clean and intact.

(5) Opening and closing status of air pressure screen valve meets operational requirements.

(6) The gas-water separator and automatic sewage discharge device have intact appearance, and pipeline connection is firm, and the wires are connected properly.

5.2.3 Items of GIS maintenance

5.2.3.1 Maintenance requirement

GIS equipment maintenance must strictly follow guidelines and process requirements of the manufacturer, special maintenance plans and work instructions shall be prepared which shall specify requirements for maintenance environment, "three measures" (organizational measures, technical measures and safety measures), processes, process quality standards, etc. The work instructions shall be fully put into practice during maintenance.

5.2.3.2 Maintenance items

(1) Circuit breaker inspection and maintenance.

(2) Inspection and maintenance of disconnectors, grounding switches and fast grounding switches.

(3) Bus inspection and maintenance.

(4) Inspection and maintenance of direct connection between GIS and cable.

(5) Inspection and maintenance of direct connection between GIS and power transformer.

(6) SF_6 gas system inspection and maintenance.

(7) Inspection and maintenance of control cabinets and secondary elements.

(8) Inspection and maintenance of housings, structural supports, foundations and grounding connections.

(9) Auxiliary system inspection and maintenance.

(10) Other elements inspection and maintenance.

5.2.3.3 Main maintenance items

1. Circuit breaker maintenance

(1) Circuit breaker body.

① Check arc striking contact terminal burn.

② Check nozzle burn.

③ Check contact finger wear.

④ Check and clean arc extinguishing chamber and its insulating parts.

⑤ Replace adsorbent and seal ring.

⑥ Check and adjust dimensions.

⑦ Check the closing resistor and its transmission parts (if any).

⑧ Check the parallel capacitor (if any).

(2) Spring operating mechanism.

① Check the opening sizes and wear of the opening and closing coils and tripping.

② Check the switching of auxiliary switch.

③ Check the fatigue of spring.

④ Check such vulnerable parts as shafts, pins and latches, and recheck the relevant dimensions of the mechanism.

⑤ Check the buffer, and replace the buffer oil (gasket) and seal.

⑥ Check the working condition and energy storage time of motor.

(3) Pneumatic operating mechanism.

① Check the opening and closing coils.

② Check the switching of auxiliary switch.

③ Check and clean the operating valve and signal cylinder, and replace the seal ring.

④ Check the pressure switch, and verify the set values of all-level pressure contacts.

⑤ Check the time of build-up pressure (from zero gauge pressure to rated pressure).

⑥ Check the buffer, and replace the buffer oil and seal.

⑦ Check the seal condition of pipeline.

⑧ Check such vulnerable parts as shafts, pins and latches, and recheck the relevant dimensions of the mechanism for the pneumatic spring operating mechanism.

⑨ Check the lubrication of rotating and driving parts.

⑩ Check the air storage tank and related valves of this interval.

(4) Hydraulic operating mechanism.

① Check the opening and closing coils.

② Check the switching of auxiliary switch.

③ Clean and inspect the operating valve, and replace the seal ring.

④ Verify the set values of all-level pressure contacts and check the pressure switch.

⑤ Check the bulge time (from zero gauge pressure to rated pressure).

⑥ Check the oil pump and safety valve for normal operation.

⑦ Check the pre-filled nitrogen pressure, and check the microswitch for the pressure accumulator in the piston rod structure. Timely processing or replacement is required in case of any nitrogen leakage or damage to the microswitch.

⑧ Check the sizes of the spring before and after energy storage for the hydraulic spring mechanism.

⑨ Clean the fuel tank, and replace the hydraulic oil before venting.

⑩ Check the anti-slow-release device for normal function.

2. Maintenance of disconnectors, grounding switches and high speed grounding switches

(1) Check the actual opening and closing positions and the wear of contact terminal.

(2) Replace adsorbent and seal ring.

(3) Operating mechanism.

① Check the service conditions of the interlocking coil and motor.

② Check the switching of auxiliary switch and microswitch.

③ Check and clean the solenoid valve and operating valve and replace the seal ring for the pneumatic operating mechanism.

④ Check such vulnerable parts as shafts, pins and latches, and recheck the relevant dimensions of the mechanism.

⑤ Check and make up the grease of the rotating and driving parts.

⑥ Check the wear of the rotor bearings and carbon brushes of the motor (DC motor).

⑦ Check the mechanical limit sizes.

⑧ Check the spring and buffer of the operating mechanism of high speed grounding switch.

3. SF_6 system

(1) Check the SF_6 density relay, pressure gauge or density meter (uninterruptible check is allowed if conditions allowed).

(2) Test the leakage of GIS air chamber and pipeline.

(3) Measure the SF_6 humidity and purity.

(4) Replace the adsorbent for the open air chamber, or replace it regularly as required by the manufacturer.

4. Inspection and maintenance of buses

(1) Check the insulating parts in the bus barrel.

(2) Replace adsorbent and seal ring.

(3) Check the connection and fixation of conductors in the bus barrel.

5. Inspection and maintenance of direct connection between GIS and cable

(1) Check the insulating parts.

(2) Replace adsorbent and seal ring.

(3) Check the connection and fixation of conductors in the cable coaming.

(4) Check the shield.

(5) Check the discharging gap.

6. Inspection and maintenance of direct connection between GIS and power transformer

(1) Check the insulating parts.

(2) Replace adsorbent and seal ring.

(3) Check the connection and fixation of conductors inside.

(4) Check the shield.

(5) Check the discharging gap.

(6) Check the Connecting expansion joint.

7. Inspection and maintenance of control cabinets (boxes) and secondary components

(1) Check the sealing condition of the control cabinet (box), and replace the aged seal ring of

the door.

(2) Check the sealing condition of the secondary cable, and replace the fireproof plugging mud aged, cracked or fallen off.

(3) Check the grounding condition of the secondary components in the box.

(4) Check the secondary components like selector switch, relay, contactor, air circuit breaker, temperature and humidity controller, heater, limit switch, terminal strip, signal indicator and rectifier module, and replace them where appropriate or regularly as required by the manufacturer.

8. Housings, supports and frames, foundations and grounding connections

(1) Check the paint layers of housing.

(2) Check the expansion joint for distortion, and check whether the stretching or compressing dimensions are within the allowed range.

(3) Check the explosion-proof film of each air chamber for the existence of corrosion and no plugging.

(4) Check the connection of three-phase busbars.

(5) Check the supports, frames and grounding connections for the existence of corrosion, deformation or damage.

(6) Check the equipment foundation for the existence of settlement.

9. Auxiliary system

(1) Centralized gas supply system of pneumatic mechanism.

① Check the valve plate, piston ring, and crankcase of air compressor, replace the seals, filter elements and vulnerable parts, clean the air inlet screen, and replace the compressor oil for oil compressor It is recommended to replace the complete compressor if there are many components to be replaced after the compressor has run for a long time.

② Check the service conditions of the oil-gas separation device and the automatic (manual) blowdown valve.

③ Check the motor, and replace the transmission belt.

④ Check the opening/closing pressures of the safety valve; Check the pressure relief valve, check valve and other valves, and replace damaged components when necessary.

⑤ Measure the bulge time; Check the compressed air pipelines for leakage; Check the primary/secondary pressures in case of the decompressor; Verify the set values of all-level pressure switches.

⑥ Check the secondary components of the gas station control system.

(2) Maintenance and inspection of online monitoring system.

① Check and maintain the sensors.

② Check and maintain the background signal processing system.

③ Check and maintain the secondary components.

④ Replace the consumables and vulnerable parts as recommended by the manufacturer.

10. Other components

(1) Current transformers, voltage transformers, lightning arresters, live displays, etc. can be checked as regulated.

(2) Check the external insulation of the outgoing bushing (regularly check the hydrophobicity of composite bushing).

(3) Check the connection with other primary equipment (oil and gas bushings connected to transformers and cable terminals connected to cables).

任务一　组合电器指示仪表检查

一、工作任务

对组合电器进行指示仪表的检查。本模块主要包含两个内容：第一，识别组合电器汇控柜的各位置显示灯；第二，检查组合电器设备断路器、隔离开关的实际位置。通过结构分析、图例展示、现场讲解、操作技能训练，掌握组合电器的基本结构；汇控柜的组成、作用和原理；并能够识别、判断组合电器汇控柜的状态与设备实际位置是否一致。

二、引用标准

（1）《气体绝缘金属封闭开关设备运行维护规程》（DL/T 603—2017）。
（2）《国家电网公司五项通用制度变电检修管理规定》（第 3 分册组合电器检修细则）。
（3）《国家电网公司变电运维管理规定》。
（4）《国家电网公司变电检修管理规定》。
（5）《国家电网公司电力安全工作规程》（变电部分）。

三、工作要求

（1）作业为室内停电作业，无天气要求。
（2）被检修间隔与其他带电间隔之间使用围栏隔离，面向通道处设置唯一出入口。
（3）作业人员精神状态良好，熟悉工作中安全措施、技术措施以及现场工作危险点。
（4）实训现场要求按生产现场规范布置安全措施，并严格执行标准化作业。
（5）作业人员应规范穿戴劳动保护用品，做好安全防护。

四、工作准备

（一）危险点及预控措施

1. 高压触电

危险点：作业中误入相邻带电间隔导致高压触电。

预控措施：作业前用围栏将被检修间隔与相邻带电间隔隔离，并且面向作业现场内悬挂适量"止步，高压危险"标识牌，在通道处设置唯一出入口，悬挂"从此进出"标识。工作时至少两人进行，一人监护一人操作，听工作负责人指挥。

2. 低压触电

危险点：断路器分合闸操作中可能造成低压触电。

预控措施：操作时，操作人员大声呼唱，得到工作班所有人员大声回应之后，方可操作。操作人操作空开时，需与带电部位保持一定安全距离。

（二）工器具及材料选择

组合电器指示仪表检查工器具及材料见表 5-1。

表 5-1 组合电器指示仪表检查工器具及材料

类别	名称	规格型号	数量	备注
专用工具	地刀操作把手		1 把	
	断路器储能把手		1 把	
	隔离开关操作把手		1 把	
	内六角扳手		1 套	
通用工具	平口钳		1 把	
	尖嘴钳		1 把	
	一字螺丝刀	150 mm	1 把	
	十字螺丝刀	150 mm	1 把	
	活动扳手	200 mm	1 把	

（三）作业人员分工

组合电器指示仪表检查人员分工见表 5-2。

表 5-2 组合电器指示仪表检查人员分工

序号	工作岗位	数量	职责
1	工作负责人	1	负责本次工作的人员分工、现场查勘、作业方案制定、召开班前会、作业过程中安全监督、工作中突发状况的处理、工作质量的监督、班后会总结
2	操作人员	1	负责组合电器指示仪表检查的主要操作
3	辅助操作人员（可无）	1	辅助操作人员进行组合电器指示仪表检查

五、作业程序

组合电器指示仪表检查作业流程见表 5-3。

表 5-3 组合电器指示仪表检查作业流程

序号	作业内容	作业步骤及标准
1	工作前准备工作	1. 检查工器具是否齐全，检查工器具外观和试验合格； 2. 工作负责人同工作许可人巡视待检修设备，确认工作票所列安全措施已经正确执行，安全措施是否完备，现场是否具备开工条件，必要时进行补充； 3. 执行工作许可手续； 4. 对工作班成员召开班前会； 5. 抄写设备铭牌参数
2	断路器分闸位置检查	检查断路器分闸指示器在分闸状态，检查时应口述并指出对应信号
3	合上交流电	打开汇控柜柜门，合上交流电源空开，储能、控制空源空开开，操作时应口述并指出对应信号

续表

序号	作业内容	作业步骤及标准
4	检查储能指示	检查汇控柜状态显示器储能指示与储能实际位置是否相对应,检查时应口述并指出对应信号
5	合闸并检查	对开关进行合闸操作检查合闸位置时汇控柜指示位置与实际位置是否对应
6	分闸并检查	对开关进行分闸操作检查合闸位置时汇控柜指示位置与实际位置是否对应
7	恢复原状	检查完毕恢复设备到检查前状态,并提交检修报告

Task 1　Indicating Instrument Inspection of GIS

1.1　Work Tasks

Check the indicating instrument of GIS. This module mainly includes two aspects: 1. Identify the position indicator lights of the GIS control cabinet; 2. Check the actual positions of the circuit breaker and disconnector of the GIS. Through structural analysis, legend presentation, on-the-spot explanation and operational skill training, master the basic structure of GIS, and the composition, functions and principle of control cabinet; and identify and judge the status of the GIS control cabinet for consistency with the actual position of the equipment.

1.2　References

(1) *Regulation of Operation and Preventive Maintenance for Gas-insulated Metal-enclosed Switchgear* (DL/T 603-2017).

(2) *Five General Regulations of State Grid Corporation of China on Substation Maintenance Management* (Volume 3: Rules for Maintenance of GIS).

(3) *Regulations of State Grid Corporation of China on Management of Substation Operation and Maintenance*.

(4) *Substation Maintenance Management Regulations of State Grid Corporation of China*.

(5) *Electric Power Safety Working Regulations (Power Transformation) of State Grid Corporation*.

1.3　Work Requirements

(1) The operation belongs to indoor power interruption operation, without weather requirements.

(2) The interval to be maintained shall be isolated by fences from other live intervals, and the only exit shall be set up at the place oriented towards the passage.

(3) Operators are to be in good mental state, and are aware of safety measures, technical measures and hazards to site operation during operation.

(4) Take safety measures on practical training site as per regulations on production site, and strictly implement standard operation.

(5) Operators are requested to wear labor protection appliances to ensure safety protection.

1.4　Preparation for Work

1.4.1　Hazards and Preventive and Control Measures

1. High-voltage electric shock

Hazard: HV electric shock which is caused by entering the adjacent charging interval by mistake during operation.

Preventive and control measures: The interval to be maintained shall be isolated by fences from adjacent live intervals before operation, and an appropriate amount of sign boards, "Stop! High Voltage, Danger!", shall be hung towards the working sites. The only exit shall be set at the passage, with the sign board "Entrance/Exit" hung. At least two persons are required during work, one for supervision and the other for operation, who both are instructed by the person in charge of work.

2. Low-voltage electric shock

Hazard: LV electric shock which may be caused by opening and closing of circuit breaker.

Preventive and control measures: During operation, the operator shall shout loudly, and start the operation only after receiving a reply from all members of the work shift. The operator shall keep a certain safe distance from the live parts during the air switch operation.

1.4.2　Tools, Equipment, and Material Selection

Refer to Table 5-1 for the inspection tools and materials of the indicating instrument of GIS.

Table 5-1　Inspection Tools and Materials of Indicating Instrument of GIS

Category	Name	Specification and model	Quantity	Remarks
Specialized tools	Grounding knife-switch operating handle		1	
	Energy storage handle of circuit breaker		1	
	Operating handle of disconnector		1	
General tools	Allen wrench		1	
	Flat-nose pliers		1	
	Long-nose pliers		1	
	Slotted screwdriver	150 mm	1	
	Cross screwdriver	150 mm	1	
	Monkey wrench	200 mm	1	

1.4.3　Division of Workforce

Refer to Table 5-2 for the division of labor of inspectors regarding the indicating instrument of GIS.

Table 5-2　Division of Labor of Inspectors for Indicating Instrument of GIS

S/N	Job	Quantity	Responsibilities
1	Person in charge of work	1	Be responsible for work division of working staffs, site survey, stipulation of operation scheme, pre-shift meeting, safety supervision during operation, handling of emergencies during work, supervision of work quality and summary of post-shift meeting
2	Operator	1	Take charge of main operations of the indicating instrument of GIS
3	Assistant operator (optional)	1	Assist the operators in inspecting the indicating instrument of GIS

1.5 Operation Procedures

Refer to Table 5-3 for the inspection procedures for indicating instrument of GIS.

Table 5-3　Inspection Procedures for Indicating Instrument of GIS

S/N	Scope of work	Operational steps and standards
1	Preparations before work	1. Check if all instruments and tools are complete, and if their appearance and test are acceptable; 2. The person in charge of work and work permitter shall make an tour inspection for equipments under maintenance, and confirm all safety measures as listed by work ticket have been properly implemented. It is also necessary to check if safety measures are complete, and if the site is provided with conditions for commencement of work, and make supplements as required; 3. Implement work permit procedures; 4. Call in toolbox meeting participated by members of work team; 5. Record parameters on the equipment nameplate
2	Inspection of open position of circuit breaker	Check that the split indicator of the circuit breaker is in the split state, which shall be verbally stated with the corresponding signal indicated
3	Closing of AC power supply air switch	Open the door of the control cabinet, close the AC power supply air switch, as well as store the energy for and control the air switch, which shall be verbally stated with the corresponding signal indicated
	Inspection of energy storage instructions	Check whether the energy storage instructions on the status display of control cabinet correspond to the actual position of energy storage, which shall be verbally stated with the corresponding signal indicated
5	Closing and inspection	Close the switch to check whether the indicated position of control cabinet corresponds to its actual position at the closed position
6	Opening and inspection	Open the switch to check whether the indicated position of control cabinet corresponds to its actual position at the closed position
7	Restoration	After inspection, restore the equipment to the state before inspection, and submit the maintenance report

参考文献

[1] 高建. 电气设备检修[M]. 成都：成都时代出版社，2019.

[2] 杨迪. 变电检修技能培训教材[M]. 北京：中国电力出版社，2019.

[3] 姜聿涵. 变压器检修技能培训教材[M]. 北京：中国电力出版社，2019.

[4] 姜聿涵，杨冰."一带一路"变电设备检修专业培训教材 变压器检修（结构及附件篇）（中英文对照）[M]. 北京：中国电力出版社，2021.

[5] 华章，雷春."一带一路"变电设备检修专业培训教材变压器检修（电气试验篇）（中英文对照[M]. 北京：中国电力出版社，2021.

[6] 邓常飞，祝捷."一带一路"变电设备检修专业培训教材变电检修（故障处理篇）（中英文对照[M]. 北京：中国电力出版社，2021.